16th Topical Conference on Gas Utilization 2016

Topical Conference at the 2016 AIChE Spring Meeting and 12th Global Congress on Process Safety

Houston, Texas, USA
10-14 April 2016

ISBN: 978-1-5108-2485-0

Printed from e-media with permission by:

Curran Associates, Inc.
57 Morehouse Lane
Red Hook, NY 12571

Printed by Curran Associates, Inc. (2016)

For permission requests, please contact AIChE
at the address below.

AIChE
120 Wall Street, FL 23
New York, NY 10005-4020

Phone: (800) 242-4363
Fax: (203) 775-5177

www.aiche.org

Additional copies of this publication are available from:

Curran Associates, Inc.
57 Morehouse Lane
Red Hook, NY 12571 USA
Phone: 845-758-0400
Fax: 845-758-2633
Email: curran@proceedings.com
Web: www.proceedings.com

16th Topical Conference on Gas Utilization 2016

Topical Conference at the 2016 AIChE Spring Meeting and 12th Global Congress on Process Safety

Houston, Texas, USA
10-14 April 2016

AIChE
The Global Home of Chemical Engineers

TABLE OF CONTENTS

444065 Flowsheet Development and Simulation on the Use of Deep Eutectic Solvents for Carbon Dioxide Capturing

Abstract:

Capturing and restoring of carbon dioxide (CO_2) is getting growing attention as a potential method for regulating greenhouse gas emissions. Carbon dioxide content in natural gas can cause corrosion and other environmental issues. Low cost ionic liquid analogues, namely, deep eutectic solvents (DESs), have attracted more attention for use in a diversity of applications. DESs exhibit many favorable properties, such as availability, non-toxicity, biodegradability, recyclability, non-flammability, and low price. The use of ionic liquids for carbon dioxide capture is considered cost-effective because it requires less energy for solvent recovery compared to other conventional processes. In this work, an attempt to study the use of ionic liquids (ILs) for CO_2 capturing in typical absorption/stripper processes is considered. The attempt is tackled through simulation of a typical treatment system where ILs are introduced and circulated. The flowsheet simulation based approach implemented in Aspen HYSYS process simulators, is used to study the effect of temperature, pressure, adsorbent concentration, energy demand and flow rate on the CO_2 removal performance using the proposed system. The CO_2 content in natural gas is within the range of local industrial companies located in Qatar. The system is simulated using the most recent Acid Gas removal technology introduced in Aspen HYSYS simulation with closed loop circulation of the absorbent. The developed system is going to be validated against conventional amine system and experimental ILs data. The preliminary results showed a promising reduction on the required energy duty compared to Amine based treatment while higher purity is achieved.

440908 Chill More with Less and Properly Control Your Mechanical Refrigeration Unit (MRU) via Dynamic Simulation

Abstract:

Use Dynamic Simulation to Validate Your Process Control of a MRU Gas Processing Plant

There are many misconceptions about the closed refrigerant loop widely used in Mechanical Refrigeration Unit (MRU) to extract Natural Gas Liquid (NGL) from natural gas. Two common mistakes are: 1) How can I control the refrigerant compressor discharge pressure? 2) Why can't I control the liquid level in the receiver? The best way to answer these kinds of questions is to model the MRU in a dynamic process simulator. The dynamic simulation allows users to operate the virtual MRU plant without worrying about shutting down the real plant. From the dynamic simulation, users quickly learn that the compressor discharge pressure is actually controlled by the ambient temperature and/or the fans speed. The reason why the liquid level can't be controlled is because, 1) the total mass (not mass flow) is fixed in the closed loop, 2) the amount of mass in liquid form is controlled in the chiller and the economizer if any; thus the liquid level in the receiver is dependent on the chiller and economizer level.

Chill More with Less by Optimizing Set Points

Further, Dynamic Simulation is able to help to extract more NGL with less energy consumption. Most operators operate the plants with the fixed setpoints given by the plant equipment provider. However, this dynamic simulation demonstrates that it is possible to achieve a lower cold separator temperature while using the same electricity consumption or less by optimizing the setpoints.

Summary

Dynamic Simulation can be used to validate the process control during the design phase and to optimize the plant output while under operation.

Energy optimisation in natural gas processing plants

Leandro Labanca, Process Integration Ltd.
Yuhang Lou, Process Integration Ltd.

Abstract

Natural gas processing plants require a series of process units to remove contaminants such as water, carbon dioxide and hydrogen sulphide. These units have different process heating and cooling demands. Proper matching of the heating and cooling requirements leads to significant operating cost reductions. Excessive heat from process units can be recovered by producing steam, which may be used for satisfying heating requirements of other units or for power generation.

Systematic procedures for energy optimisation in natural gas processing plants are introduced and discussed during the paper. Novel software design tools with embedded state-of-the-art process integration technologies are presented for the simulation and optimisation of natural gas processing plants.

An industrial case study is presented to illustrate the benefits of performance optimisation both on the individual processes and on the overall steam system simultaneously. Eleven cost-effective projects were identified to save energy consumption in individual process units through enhanced heat recovery and process modifications. The influence and impact of the individual energy saving projects on the utility system was evaluated effectively with an integrated process / utility system model. Such a systematic and integrated energy optimisation approach significantly reduced utility consumption and improved the overall economics of the natural processing plant. Typical range of Internal Rate of Return (IRR) achieved for these energy saving projects were found to be between 25% - 180%.

The paper is extracted from a publication in GPA Europe's AGM and Technical Meeting, 2015 [7].

Introduction

Natural gas has been playing an increasingly important role in the past few decades as a clean and safe source of energy. Natural gas from underground has to be treated to remove impurities such as water, carbon dioxide and hydrogen sulphide before it is transported to end consumers. Separation of heavy hydrocarbons and sulphur recovery are often employed in natural gas processing plants to take advantage of the commercial value of by-products.

A typical natural gas processing plant consists of inlet facilities, an acid gas removal unit, a dehydration unit and other supporting process units. In all these process units, varieties of process heating and cooling are required. For example, in the acid gas removal unit, process heating is required at the bottom of the regeneration column, whilst the lean amine stream from the reboiler requires significant cooling before being sent to the absorption column. These process heating and cooling requirements are met by process-to-process heat exchangers or by heaters and coolers connected to site utilities.

If excess heat is available in an individual unit, steam could be produced to move heat across different process units and used for process heating or power generation purposes. In order to identify the potential for indirect heat recovery in the site, energy performance evaluation on a site-wide level, Total Site Analysis, is useful to review the steam balance and estimate the potential for power generation. Selection of mechanical drivers, including steam turbines, direct drive gas turbines or electric motors, has a direct impact on the ratio of heat and power supply. All these areas for utility system design should be carefully reviewed and evaluated systematically to achieve the best energy performance.

Process Integration Ltd. (PIL) has developed state-of-the-art software design tools, i-Heat™ and i-Steam™, for systematic energy optimisation of Heat Exchanger Networks (HENs) and utility systems, respectively. Advanced process integration technologies embedded in both software tools can be applied to effectively and efficiently synthesise a new process design or to identify and evaluate cost-effective retrofit options for an existing process. A recent heat integration and energy optimisation study for a natural gas processing plant using these novel tools is presented in this paper for demonstration purposes.

HEN design and optimisation with i-Heat™

Environmental policies and high energy prices have led companies to search for options to reduce their energy consumption as well as their carbon footprint. The optimal design of HENs may improve heat recovery and, hence, reduce the operating costs. Increasing heat recovery often requires an increase in the heat transfer area. Generally, there are economic trade-offs between the operating costs and the capital costs when designing and retrofitting HENs [1].

It is common to find options for process changes in a production process to unlock opportunities to reduce the operating and capital costs of the HEN. At this point, the user might consider reviewing the design and operation of other sections of the process before considering the HEN design further.

Across the year, the process is subjected to different operating schemes due to variations in the operation of other units or fluctuations in the ambient conditions. The HEN has to be designed to handle these variations and guarantee continuous operation of the unit [2].

To summarise, a good HEN design must achieve high energy recovery whilst balancing its capital investment. In addition, the HEN has to present flexibility to adapt to different operating scenarios [3]. Therefore, a HEN model needs to be set up to analyse different configurations and accurately evaluate its operating and capital costs.

1. Standard procedure for HEN design and optimisation

The following systematic procedure is used by PIL to design or retrofit a HEN.

- Step 1: Data collection

Base case and detailed scope of work are defined and agreed with the client. The relevant design or operating data from the process as well as the economic information for the study are collected. Different operating scenarios to be considered during the study are also agreed.

- Step 2: Stream data extraction

Stream's heating and cooling demands (duties, supply and target temperatures) are extracted from the collected documents for the process and entered in i-Heat™.

- Step 3: Establish HEN

Unit models such as heat exchanger, splitters and mixers are used construct new or existing HEN in i-Heat™.

- Step 4: Pinch Analysis

i-Heat™ performs pinch analysis [4] to identify energy targets for maximum heat recovery and minimum utility requirements at the selected minimum approach temperature for existing or new HENs.

- Step 5: Identify sources of inefficiency in the HEN

A suite of analysis tools are available in i-Heat™ to evaluate the energy performance of a HEN and identify the bottlenecks in the network. Sources of thermal inefficiencies, such as cross-pinch heat transfer, are identified and quantified automatically by the software tool.

- Step 6: HEN Optimisation

Operational optimisation of an existing HEN is carried out by adjusting operational variables, such as flows in stream-splits and duties in the exchangers, to achieve the minimum energy consumption without additional capital investment.

i-Heat™ also considers a combination of operational changes and structural modifications to automatically resolve the performance bottlenecks which limit energy recovery. Different retrofit designs are obtained based on maximum additional area and allowed locations of new heat exchangers defined by the user.

For a new design, i-Heat™ automatically synthesises cost-optimal heat exchanger networks whilst satisfying design, practical and operational constraints set by the user.

- Step 7: Economic analysis of retrofit options

Economic analysis is performed by i-Heat™ to quantify the trade-offs between the reduced energy costs and the additional capital investment. The user can compare the alternative retrofit designs based on overall costs to identify the best HEN configuration for the process.

Utility system design and optimisation with i-Steam™

Most production industrial sites possess significant heating and cooling requirements for their normal operation [5]. Most of the heating needs of various processes is satisfied by the same utility system which distributes steam around the site at several pressure levels. The steam system is also vital in meeting the power demands within the site [3]. Thus, the main aim of utility systems is to convert fuels to steam and electricity and satisfy process energy requirements within a production site.

There are two main scenarios to study and optimise site utility systems during the design stage: new design and retrofit design. A grassroots utility system may be needed for a site. Modifications to the configuration of an existing utility system might be required to reduce operating costs or to deal with changes in heat and power demands. At this stage, several decisions need to be addressed that will have a direct impact on the performance of the final design.

A number of configurations can be considered to satisfy the steam and power requirements of a site. For instance, boilers may be considered to generate high pressure steam, whilst steam turbines may be used to generate power through steam expansion. Alternatively, gas turbines may be employed to produce power whilst the exhaust gases could be sent through a Heat Recovery Steam Generator (HRSG) for steam production. In general, the efficiency of the utility systems may be increased with higher capital investment being considered. The user will need to identify the optimal configuration to satisfy the site requirements within the client's budget.

Process operating conditions change throughout the year for various reasons (e.g. start-up and shut-down of different units and planned maintenance cycles). In some regions, ambient conditions vary across the year or even across the day, which influences the operating conditions of processing units. Utility systems have to be prepared to cope with these variations to guarantee continuous operation of the site. The penalty for ensuring the integrity of the site operation is that the utility systems are generally over-designed and present significant redundancy in place. The redundancy also provides a number of degrees of freedom to improve the performance of the system [3].

Large and small process machines on a site need a mechanical driver (electric motor, steam turbine or other driving device) for their operation. The user should evaluate which is the best option for each particular driver taking into consideration that the selection of drivers will have a direct effect on the site requirements of steam and power.

To summarise, utility systems must be designed to cope with steam and power demands of the site whilst achieving the highest efficiencies with the lowest possible economic cost. To obtain the best configuration to accommodate the site needs and carry out a detailed economic evaluation of the diverse designs, an accurate optimised model should be developed [3].

1. Standard procedure for utility system design and optimisation

The following is the systematic procedure used by PIL to design a new utility system. A similar procedure is followed when analysing and optimising an existing utility system.

- Step 1: Data Collection

Detailed scope of work is defined and agreed with the client. A study basis is set which includes the definition of the base case. The relevant design or operating data and the economic information for the study are collected from the site.

- Step 2: Identify and understand utility demands

Hot and cold utility requirements are identified by reviewing the design documents of individual processes. Utility duties as well as the required temperatures for each utility requirement are extracted for energy performance analysis.

- Step 3: Definition of multiple scenarios

A number of scenarios considering various process operating conditions and seasonal ambient variations are defined to ensure the utility system can cope with different energy demands across the year.

- Step 4: Total Site Analysis

Utility demands are extracted from the processes and are used to construct the total site composite curves in i-Steam™. i-Steam™ contains sophisticated targeting tools to define the best energy performance for a site under a range of different operating scenarios. Total Site Analysis helps to identify heat and power co-generation targets and optimal steam header conditions.

- Step 5: Utility Systems Modelling

A comprehensive steam system model is produced in i-Steam™ to represent the major equipment items, such as boilers, steam turbines and gas turbines, and their interactions. Modelling of the steam network and the individual equipment items is based on thermodynamic principles and the overall heat and material balance of the system. Part-load performance models for key equipment are established [6].

- Step 6: Utility Systems Optimisation

Considering the multiple scenarios defined, operational optimisation is carried out by adjusting operational variables, such as steam distribution or steam header pressures, to explore the potential for performance improvement of the existing utility system. i-Steam™

automatically selects between electric motors and steam turbine drivers for all power demands within the utility system taking into consideration operating conditions and market prices.

Structural changes are also analysed in order to improve the flexibility of the system as well as to adjust site heat to power demand ratio. Finally, the overall steam network is verified for performance, flexibility and economics.

Energy optimisation studies in natural gas processing plants

PIL follows standard procedures to ensure high quality designs are produced in its energy optimisation studies. These procedures cover every aspect of the study from data extraction to process modelling and optimisation. Multiple layers of checking ensure that the energy study meets PIL's quality standards and every potential design improvement is captured during the heat integration and energy optimisation study.

- Step 1: Data collection

Obtain process design information from the client, including Basis of Design document, process description, Process Flow Diagrams (PFDs) and Heat and Material Balance (H&MB).

- Step 2: Data review and reconciliation

Review and cross check process design information from various sources. Discuss with the client any inconsistent data items and agree on priority for different data sources.

- Step 3: Process unit screening

Review and select those process units containing heat exchangers and therefore providing opportunities for design improvement.

- Step 4: Heat integration study for individual process units

Standard procedure for HEN design and optimisation is followed in the heat integration study for individual process units. As discussed in the previous sections, the key steps include stream data extraction, establish HEN in i-Heat™, pinch analysis, identify sources of inefficiency in the HEN, HEN optimisation and economic analysis of retrofit options.

- Step 5: Evaluate the potential for cross-process heat recovery

Stream data from multiple process units is analysed simultaneously to exploit the potential for cross-process heat recovery.

- Step 6: Energy optimisation for site-wide utility system

Standard procedure for utility system design and optimisation is followed in the energy optimisation for site-wide utility system. As discussed in the previous sections, the key steps include data collection, identify and understand utility demands, definition of multiple scenarios, total site analysis, utility systems modelling and optimisation in i-Steam™.

- Step 7: Economic evaluation of combined improvements

1) Verify the recommended retrofit option from each heat integration study for practicality and feasibility.
2) Establish new steam balance in i-Steam™ taking into consideration the recommended retrofit options from the heat integration studies.
3) Adjust the operation of the utility system to meet the new steam supply and demand.
4) Verify the operation of the utility system under different operating scenarios to guarantee the reliability of the site.
5) Evaluate the economic benefits with all recommendations combined.

Case study

PIL completed a heat integration and energy optimisation study for a new design of a natural gas processing plant. The natural gas processing plant imports steam at various pressure levels from a co-generation plant nearby. PIL was asked to identify opportunities to improve energy efficiency of the existing process design and exploit the interactions between the natural gas processing plant and the co-generation plant to minimise the overall life cycle cost.

- Step 1: Data collection

Various design information was provided by the client to support the heat integration and energy optimisation study.

- Step 2: Data review and reconciliation

Since data discrepancies were observed in some cases, a hierarchy of data sources was agreed between PIL and the client for the project to proceed further.

- Step 3: Process unit screening

PIL received the requested design information for twenty-three process units in the natural gas processing plant. After reviewing the process design data, PIL identified those process units containing heat exchangers and, therefore, providing opportunities for heat integration. Thirteen out of twenty-three units were selected for the energy optimisation study.

Three out of thirteen process units are discussed to demonstrate the benefits of heat integration for design and performance improvement using i-Heat™.

- Step 4: Heat integration study for individual process units

Unit 1

Unit 1 is used as an example to illustrate the standard procedure for a heat integration study. This unit has three hot streams and four cold streams. Process stream conditions and heat exchange duties were extracted from the design information and were entered in i-Heat™, as shown in Table 1. It was agreed with the client that heat loss during heat exchange was not considered as part of the study.

Table 1. Stream data summary for Unit 1

	No.	Stream	Segment	TS [°C]	TT [°C]	MCp [kW/°C]	DH [kW]	DTmin [°C]	HLoss [%]
1	1	Hot Stream 1	1	139.4	37.8	26.22	2666.1	Global	0.0000
2	2	Cold Stream 1	1	4.4	50	58.52	2666.1	Global	-
3	3	Cold Stream 2	1	80	90	186.54	1865.4	Global	-
4	4	Cold Stream 3	1	86.7	139.4	396.95	20950.2	Global	-
5	5	Hot Stream 2	1	110.6	60	385.51	19489.6	19	0.0000
6	6	Hot Stream 3	1	136.9	60	39.18	3012.2	19	0.0000
7	7	Cold Stream 4	1	137.2	147.2	245.2	2452	Global	-

Figure 1 shows the existing heat exchanger network design of Unit 1 in i-Heat™.

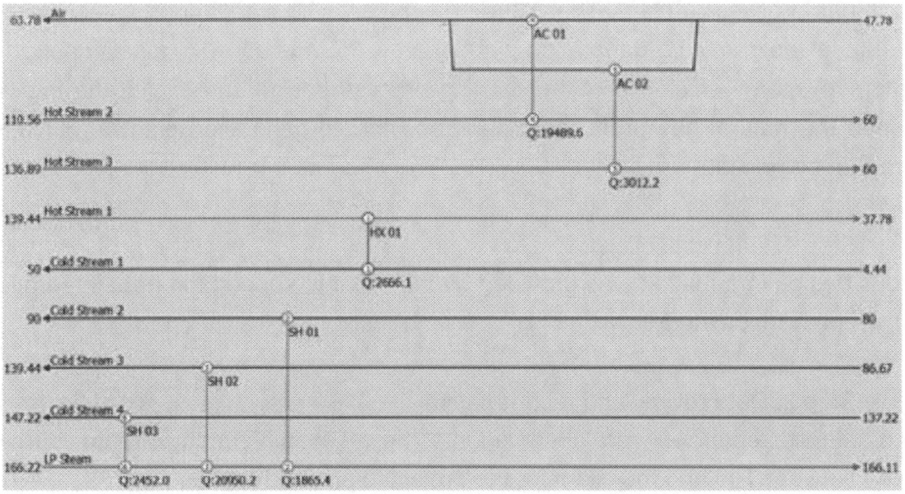

Figure 1. Existing heat exchanger network design for Unit 1.
Duties (Q) are in kW and temperatures are in °C

There is one heat exchanger HX 01 for process heat recovery in this unit. Three cold process streams are heated by LP Steam. Two hot process streams are cooled down in air coolers to their target temperatures. The utility consumption of the existing design is summarised in Table 2.

Table 2. Utility consumption of the existing design (Unit 1)

Utility Type	Utility	Heat Exchanger	Duty (kW)
Hot	LP Steam	SH 01	1,865
		SH 02	20,950
		SH 03	2,452
		Total	25,267
Cold	Air	AC 01	19,490
		AC 02	3,012
		Total	22,502

Pinch analysis was carried out in i-Heat™ to identify the minimum energy targets for both hot and cold utilities. Temperature vs. Enthalpy segments for hot and cold process streams were combined to generate the hot and cold composite curves, as shown in Figure 2. The hot and cold temperature pinch points are represented by the red and blue dots in the diagram.

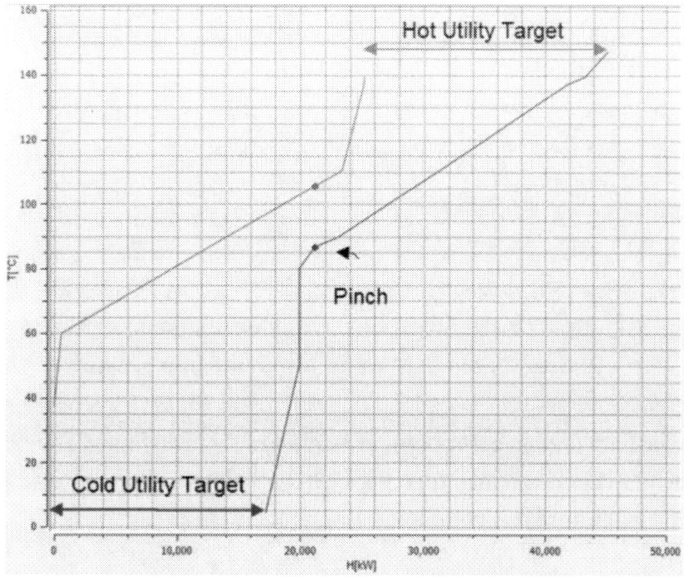

Figure 2. Composite curves for Unit 1

The global minimum approach temperature (minimum DT) at the thermodynamic pinch is set to 19°C. The corresponding energy targets were obtained from i-Heat™ and compared with current consumption, as shown in Table 3.

Table 3. Energy consumption and target for Unit 1 with minimum DT = 19°C

	Current Consumption	Energy Targets	Difference
Hot Utility Duty [kW[	**25,267**	**20,030**	**5,237**
Cold Utility Duty [kW]	**22,502**	**17,265**	**5,237**

By comparing the utility consumption of the existing design against the minimum energy target, thermodynamically (without any constraints), there is the opportunity to reduce hot and cold utility consumption by 5,237 kW.

Figure 3 shows the analysis of cross pinch heat transfer on the existing design. Heat exchangers that have cross pinch heat transfer are represented with an oblique line that crosses the pinch. Four heat exchangers were found to have cross pinch heat transfer, as shown in Figure 3. These heat exchangers are the source of thermodynamic inefficiency in the unit.

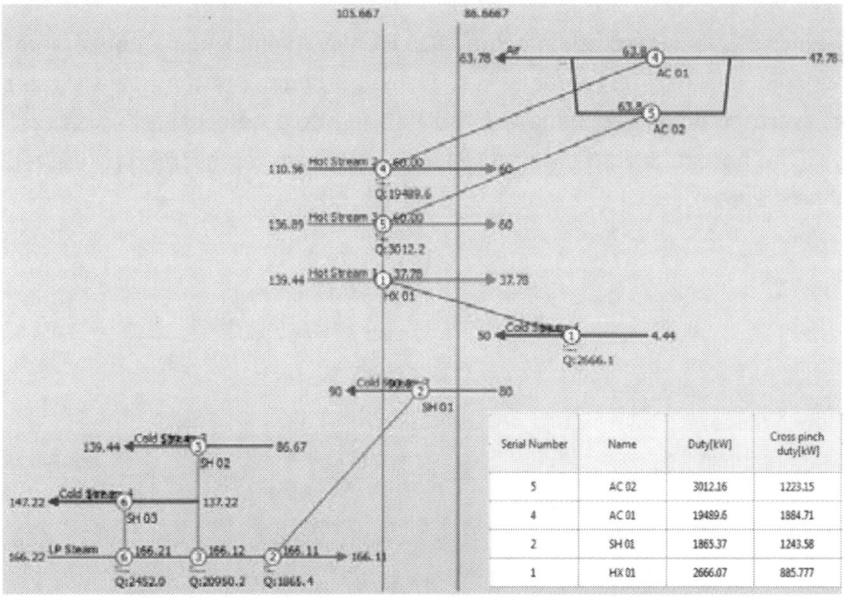

Serial Number	Name	Duty[kW]	Cross pinch duty[kW]
5	AC 02	3012.16	1223.15
4	AC 01	19489.6	1884.71
2	SH 01	1865.37	1243.58
1	HX 01	2666.07	885.777

Figure 3. Cross pinch heat transfer analysis for Unit 1
Duties (Q) are in kW and temperatures are in °C

Design improvements on this unit could be achieved by reducing cross pinch duties within the existing HEN. An energy saving project, PROJECT1, was proposed by PIL to reduce the cross pinch heat transfer duty in SH 01.

As shown in Figure 4, a new heat exchanger named 'New HX 01' was added in the design to recover heat between Hot Stream 2 and Cold Stream 2 which results in a reduction of LP Steam consumption by 1,865 kW. As a result, the LP steam heater SH 01 is no longer required and the duty of the air cooler AC 01 is reduced by the same amount.

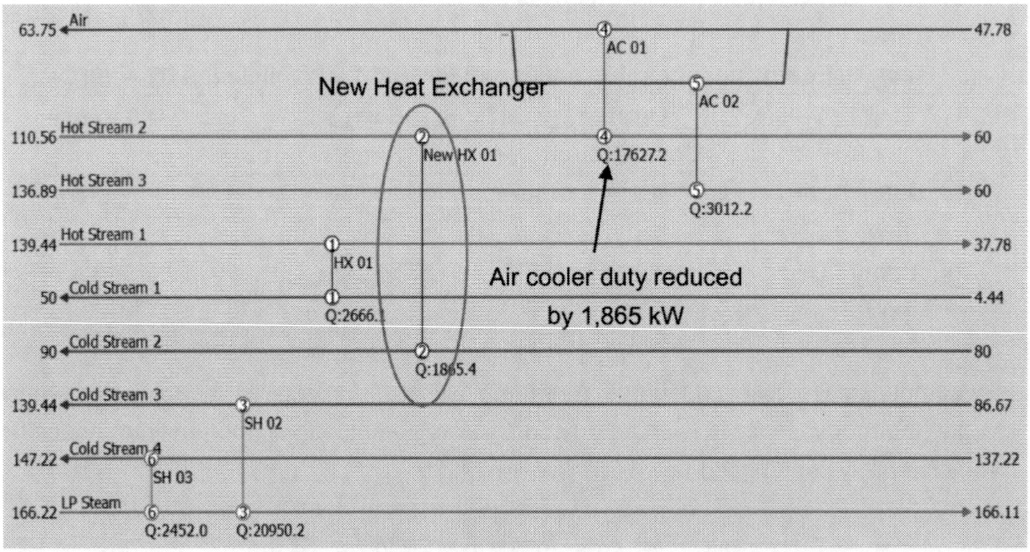

Figure 4. New heat exchanger network for Unit 1
Duties (Q) are in kW and temperatures are in °C

Process economic analysis was carried out to explore the trade-offs between the reduced operating cost and the extra capital investment for the additional heat exchanger. The changes in overall capital and operating costs between existing design and new design are summarised in Table 4. The addition of a new heat exchanger 'New HX 01' and removal of existing exchanger SH 01 require an increase in investment of 2.46 MM$. Energy savings of 1,865 kW of LP steam and air cooling duties are worth 0.62 MM$/yr. The project was found economically attractive with 20-year Net Present Value (NPV) of 6.51 MM$ and Internal Rate of Return (IRR) of 25.8%. PROJECT1 was recommended to the client for further investigation.

Table 4. Process economics comparison for PROJECT1

	Change in Installed Capital Costs (MM$)	Change in Operating Costs [1] (MM$/yr)
Single Train	+ 0.82	- 0.31
All Trains	+ 2.46	- 0.62

[1] Annual operating hours of 8760

Unit 2

A similar heat integration study was carried out for Unit 2. The existing heat exchanger network design is composed of 20 heat exchangers. Practical constraints, such as forbidden matches between process streams, were considered when looking for opportunities for further heat integration. The process streams and heat exchangers that do not allow structural changes were removed from the HEN model in i-Heat™. Figure 5 illustrates the simplified HEN composed of 6 heat exchangers.

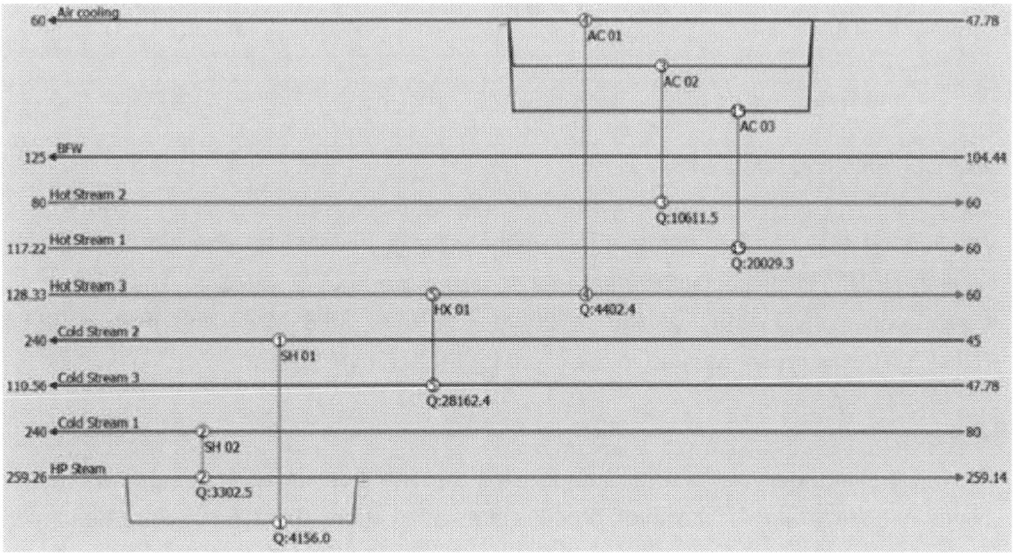

Figure 5. Simplified existing heat exchanger network design for Unit 2
Duties (Q) are in kW and temperatures are in °C

Cross pinch heat transfer analysis showed that steam heaters SH 01 and SH 02 have a cross pinch duty of 1,409 kW and 642 kW, respectively. In order to eliminate the cross pinch heat transfer in the existing design, two heat exchangers "New HX 01" and "New HX 02" were added in the heat exchanger network, as shown in Figure 6.

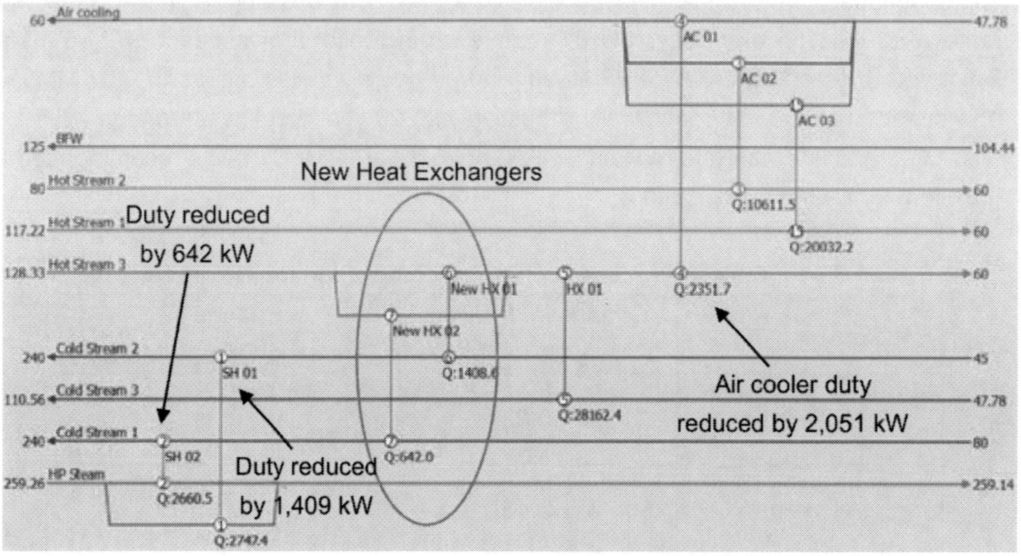

Figure 6. New heat exchanger network design for Unit 2
Duties (Q) are in kW and temperatures are in °C

The results summary of process economic analysis for the new heat exchanger network is summarised in Table 5.

Table 5. Process economics comparison for PROJECT2

	Change in Installed Capital Costs (MM$)	Change in Operating Costs [1] (MM$/yr)
Single Train	+ 0.65	- 0.554
All Trains	+ 3.92	- 2.77

[1] Annual operating hours of 8760.

Considering all trains, an increase in investment of 3.92 MM$ is required to modify the existing HEN. The associated energy savings are worth 2.77 MM$/yr. The project was found economically attractive with a 20-year NPV of 38.8 MM$ and IRR of 72.8%. PROJECT2 was recommended to the client for further investigation.

Unit 3

The initial pinch analysis for Unit 3 suggested that the existing HEN design had met the minimum energy targets. However, process simulation showed that further pre-heating the feed to the column could potentially reduce the energy consumption of the existing design. By using heat recovery to pre-heat the stream, LP steam consumption in the column reboiler was successfully reduced by 2,198 kW. The overall economic benefit of this design change is a reduction in operating costs of 0.37 MM$/yr. The required capital investment is increased by 0.20 MM$. Therefore, the project was economically attractive with a 20-year NPV of 6.0 MM$ and IRR of 188.7%. PROJECT3 was recommended to the client for further investigation.

The previous examples illustrate how heat integration analysis and optimisation can be carried out easily and efficiently with i-Heat™. Cost-effective energy saving projects can be quickly identified and evaluated. In total, eleven energy saving projects were identified and evaluated by PIL using i-Heat™. Seven projects were found to be cost-effective and recommended to the client for further feasibility evaluation.

- Step 5: Evaluate the potential for cross-process heat recovery in i-Heat ™.

Following the heat integration studies of the individual process units, the potential for heat integration between process units was also considered in the energy optimisation study. Although it was possible to match heat sources and heat sinks across process units in the thermodynamic analysis, practical constraints ruled out options for feasible heat recovery.

- Step 6: Energy optimisation for site-wide utility system

The steam system and the nearby co-generation plant were analysed for energy optimisation. The steam system was designed to provide process heating and power for the natural gas processing plant. The existing steam system design was modelled in i-Steam™ for performance analysis. Figure 7 illustrates the various steam headers, process steam generators, steam turbine generator and steam consumers present in the system.

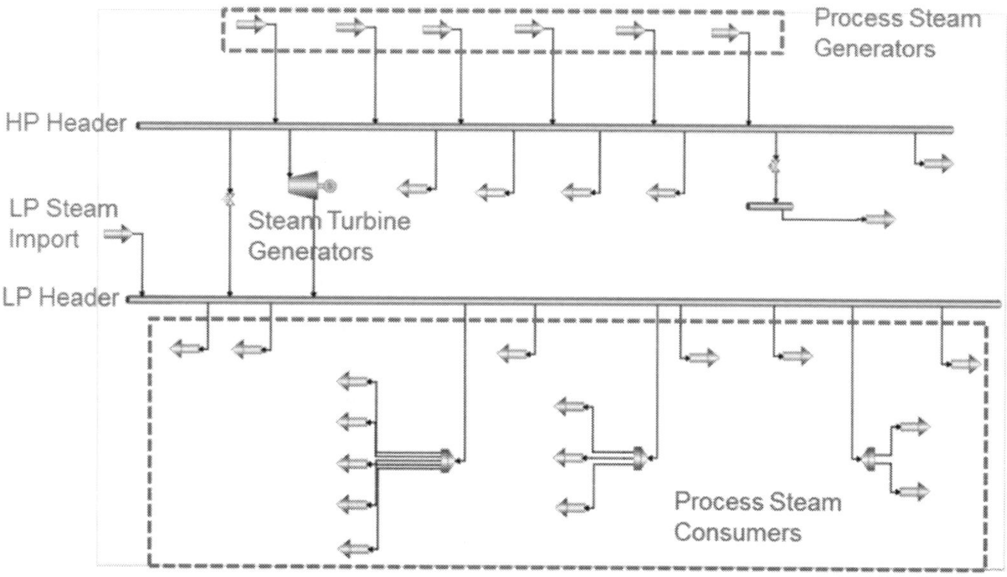

Figure 7. Steam system model in i-Steam™

The equipment items present in the nearby co-generation plant were also modelled in i-Steam™ to represent the power generation and LP steam export to the natural gas processing plant. The co-generation plant model was used to estimate the price of LP steam, as shown in Figure 8. Considering fixed steam flow rate through the HP turbine, if the LP steam export is reduced by 1t/h, the steam flow rate going through the condensing turbine is increased by 1t/h to generate additional power. The value of 1t/h LP steam was estimated as the economic benefit from extra power generation. Based on an electricity price of 0.07$/kWh, the value of LP steam was estimated at 10.87$/t. This LP steam price served as the basis for analysis and evaluation of process economics.

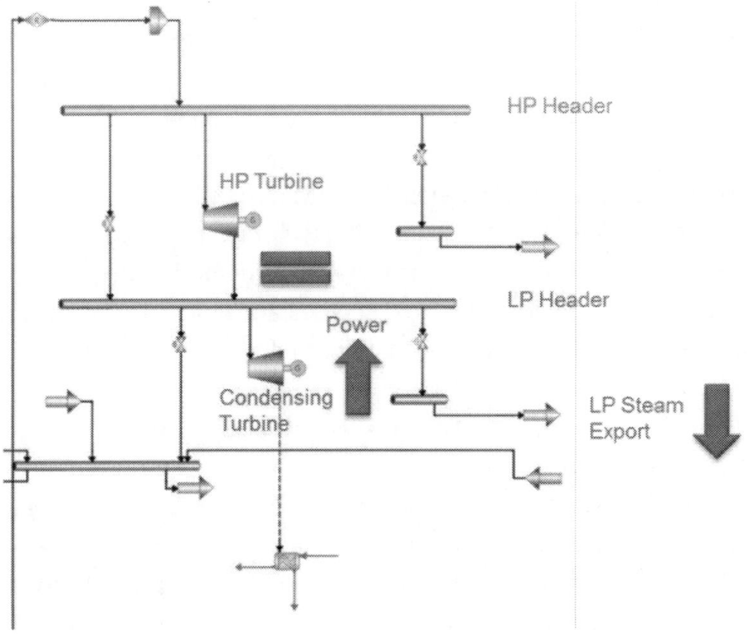

Figure 8. Co-generation plant model in i-Steam™

During the energy optimisation study, PIL also considered other options to improve the overall performance and economics of the natural gas processing plant. These options included driver selection as well as the optimisation of refrigeration systems. The integration of a chilled water system in the overall process was found to improve the process economics significantly.

- Step 7: Economic evaluation of combined improvements

Every cost-effective energy saving project was discussed with the client for further evaluation. A combined overall process model was developed to evaluate how the operation and configuration of the steam system and the co-generation plant were affected with the design improvements. The process economics of the combined process model was represented by the improvement of the NPV over the plant life of 20 years.

When considering all the energy saving projects that were accepted by the client, 20-year NPV of the natural gas processing plant is increased by 284.1 MM$. The IRR is 137.8%.

Conclusion

The case study illustrates how the advanced design tools for heat integration and energy optimisation, i-Heat™ and i-Steam™, can significantly improve the energy performance and process economics of a natural gas processing plant. Cost-effective projects can be identified by i-Heat™ to save energy consumption in individual process units through enhanced heat recovery. The influence and impact of the individual energy saving projects on the utility system can be evaluated effectively with the combined overall process model in i-Steam™. Such a systematic and integrated energy optimisation approach significantly reduces utility consumption and improves the overall economics of the processing plants.

The paper is extracted from a publication in GPA Europe's AGM and Technical Meeting, 2015 [7].

References

1. Douglas, J., 1988, Conceptual Design of Chemical Processes, McGraw-Hill.

2. Chen. L., Heat-Integrated Crude Oil Distillation Design, PhD thesis, Centre for Process Integration, University of Manchester, UK, 2008.

3. Smith, R., 2005, Chemical Process Design and Integration, John Wiley & Sons, Ltd., Chapter 23.

4. Kemp, I.C., 2007, Pinch Analysis and Process Integration: A User Guide on Process Integration for the Efficient Use of Energy, 2nd edition, Elsevier/Butterworth-Heinemann.

5. Varbanov, P., 2004, Optimisation and Synthesis of Process Utility Systems, University of Manchester Institute of Science and Technology (UMIST), Manchester.

6. Lou, Y., Comprehensive Modelling and Optimisation of Industrial Steam Systems, AIChE Spring Meeting, Austin, USA, 2015.

7. Zheng, X., Labanca, L., Energy Optimisation in Natural Gas Processing Plants, GPA Europe's AGM and Technical Meeting, London, UK, 2015.

Modeling of LNG Spreading and Vaporization Using Smoothed Particle Hydrodynamics Method

Anh Viet Bui
AnhBui@flowsim3d.com

Morgan Reed, Tony Liu, Eric Peterson
MMI Engineering Inc.
11490 Westheimer Road, Suite 150, Houston, Texas 77077
MReed@MMIEngineering.com
TLiu@MMIEngineering.com
EPeterson@MMIEngineering.com

Prepared for Presentation at
American Institute of Chemical Engineers
2016Spring Meeting
12thGlobal Congress on Process Safety
Houston, Texas
April 11-23, 2016

Modeling of LNG Spreading and Vaporization Using Smoothed Particle Hydrodynamics Method

Anh Viet Bui
AnhBui@flowsim3d.com

Morgan Reed, Tony Liu, Eric Peterson
MMI Engineering Inc.
11490 Westheimer Road, Suite 150, Houston, Texas 77077

Keywords: LNG, vaporization, CFD, Smoothed Particle Hydrodynamic (SPH)

Abstract

Smoothed Particle Hydrodynamics (SPH) has been used for numerical simulation of the release of LNG and the subsequent spreading of the liquid. Although the SPH method has been used for years in the astrophysics industry, its application to the LNG industry is quite novel. The spreading and vaporization of LNG creates the potential for a large vapor cloud with the risk of an explosion or flash fire. The SPH method offers excellent mass conservation, especially when modeling small scale changing fluid interfaces (i.e. thin film or small vapor bubbles). Based on LNG consequence hazards, many regulations (i.e. NFPA 59A) restrict vapor cloud dispersion extent to the facility property perimeter; less expensive (time and money) calculation methods are sought.

Vaporization of LNG during its spreading over water or solid substrate involves either evaporation on the air-LNG interface or nucleate/film boiling on the contact of cold LNG with relatively high-temperature substrate surface. For the latter vaporization mechanism to be correctly modeled, surface temperature and its subsequent change due to contact with LNG needs to be described. Consequently, a conjugate heat transfer model which includes wall heat conduction is necessary.

This paper discusses the continued development of a multiphase flow model with heat transfer and phase change based on the SPH model available in the LAMMPS Molecular Dynamics Simulator software. A mechanistic film boiling model is implemented and incorporated into our SPH flow model. Limited validation of the SPH prediction of the LNG spreading will be conducted and examples showing the enhanced boiling and heat transfer capabilities of the proposed LNG spreading model are outlined and discussed.

1 Introduction

LNG release, spreading, and vaporization and the formation of a flammable vapor cloud present a significant fire and explosion hazard to oil & gas facilities. NFPA 59A Standard for the Production, Storage and Handling of Liquefied Natural Gas, requires that the periphery of a vapor cloud (where fuel gas concentration becomes 50% of LFL) formed as a result of LNG release be within the property limits of LNG facilities.

Gravity-driven spread of LNG is a non-trivial process due to the intensive heat exchange of cryogenic LNG with the surrounding environment (soil, water, wind) which involves a rapid thermal exchange and evaporation of LNG. Various boiling regimes, such as nucleate, film, and/or explosive, may happen depending on the thermal and mechanical interaction of LNG and the substrate. In a scenario of LNG release to water, for instance, mechanical mixing of very cold LNG and relatively "hot" water may occur leading to a chain reaction of explosive increase of LNG-water interphase area and fiery LNG boiling dynamics (i.e. rapid phase transition).

The presence of a large free surface of spreading LNG and possible thermal and mechanical interactions of LNG with complex terrain and structural geometries pose significant difficulty to the modeling of LNG spreading using traditional grid-based CFD approaches. Such CFD approaches usually require a refined grid to resolve the physics happening inside an increasingly thinning layer of LNG when it spreads further away from the release origin, which is computationally expensive. Complex geometry of and abrupt change of material properties across the gas-liquid interfaces may also pose a significant problem to numerical methods used to solve the system of transport equations.

The Smoothed Particle Hydrodynamics (SPH) method has been proposed to model the spreading of LNG and the mechanical and thermal interactions with the surrounding environment and structures [1]. In comparison to the traditional grid-based CFD method, SPH can naturally and efficiently deal with complex (moving) fluid interfaces and terrains. This paper discusses the continued development of the SPH LNG spreading model with heat transfer and phase change [1] which is based on a SPH model available in the LAMMPS Molecular Dynamics Simulator software [2]. The model has been revised to enable the simulation of LNG spreading over complex solid boundaries represented by the STL (STereoLithography) geometries. The model of thermal interaction of LNG with such complex boundaries has been implemented which may include a mechanistic LNG boiling model. Limited testing and validation of the model will be conducted and examples showing the enhanced capabilities of the proposed LNG spreading model will be outlined and discussed.

2 Methodology

2.1 A brief description of the SPH method

Smoothed Particle Hydrodynamics method is based on a Lagrangian description of a fluid domain in which a limited number of particles with their own individual masses, momenta, and energies are employed for the modeling of fluid motion. The governing equations of fluid motion, i.e. Navier-Stokes equations, are solved with use of discrete (particle) approximations of field variables, i.e. density, velocity and energy, as well as their derivatives and integrals.

In the SPH method, the system of Navier-Stokes equations is expressed using particle approximations of the field variables and their derivatives as follows

$$\frac{D\rho_i}{Dt} = -\sum_{j=1}^{N} m_j W_{ij},$$

$$m_i \frac{D\mathbf{v}_i}{Dt} = -\sum_{j=1}^{N} m_i m_j \left(\frac{P_i}{\rho_i^2} + \frac{P_j}{\rho_j^2}\right) \nabla_j W_{ij} + \sum_{j=1}^{N} \frac{m_i m_j}{\rho_i \rho_j} (\mu_i + \mu_j) \mathbf{v}_{ij} \left(\frac{1}{r_{ij}} \frac{\partial W_{ij}}{\partial r_{ij}}\right) + \boldsymbol{F}_i,$$

[Eq.1]

$$m_i \frac{De}{Dt} = -\frac{1}{2} \sum_{j=1}^{N} m_i m_j \left(\frac{P_i}{\rho_i^2} + \frac{P_j}{\rho_j^2}\right) : \mathbf{v}_{ij} \nabla_j W_{ij} - \sum_{j=1}^{N} \frac{m_i m_j}{\rho_i \rho_j} \frac{(k_i + k_j)(T_i - T_j)}{r_{ij}^2} \boldsymbol{r}_{ij} \cdot \nabla_j W_{ij}.$$

where $\boldsymbol{r}_{ij} = \boldsymbol{x}_i - \boldsymbol{x}_j$, $r_{ij} = |\boldsymbol{r}_{ij}|$, $\mathbf{v}_{ij} = \mathbf{v}_i - \mathbf{v}_j$; e, P, μ and k are the particle energy, pressure, dynamic viscosity and thermal conductivity, respectively; and m_i, ρ_I and N denote the particle i mass, density, and the total number of particles located within the influence area. W_{ij} or $W(\boldsymbol{x}_i - \boldsymbol{x}_j, h)$ is a smoothing function which defines the influence or support area around each particle. This function determines the accuracy of the SPH approximations and efficiency of the computation. A *cubic spline smoothing function* is used in this work.

A weakly-compressible approach to simulate incompressible fluids is applied which is based on the following equation of state (EOS):

$$p(\rho) = \frac{\rho_0 c^2}{\gamma} \left[\left(\frac{\rho}{\rho_0}\right)^{\gamma} - 1\right] \qquad \text{[Eq.2]}$$

where c is the sound speed, ρ_0 is the reference density, and γ is a factor defining fluid stiffness. The time integration of Eqn.2 is conducted using a Velocity-Verlet scheme [3].

2.2 *Treatment of complex wall boundary conditions*

Most SPH models in the literature employ the so-called ghost (or dummy) particles technique to represent solid wall boundaries. In this approach, two or more layers of ghost particles are placed around a boundary (Figure 1). These ghost particles interact with the fluid particles confined inside the boundary, but are stationary during simulation. For complex boundary geometry with curved walls, appropriate generation and placement of ghost particles can be a challenging task.

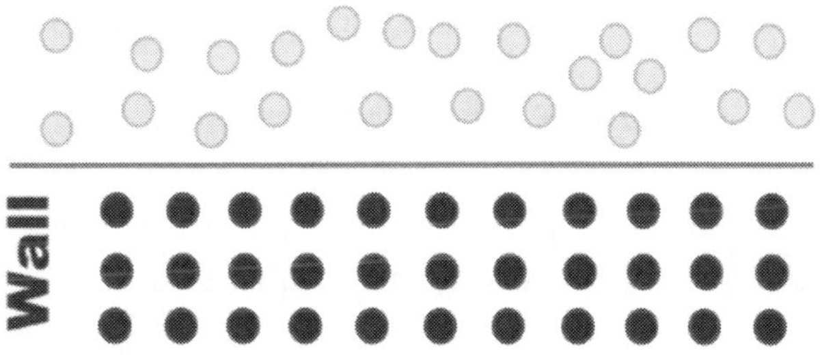

Figure 1: SPH particles – fluid (yellow) and boundary (purple) particles.

In this work, complex wall geometry is treated with use of a repulsive force boundary condition which prevents the penetration of interior fluid particles through the wall [4]. The wall is represented by STL geometries which approximate complex surfaces with triangular meshes. The repulsive force is calculated for every particle which is in the proximity of a boundary triangle. This force acts normal to the wall triangle and is given by

$$F_{rep,i}(d) = \begin{cases} D\left[\left(\frac{r_0}{d}\right)^a - \left(\frac{r_0}{d}\right)^b\right], & d < r_0 \\ 0, & d \geq r_0 \end{cases} \quad \text{[Eq.3]}$$

where d is the distance from a particle to a wall triangle and D, r_0, a, and b ($a > b$) are some regulation parameters.

The viscous forces acting on the directions tangential to the wall surface are accounted for by a modified viscous term which has the following form [4]:

$$\boldsymbol{F}_{vis,i}(d) = \frac{2m_i\mu_i\mathbf{v}_{i,tan}}{\rho_i^2}\left(\frac{1}{d}\frac{\partial W_d}{\partial d}\right) \quad \text{[Eq.4]}$$

with $\mathbf{v}_{i,tan}$ being the projection of the velocity vector of particle $i$on the wall surface plane.

2.3 Models of surface tension and cohesion forces

The surface tension force is defined simply by [5]

$$\boldsymbol{F}_{s,i} = -\sigma_s \sum_j m_j \boldsymbol{r}_{ij} W_{ij} \quad \text{[Eq.5]}$$

with σ_s being the surface tension coefficient.

A cohesion force can also be added which attracts nearby small bubbles (vapor particles) to form a large one. This force is given by [5]

$$\boldsymbol{F}_{cohesion,i} = -k_c m_i \sum_j \rho_j \boldsymbol{r}_{ij} \quad \text{[Eq.6]}$$

with k_c being a parameter that controls the strength of the cohesion force.

2.4 Modeling of wall heat transfer

Constant temperature or heat flux can be specified on the solid wall boundaries. When a constant wall temperature is set, the heat flux from the wall to the fluid particle $i$is defined by

$$Q_i = h(T_w - T_i) \quad \text{[Eq.7]}$$

where T_w and T_i are the wall and particle temperatures, respectively, and h is the convective heat transfer coefficient.

This wall heat flux can be added to the energy equation using the following term:

$$Q_i = \frac{2m_i^2 h(T_w - T_i)}{\rho_i^2}\left(\frac{\partial W_d}{\partial d}\right). \quad \text{[Eq.8]}$$

Upon a transition from convective heat transfer to nucleate or film boiling heat transfer, an appropriate boiling heat transfer coefficient should be utilized for h.

2.5 Modeling of LNG-vapor coexistence and evaporation

The flow of LNG and its vapor can be modeled with use of a *multiphase* or *multi-fluid* SPH model where separate transport equations and equations of state (EOS) are utilized for liquid and vapor LNG [1,6,7]. The mechanical and thermal interactions between phases including phase transition (LNG evaporation) are described by the interphase exchange terms. Different particle types are used to represent LNG and its vapor and the mass source or sink terms define the phase change rates.

Alternatively, the coexisted LNG liquid and vapor phases can be described by a single set of transport equations and a common EOS. The van der Waals equation of state is normally used in such a model, although it may impose some stability constraint on the numerical solutions.

Modeling of the near-wall boiling is complicated due to small scale physics involved. The physics of boiling and the transition from nucleate to film boiling are still not well understood and the split of wall heat flux (to convective and boiling heat transfers) and the wall evaporation rate are still mostly defined based on experimental data.

Phase change, i.e. evaporation or condensation, happens as a result of heat transfer between heating wall and flow or thermal exchange across the fluid-vapor interface. In this work, only evaporation near a heating wall is considered with the wall heat flux (Eq.7) assumed to be decomposed into a fluid heating and an evaporation components [6,7]. The evaporation heat flux Q_{evap} results in an evaporation mass flow rate of $\dot{\boldsymbol{m}}_{\boldsymbol{evap}} = \boldsymbol{Q}_{\boldsymbol{evap}}/h_{evap}$ (h_{evap}– the latent heat of evaporation). A corresponding number of vapor particles needs to be generated and inserted into the computational domain. In the current model, a simple evaporation model based on simple phase/particle type switching is employed.

3 Simulations of Fluid Spreading and Phase Change

Limited testing of the SPH method in modeling the spread of water over dry and wet flat surfaces has been conducted as described in a previous work [1]. In this study, new capabilities of the further developed SPH model are tested.

3.1 Simulation of fluid spreading over STL boundaries

The latest approach tracks the liquid running pool as it transverses a terrain other than a flat surface. Figure 2 shows an example of what can be considered a trench system common at many LNG terminals throughout the United States. The fluid temperature is set at 20 °C, whereas the surface temperature is at 100 °C. While the problem is essentially 2D, the simulation is 3D due to the presence of 2D STL boundary. The initial water column has the dimensions of 0.1m × 0.1m × 0.2m, and is described by approximately 3000 fluid particles.

Figure 2: A simple STL surface.

The simulation results are show in Figure 3 which indicates that the SPH model can successfully capture the dynamics of fluid spreading over a relatively complex boundary described by a STL surface meshe. The SPH method is seen to be able to capture the void space developed under the fluid jet in the beginning stage of the spreading (which can hardly be described by a shallow water model relying on a fluid height equation). Heat transfer between the wall and the fluid is also found to be reasonably simulated. It is worth noting that only constant temperature or constant heat flux boundary condition can be specified for the STL walls. If heat conduction inside the wall needs to be considered, the traditional approach of using stationary ghost particles needs to be applied.

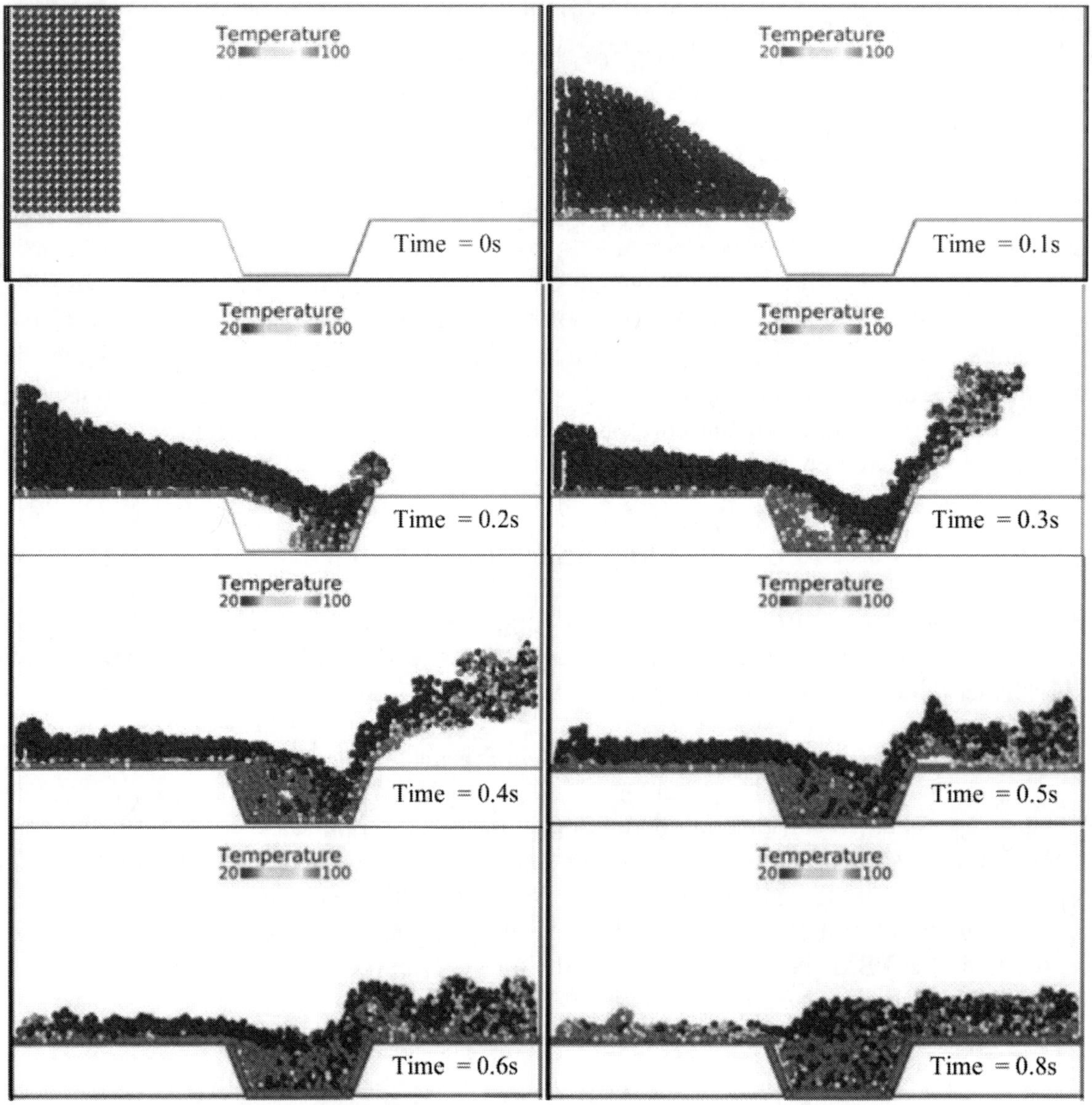

Figure 3: Spreading and heating of fluid over a STL surface.

3.2 Simulation of fluid evaporation during spreading

The SPH has been tested for its capability to correctly simulate the evaporation of fluid during spreading and the dynamics of the vapor phase. The fluid spreading is initiated by the collapse of a fluid column, which had the dimensions of 0.5m width by 1m height. The fluid and vapor densities are set at 450 kg/m^3 and 2 kg/m^3, respectively. The temperature difference between the spreading surface and the fluid is set at 900°C to facilitate faster heat transfer and evaporation of fluid.

The simulation results are presented in Figure 4, which shows the creation of an unstable vapor layer along the hot spreading surface. In comparison to previous simulations with similar settings and conditions [1], predictions of the wall evaporation dynamics and the rising of vapor phase across the fluid layer are more accurately predicted in Figure 4. The progressive thinning of the fluid layer is also seen.

To simulate the atmospheric transport and possible combustion of the vapor phase, the open-sourced CFD software OpenFOAM will be used. A coupling of OpenFOAM and this SPH spreading model is currently under development.

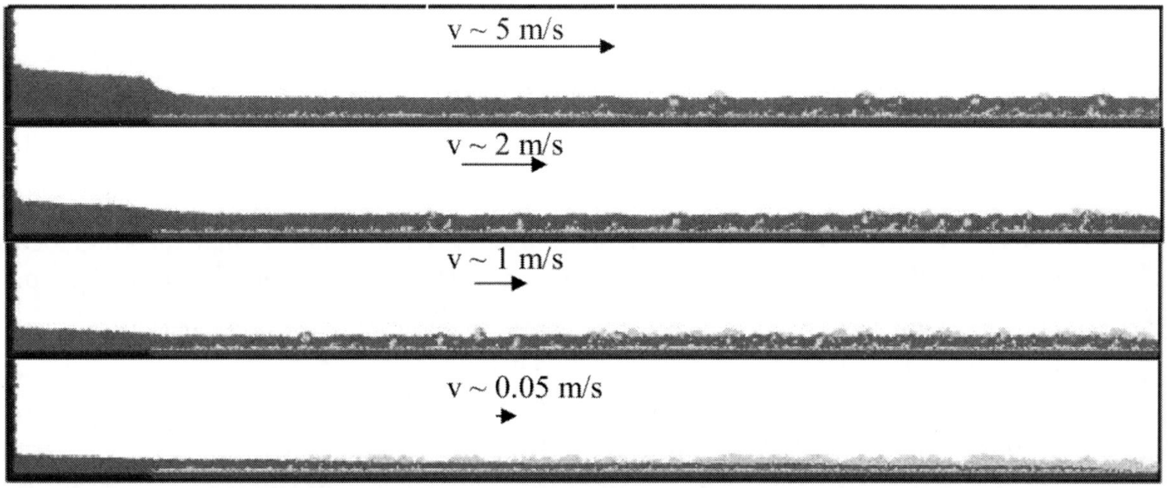

Figure 4: Fluid spreading, evaporation and dryout.

4 Conclusion

A SPH model of LNG spreading and evaporation [1] has been further developed in this work to include the capabilities of

- Modeling the mechanical and thermal interactions of the spreading fluid with terrain and objects of arbitrarily complex 3D geometries represented by STL surface meshes,
- Taking into consideration the effect of interactions between phases, i.e. surface tension and cohesion forces.

The wall evaporation model has also been improved to account for the effect of subcooled boiling.

Limited testing of the model has been conducted, but further model verification and validation are needed before it can be applied to the simulation of realistic industrial scenarios of LNG spreading.

Using CFDEM®coupling simulation engine [9], the proposed LNG spreading model can easily be coupled with the OpenFOAM software to enable 3D CFD modeling of LNG vapor dispersion and fire.

5 References

[1] A.V. Bui, T. Liu, M. Reed and E. Peterson. "CFD Modeling of LNG Spreading and Vaporization," Proceedings of the 11thAIChEGlobal Congress on Process Safety, Austin, Texas, April 2015.

[2] LAMMPS Molecular Dynamics Simulator, Available at http://lammps.sandia.gov, Assessed on 02/15/2016.

[3] G.C. Ganzenmuller, M.O. Steinhauser, and P.V. Liedekerke, The implementation of SPH in LAMMPS – A guide to the SPH-USER package, Germany, 2011.

[4] A. Eitzlmayr, G. Koscher, and J. Khinast, "A novel method for modeling of complex wall geometries in smoothed particle hydrodynamics," Comput. Phys. Comm., 185, pp.2436-2448, 2014.

[5] M. Ihmsen, J. Bader, G. Akinci, and M. Teschner, "Animation of air bubbles with SPH," Proceedings of International Conference on Computer Graphics Theory and Application, 2011.

[6] N. Shirakawa, Y. Yamamoto, H. Horie, and S. Tsunoyama, "Analysis of subcooled boiling with the two-fluid particle interaction method," J NuclSci Tech, 40(3), pp.125-135, 2003.

[7] A.K. Das and P.K. Das, "Modeling of liquid-vapor phase change using smoothed particle hydrodynamics," J Comput Phys, 303, pp.125-145, 2015.

[8] J.L. Woodward and R.M. Pitblado, "LNG Risk Based Safety – Modeling and Consequence Analysis," John Wiley & Sons, Inc., Hoboken, New Jersey, 2010.

[9] CFDEM®project, Available at http://www.cfdem.com, Assessed on 02/15/2016.

12TH GLOBAL CONGRESS ON PROCESS SAFETY

50th LOSS PREVENTION SYMPOSIUM

LNG Facility Siting Regulations – Home (USA) and Away (International)

Chaitanya Gottimukkala
RISKbytes Inc
1318 Nance St Houston TX 77002
Chaitanya@riskbytes.com

Prepared for Presentation at
American Institute of Chemical Engineers
2016 Spring Meeting
12th Global Congress on Process Safety
Houston, Texas
April 11-23, 2016

LNG Facility Siting Regulations –

Home (USA) and Away (International)

Chaitanya Gottimukkala
RISKbytes Inc
1318 Nance St Houston TX-77002
Chaitanya@riskbytes.com

Keywords: LNG, Facility Siting, Regulations, USA, International, Risk-based, Prescriptive

Abstract

Several LNG facilities are being proposed, designed and built around the world. These facilities will have to be designed, built and operated according to their local and national codes and standards. This paper provides an overview of the various codes, regulations and standards that govern the siting of the onshore LNG Facilities in the U.S.A and around the world. This paper also highlights some of the important siting requirements including risk criteria, vapor dispersion and overpressure criteria, acceptable thermal radiation levels found in these regulations and standards.

The primary siting regulations for U.S. onshore LNG facilities are 49 Code of Federal Regulations (CFR), Part 193 and NFPA 59A, "Standard for the Production, Storage, and Handling of Liquefied Natural Gas (LNG)". The design and siting standard for Europe and some facilities internationally is the European Standard, EN 1473, "Installation and equipment for liquefied natural gas - Design of onshore installations".

1 Introduction

With the demand for cleaner fuels rapidly growing around the world, Liquefied Natural Gas (LNG) is playing a vital role in this market. In order to meet the growing demand, LNG has to be transported from places of availability to places of demand. Therefore, a number of LNG Liquefaction and Regasification facilities are in proposal, design or construction around the world.

These LNG facilities must follow the industry codes, local and national regulations, and standards applicable to their location. The next section of this paper provides an overview of these required codes, regulations and standards for siting of LNG facilities both in the USA (home) and internationally (away). This paper's major focus is on land-based large scale LNG facilities; however, some of the standards for offshore based LNG facilities are also listed.

The major difference between the U.S. and international (for example - Europe & Australia) LNG siting requirements is that the United States (U.S) generally follows a "Prescriptive" approach and majority of the rest of the world follows a "Performance or Risk-based" approach. These approaches are consistent with the general approaches the onshore industries follow in these countries.

Prescriptive approaches in the U.S must follow a defined set of spacing criteria as outlined in NFPA 59A [3] and 49 CFR Part 193 [4]. In the risk-based approach, the proposed facility's spacing needs to demonstrate that it meets the tolerable risk criteria established by the host Nation, company or as outlined in codes such as EN 1473 [5], whichever is most stringent.

The subsequent sections of this paper highlight some of the important requirements with respect to the facility siting process including determination of design spills, risk criteria, allowable vapor dispersion and overpressure criteria, allowable thermal radiation levels.

2 Regulations, Codes, and Standards

The regulations, codes and standards applicable to facility siting in the US and internationally are summarized in the subsections below.

Local regulations, codes, and standards exist for countries such as Australia, United Kingdom, Canada, China, Europe, India, Japan and Mexico. Both the NFPA 59A (Prescriptive) and EN 1473 (Risk-based) have been adopted globally either individually or as a combination of both. Table 1 below summarizes the approaches followed by some of the countries.

Table 1. Summary of Some of the Countries with LNG Facilities and their Approaches

Country	Approach
U.S.A - NFPA 59A 2001 edition	Prescriptive
Canada - CSA Z276	Prescriptive
Japan - JGA-102	Prescriptive
Mexico - NOM-013-SECRE-2004	Combination of Prescriptive and Risk-based
India - OISD-STD-194	Combination of Prescriptive and Risk-based
Europe (U.K, Netherlands, France) - EN 1473	Risk-based
Australia - AS3961	Risk-based
Singapore	Risk-based

For example, the Canadian standard, CSA Z276, adopted NFPA 59A's prescriptive approach and the Australian standard, AS3961, adopted a risk-based approach similar to EN1473.

The primary focus of this paper is to compare the Prescriptive approach in the U.S (home) versus the Risk-based approach followed in Europe and other parts of the world (away). For purposes of this paper, when "U.S" or "Home" is mentioned it refers to the countries following Prescriptive approach and when "International", "EN1473" or "Away" is mentioned it refers to countries following Risk-based approach.

Offshore LNG facility siting process is outside the scope of this paper. However, some of the industry standards applicable to the offshore LNG facilities are listed below:

- DNV-OSS-103, Rules for Classification of Floating LNG/LPG Production, Storage and Loading Units
- ABS Guide for Building and Classing Floating Offshore Liquefied Gas Terminals

2.1 U.S (Home) Regulations, Codes and Standards – Onshore LNG Facility Siting

The following are the facility siting regulations and standards applicable to the U.S land based LNG facilities.

- 49 CFR Part 193, Liquefied Natural Gas Facilities: Federal Safety Standards
- 33 CFR 127 Subpart B, U.S Coast Guard - LNG Facility Inspection
- PHMSA, U.S Department of Transportation Pipeline and Hazardous Materials Administration
- FERC, Federal Energy Regulatory Commission
- NFPA 59A (2001/2009) – Standard for the Production, Storage, and Handling of Liquefied Natural Gas (LNG)
- API RP 752, Management of Hazards Associated with Location of Process Plant Permanent Buildings
- API RP 753, Management of Hazards Associated with Location of Process Plant Portable Buildings
- API RP 500, Recommended Practice for Classification of Locations for Electrical Installations at Petroleum Facilities
- API RP 505, Recommended Practice for Classification of Locations for Electrical Installations at Petroleum Facilities Classified as Class I, Zone 0, Zone 1, and Zone 2
- API STD 2510, Design and Construction of LPG Installations

2.2 International (Away - European Norm) Regulations, Codes and Standards – Onshore LNG Facility Siting

The following are the facility siting regulations and standards applicable to the land based LNG facilities following Risk-based approach, specifically, countries adopting European Standard - EN 1473:

- EN 1473 (2007), Installation and equipment for liquefied natural gas - Design of onshore installations

- EN 1160, Installations and equipment for liquefied natural gas - General characteristics of liquefied natural gas
- GAP 2.5.2, Oil and chemical plant layout and spacing
- GAP 2.5.2 A : Hazard Classification of Process operations for spacing requirements
- IP15, Area Classification code for installation handling flammable fluids Part 15, 2d edition (2002), by the Institute of Petroleum
- Site selection and design for LNG Ports and jetties: SIGTTO

The Host Nations have specific standards and regulations for their LNG facilities in addition to the above standards, which have not been included in this paper.

3 Facility Siting Process

U.S (Home) LNG facilities follow a prescriptive approach to siting as described in US 49CFR193 and NFPA 59A (2001) requiring strict compliance to the prescribed thermal radiation levels, vapor dispersion levels and overpressure levels at the facility's property line that can be built upon. The primary focus of the NFPA 59A and US 49CFR193 regulations is offsite risk to public. Other standards and regulations such as API, Environmental Protection Agency (EPA), Occupational Safety and Health Administration (OSHA) provide guidance for onsite risk criteria.

Some of the International (Away) LNG facilities follow a risk-based approach as described in EN 1473. Three categories of are risk listed in EN 1473 [5]:

- Level 3: Situation which is undesirable and cannot be tolerated. Remedial action required (Not Acceptable);
- Level 2: Situation which shall be improved. A level at which it shall be demonstrated that the risk is made As Low As Reasonably Practical (ALARP);
- Level 1: Normal situation (Acceptable).

These facilities should meet the risk acceptance criteria i.e risks in the "Acceptable" and "As Low As Reasonably Practicable (ALARP)" region for both public and on-site personnel either set by the Host Nation, Company or the EN 1473 standard, whichever is more stringent. If the quantified risk estimate lies in the Acceptable and ALARP regions, then the risk is "tolerable". The acceptability criteria are more stringent for hazards to the public.

The first step for facility siting in both the Home and Away approaches is hazard analysis to determine the maximum design spillage. The criteria for calculation of design spills are different in both approaches. Figure 1 provides a flow chart for facility siting process for both the Prescriptive (US) and Risk-based (International) approaches.

Home (US) Regulations and codes require the determination of design spill, referred to as Single Accidental Leakage Source (SALS) in NFPA 59A based on a maximum 10 minute release. Recent Nominal Failure Rate data published by FERC [2] and PHMSA [1] should be utilized to determine the failure rate criteria of equipment, piping and other appurtenances. Approved modeling tools should be used for evaluating the prescribed criteria for vapor dispersion, thermal radiation and overpressure at the property line that can be built upon. In case the prescribed siting criteria is not satisfied, passive mitigation measures can be specified. Some of the passive mitigation measures are rerouting of piping and relocation of hazardous equipment / tanks; modifying design conditions such as changing flow rates, pipe sizes, and capacities; constructing vapor barriers; or buying more land to extend the property line.

Away (EN 1473) requires the determination of the design spill either following a deterministic approach and/or probabilistic approach as described in Section 4.4.2 of the EN 1473 standard and Figure 1 below. Deterministic approach requires identification of hazards, establishment of credible hazards, determination of consequences, and mitigation measures to reduce consequences if risk is not tolerable. Probabilistic approach requires identification of hazards, determination of consequences, application of failure rate data, determination of frequency of hazards, and quantifying risk level from these data. If risk level is not tolerable, it can be reduced by active or passive mitigation measures including water deluge, fire and gas detection, vapor barriers and thermal shelters. The facility layout can be optimized and finalized following careful evaluation of various "tolerable" options.

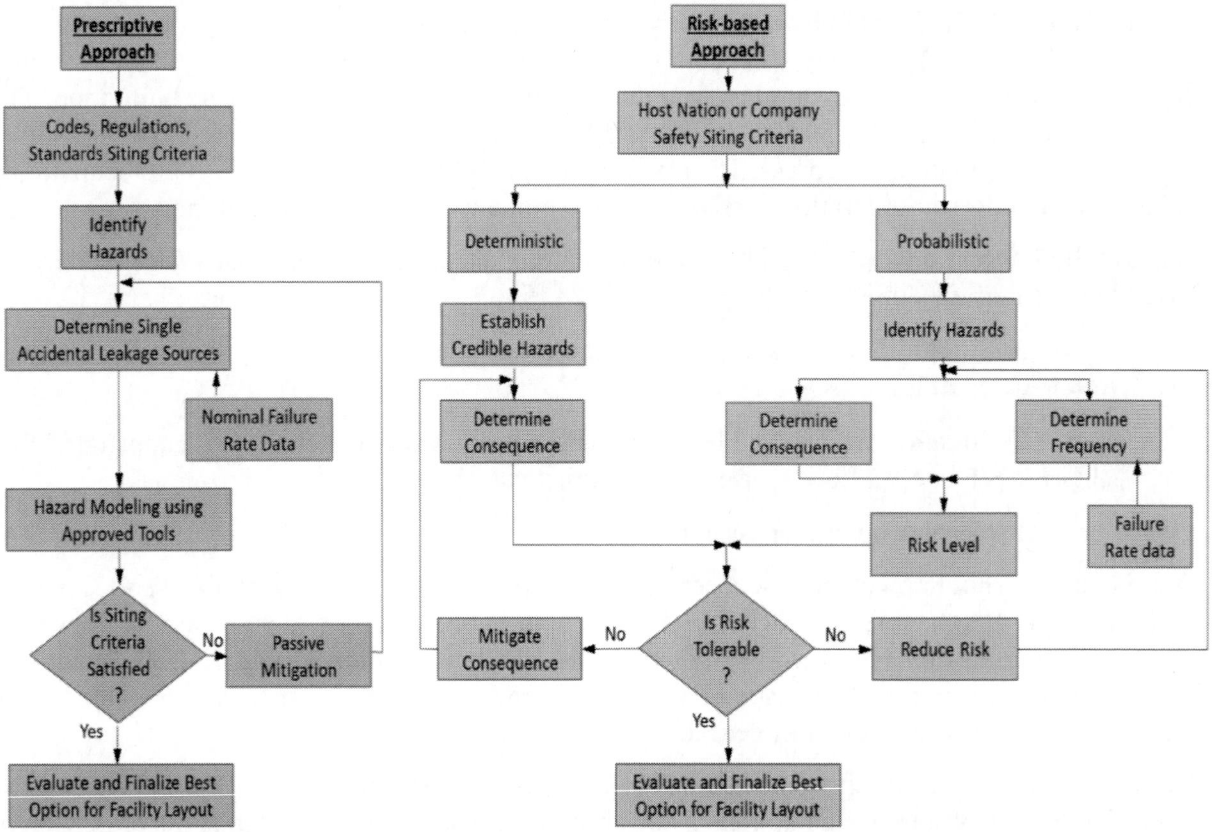

Figure 1. Facility Siting Process – Prescriptive (U.S) vs. Risk-based (International – EN1473) Approaches

The following subsections describe and compare the ambient conditions, required thermal radiation levels, vapor dispersion and overpressure criteria for both the Home (U.S) and Away (European) facilities.

3.1 Ambient Conditions

Table 2 describes the ambient conditions used for both the Prescriptive (U.S) and Risk-based (International) LNG facilities.

Table 2. Ambient Conditions for Home (49CFR193 / NFPA 59A, 2001) and Away (EN 1473, 2007)

Parameter	Home (49CFR193 / NFPA 59A, 2001)		Away (EN 1473, 2007)	
	Thermal Radiation	Vapor Dispersion	Thermal Radiation	Vapor Dispersion
Wind Speed	Wind speed producing the maximum exclusion distances shall be used. [1]	Wind speed producing longer predicted downwind dispersion distances than other weather conditions at the site at least 90 percent of the time. Alternatively, 4.5 miles per hour (2.01 meters/sec)	Combination of wind speed and atmospheric stability that can occur simultaneously and result in the longest predictable downwind dispersion distance that is exceeded less than 10 % of the time. Alternatively, 10 meters/sec	Combination of wind speed and atmospheric stability that can occur simultaneously and result in the longest predictable downwind dispersion distance that is exceeded less than 10 % of the time. Alternatively, 2 meters/sec
Temperature	Ambient temperature that produces the maximum exclusion distances shall be used.[2]	Average ambient temperature in the region	Ambient temperature in the region	Average ambient temperature in the region
Relative Humidity	The ambient relative humidity that produces the maximum exclusion distances shall be used.[2]	50%	50%	50%
Pasquill Stability Class	N/A	F	N/A	Combination of wind speed and atmospheric

Parameter	Home (49CFR193 / NFPA 59A, 2001)		Away (EN 1473, 2007)	
	Thermal Radiation	**Vapor Dispersion**	**Thermal Radiation**	**Vapor Dispersion**
				stability that can occur simultaneously and result in the longest predictable downwind dispersion distance that is exceeded less than 10 % of the time. Alternatively, Stability Class "F" can be used.
Surface roughness	N/A	0.03 m, higher values may be used based on the nature of terrain	N/A	Not specified. Industry standard values can be used.
Average gas concentration in air	N/A	½ of Lower Flammability Level (LFL) or 2.5 %	N/A	Not specified. Industry standard values can be used.

1. Per 49 CFR 193.2057, 'The wind speed producing the maximum exclusion distances shall be used except for wind speeds that occur less than 5 percent of the time based on recorded data for the area."
2. Per 49 CFR 193.2059, "in calculating exclusion distances, the ambient temperature and relative humidity that produce the maximum exclusion distances shall be used except for values that occur less than five percent of the time based on recorded data for the area."

3.2 Allowable Thermal Radiation Levels

The allowable thermal radiation levels are described in both the standards, U.S and International.

3.2.1 U.S (Prescriptive) Thermal Radiation Criteria

U.S 49CFR193 and NFPA 59A specify the thermal radiation flux levels and the exclusion zones for impounding areas containing LNG spills as listed in Table 3. U.S standards do not account for mitigation measures such as water deluge, shelters in reducing the heat flux level to the required limit. This often results in conservative estimates for thermal radiation hazards leading to larger exclusion zones. LNG spills from loading/unloading arms and LNG carriers should also be included when calculating the incident thermal radiation flux. Details on thermal radiation and impounding area design and siting requirements can be found in Section 2.2 of NFPA 59A and 49CFR193.2057 Subpart B. Thermal Radiation calculations should be calculated using the approved computer model LNGFIRE3.

Table 3. Allowable Thermal Radiation Flux Levels per Section 2.2.3 NFPA 59A

Exposure at Exclusion Zone	Thermal Radiation Flux	
	BTU/hr/ft^2	KW/m^2
At a property line that can be built upon upon ignition of a design spill	1600	5
At the nearest point located outside the owner's property line that, at the time of plant siting, used for outdoor assembly by groups of 50 or more persons for a fire over an impounding area containing a volume, V, of LNG	1600	5
At the nearest point of the building or structure outside the owner's property line that is in existence at the time of plant siting and used for occupancies	3000	9
At a property line that can be built upon for a fire over an impounding area containing a volume, V, of LNG	10,000	30

3.2.2 International (Risk-based: EN 1473) Thermal Radiation Criteria

EN 1473 specifies the thermal radiation flux levels excluding solar radiation for equipment inside the boundary and for areas outside the plant boundary as listed in Table 4. There is a provision for reducing the heat flux to the required limit by means of spacing, water sprays, fireproofing and radiation barriers. The main advantage with this approach over prescriptive standards is that overly conservative thermal exclusion zones will not be required after application of mitigation measures to reduce the risk to be in the "Tolerable" risk region.

LNG spills from loading/unloading trucks or LNG carriers should also be included when calculating the incident thermal radiation flux. It also specifies the allowable thermal radiation flux from flares or ignited vent stacks on equipment inside the plant boundary and on areas outside the plant boundary. This standard does not specify a required computer model to estimate thermal radiation and specifies that the radiation flux from an LNG fire shall be calculated using appropriate and validated models.

Table 4. Allowable Thermal Radiation Flux Levels per Table A.1 & A.2 of EN 1473

Exposure at Equipment Inside Boundary or Area Outside Boundary	Maximum Thermal Radiation Flux KW/m^2
Equipment Inside Boundary	
Concrete outer surface of adjacent storage tanks [1]	32
Metal outer surface of adjacent storage tanks	15
The outer surfaces of adjacent pressure storage vessels and process facilities	15
Control rooms, Maintenance workshops, laboratories, warehouses	8
Administrative buildings	5

Exposure at Equipment Inside Boundary or Area Outside Boundary	Maximum Thermal Radiation Flux KW/m^2
Area Outside Boundary	
Remote area [2]	8
Critical area [3]	1.5
Other areas [4]	5

1. For pre-stressed concrete tanks, maximum radiation fluxes may be determined by the requirements given in A.1.1 of EN 1473.
2. An area only infrequently occupied by small numbers of persons, e.g. moor land, farmland and desert.
3. This is either an unshielded area of critical importance where people without protective clothing can be required at all times including during emergencies or urban area (defined as an area with more than 20 persons per square kilometre) or a place difficult or dangerous to evacuate at short notice (e.g. hospital, retirement house, sports stadium, school, outdoor theatre).
4. Other areas typically include industrial areas not under control of the operator/occupier of the LNG facilities.

3.3 Vapor Dispersion Criteria

The vapor dispersion criteria have been specified in both the standards, U.S and International.

3.3.1 U.S (Prescriptive) Vapor Dispersion Criteria

In the U.S, the regulation 49CFR193, and code NFPA59A, prescribe the criteria. Clarifications and additional details for these criteria can be found on the PHMSA website [1]. These regulations require assessment of vapor dispersion hazards resulting from design spills of LNG and other flammable substances. Flammable vapor dispersion distances should be calculated at ½ Lower Flammability Limit (LFL) concentrations for both LNG and other flammable fluids.

FERC and PHMSA provide guidance on determining the single accidental leakage source to assess the flammable vapor dispersion hazards. Process and ambient conditions, among other information is used to determine the design spills. The potential release source is determined using nominal failure rate criteria recently published by FERC and PHMSA. However, these failure rates do not account for the consequence part of risk in terms of injury or fatality onsite and/or offsite.

The approved computer models for vapor dispersion exclusion zones calculations are DEGADIS 2.3, FEM3A, FLACS 9.1 Release 2 (Oct. 7, 2011) and PHAST-UDM Version 6.6 and 6. 7 (Oct. 7, 2011).

In addition to the flammable vapor hazard analysis, U.S regulations also require the toxic gas (for example - benzene) hazard analysis, as applicable.

3.3.2 International (Risk-based: EN 1473) Vapor Dispersion Criteria

EN1473 specifies a risk-based approach for vapor hazard analysis. The determination of consequence, probability and associated risk level of a hazardous event is specified in the standard. If the local, country or company criteria are more stringent, then that criteria should be utilized. The criteria for estimation of consequences of LNG and other hazardous fluids is either detailed in EN 1473 or a reference is cited. More information is available in Section 4.4 and Annexures J, K and L of EN 1473.

Risk to personnel and public should be calculated and it should be within the "Tolerable" range of the risk criteria.

3.4 Overpressure Criteria

3.4.1 U.S (Prescriptive) Overpressure Criteria

There is no prescriptive requirement for overpressure and explosion in U.S 49CFR193 but Chapter 2.1.1(d) of NFPA59A requires that all hazards having an impact on public or plant personnel should be evaluated. FERC and PHMSA require the use of the 1.0 psi to determine the offsite impact, and are consistent with the EPA's "Risk Management Program for Chemical Accidental Release Prevention" (RMP, 40 CFR 68).

FLACS 9.1 Release 2 (Oct. 7, 2011) and PHAST-UDM Version 6.6 and 6. 7 (Oct. 7, 2011) are accepted models for use in determination of the overpressure hazards.

3.4.2 International (Risk-based: EN 1473) Overpressure Criteria

EN 1473 describes the methodology for overpressure analysis in Section 4.4.2.5 d. It also specifies a risk-based approach for overpressure analysis. The determination of consequence, probability and associated risk level of a hazardous event is specified in the standard. If the local, country or company criteria are more stringent, then that criteria should be utilized.

EN 1473 refers to EN 1160 for flammability range of mixtures of gas and air. The standard indicates that some of the recognized methods and models such as the multi energy method and/or deflagration at constant speed method can be used to calculate the over-pressure.

Risk to personnel and public should be calculated and it should be within the "Tolerable" range of the risk criteria.

3.5 Other Siting Criteria

3.5.1 U.S (Prescriptive) – Other Siting Criteria

U.S Federal regulations prescribe additional siting criteria for impoundments, equipment, buildings and structures. Spacing criteria within the facility as well as with respect to the property line is described in detail. Details of these prescriptive requirements are in U.S 49 CFR 193 and NFPA 59A.

Consequences from Boiling Liquid Expanding Vapor Explosions (BLEVE) should be considered during siting and design.

Permanent and portable buildings siting shall be in accordance with API 752 and 753.

Escape routes, SIMOPS and construction activities should also be considered during siting.

3.5.2 International (Risk-based: EN 1473) – Other Siting Criteria

EN 1473 specifies a risk-based approach for facility siting based on hazard assessment. Separation distances should include radiation flux levels, LFL contours, noise, blast effects. It also describes the siting of buildings, equipment and tanks. It refers to industry standards such as GAP 2.5.2 and GAP 2.5.2.a for additional guidance on facility siting.

Escape routes, SIMOPS and construction activities should also be considered during facility siting.

4 Prescriptive (US) vs Risk-based (International) Approaches – Pros and Cons

Based on the defined criteria, the siting of facilities using Prescriptive and Risk-based approaches have their share of pros and cons. Table 4 below describes a few pros and cons for both these approaches.

Prescriptive approach warrants the LNG facilities to follow the facility siting requirements without specific consideration for their location with respect to population density in that area, while the Risk-based approach takes into consideration the overall risk to public outside the fence based on the quantitative risk estimates included during modeling. The prescriptive approach does not take into account active mitigation measures for risk reduction while in the risk-based approach risk levels may be reduced using appropriate mitigation measures.

For example, U.S 49 CFR 193 and NFPA 59A require that the LNG facilities should adhere to the thermal radiation flux of 1600 BTU/hr-ft^2 and vapor dispersion exclusion criteria of ½ LFL at the property line that can be built upon. These criteria have to be followed without consideration of the population density in that area which may result in the LNG facility needing to own more acreage than required to satisfy the facility siting requirements. International standard, EN1473, the facility siting requirements are satisfied as long as the thermal radiation, vapor dispersion and other hazards associated with the LNG facility comply with the Company or Country required "Tolerable" risk levels.

Table 5. Prescriptive (Home) vs Risk-based (Away) Approaches – Pros & Cons

Prescriptive (Home - US)	Risk-based (Away - International)
Pros	
Application of Defined criteria is easy to follow	Flexibility in facility siting and layout
Defined criteria provide consistency across LNG industry based on industry knowledge and best practices.	Takes Location's population density into account
Less time consuming during design phases	Active and passive mitigation measure accepted to demonstrate tolerability
Cons	
Strict compliance irrespective of Location's population density	Definition of Tolerability Criteria
Limited mitigation measures (passive)	Requires complex calculations and assumptions
Special cases require longer processing time	Longer Project duration for finalizing Facility layout to demonstrate tolerable risk.

5 Conclusion

The growing demand for LNG requires proposal, design and construction of LNG facilities around the world, U.S (home) and International (away). This paper highlighted some of the regulations, codes and standards distinguishing between U.S (Prescriptive approach) and International (Risk-based approach: EN 1473) requirements.

U.S facilities follow a "Prescriptive" approach (U.S 49CFR Part 193 and NFPA 59A, 2001) and some of the International facilities follow a "Risk-based" approach (EN 1473). This paper highlighted the various requirements between these two approaches with respect to facility siting.

Criteria for ambient conditions, allowable thermal radiation levels, vapor dispersion, overpressure and additional spacing requirements for both approaches has been described.

The pros and cons of the Prescriptive and Risk-based approaches have been discussed.

In the U.S, the newer editions of NFPA 59A such as the 2013 edition include QRA (Quantitative Risk Assessment) methodology for facility siting. However, US 49 CFR 193 and FERC still refer to the 2001 edition and have not adopted the latest edition. Future adoption of the risk-based approach into the U.S regulatory ecosystem such as the U.S 49 CFR 193 and FERC will have its own challenges from public, judicial and regulatory systems.

Acknowledgment:

The author would like to thank his supervisor, Robert McDonnell, for his valuable guidance and review of the manuscript.

6 References

[1] U.S Department Of Transportation Pipeline Transportation Pipeline and Hazardous Materials Administration, Available at http://primis.phmsa.dot.gov/ptr.htm, Accessed on March 15, 2016

[2] Federal Energy Regulatory Commission, Available at http://www.ferc.gov/, Accessed on March 15, 2016

[3] National Fire Protection Agency 59A, "Standard for the Production, Storage, and Handling of Liquefied Natural Gas (LNG)", 2001

[4] Code of Federal Regulations, 49 CFR Part 193, Subpart B, Available at http://www.ecfr.gov/, Accessed on March 15, 2016

[5] European Standard, EN 1473, "Installation and Equipment for Liquefied Natural Gas - Design of Onshore Installations", 2007

[6] Australian Standard, AS 3961, "The Storage and Handling of Liquefied Natural Gas", 2005

[7] Canadian Standard, CSA Z276, "Liquefied Natural Gas (LNG) - Production, Storage, and Handling", 2011

[8] International Association of Oil and Gas Producers, OGP Draft 116901, "Guidance on Performing Risk Assessment in the Design of Onshore LNG Installations including the Ship/Shore Interface", 2013

[9] TNO, "Guidelines for Quantitative Risk Assessment, 'Purple Book,'" CPR 18E, Publication Series on Dangerous Substances (PGS 3), 2005

[10] T.A. Williams and N. Assing, " Risk-Based LNG Facility Siting and Safety Analysis in the U.S.: Recent Developments", LNG 17 Proceedings, Available at http://www.gastechnology.org/, Accessed on March 15, 2016

[11] J.L. Woodward and R.M. Pitblado, "LNG Risk based Safety Modeling and Consequence Analysis", John Wiley & Sons, Hoboken, New Jersey, 2010.

Flare Reduction Strategies for LNG Plant Design

Meghana Cherukuri
Process Engineering Professional
Houston, TX - 77054
Meghana.Cherukuri@yahoo.com

Sumit Sharma
Senior Technical Professional – Process
KBR
601 Jefferson St., Houston, TX - 77002
Sumit.Sharma@kbr.com

Kristina Martinez
Principal Technical Professional – Process
KBR
601 Jefferson St., Houston, TX - 77002
Kristina.Martinez@kbr.com

Prepared for Presentation at
American Institute of Chemical Engineers
2016 Spring Meeting
12th Global Congress on Process Safety
Houston, Texas
April 11-23, 2016

Flare Reduction Strategies for LNG Plant Design

Meghana Cherukuri
Process Engineering Professional
Houston, TX - 77054
Meghana.Cherukuri@yahoo.com

Sumit Sharma
Senior Technical Professional – Process
KBR, 601 Jefferson St., Houston, TX - 77002
Sumit.Sharma@kbr.com

Kristina Martinez
Principal Technical Professional – Process
KBR, 601 Jefferson St., Houston, TX - 77002
Kristina.Martinez@kbr.com

Keywords: Flare system, flare optimization, flare minimization, LNG, Plant design.

Abstract

In liquefied natural gas (LNG) plants, the flare system provides a reliable and safe method for vapor and liquid hydrocarbon disposal. The flare system enables controlled handling of off-spec streams during plant upsets, emergencies, and short term operations such as start-up, shutdown, or equipment maintenance activities. Although the flare system is essential for the safe operation of the plant, proper design and optimization is necessary to control capital cost.

A typical LNG flare system includes Warm (or Wet), Cold (or Dry), and Low Pressure (or LP/Marine) flares. The disposal of streams with potentially freezable components originating from plant inlet and gas treating sections is achieved through the Warm flare. The Cold flare is utilized to dispose of streams with non-freezable components and very low operating temperatures. Low pressure streams from the storage or loading area of the LNG plant are routed through the LP flare for safe disposal. Based on project design criteria, the flare system can require an elevated, ground, enclosed or smokeless flare design.

Addressing flare optimization issues provides opportunities for controlling material loss during plant operation. In this paper, an overview of factors that influence flare type selection and determine flare size are discussed. Significance of flare type and size for overall plot space requirements is also addressed. Based on typical major relieving scenarios, flare load reduction strategies for each of the LNG plant flares are presented.

LNG plant design configurations for multiple trains, refrigerant compressors, and process recycle streams can significantly affect design flare loads. High design flare loads increase flare size and require large plot space due to flare radiation limits. Proper selection of design options for major LNG plant sections, guided by flare reduction strategies, can help in reducing flare design flow rate, flare size, and plot space requirements.

1. Introduction

The flare system in Liquefied Natural Gas (LNG) plants provides a reliable and safe disposal method for hydrocarbon vapor and liquid streams. These streams may result from a deviation from normal plant operation or from intermittent operations such as start-up, shutdown, venting, draining, or purging. Typical plant safety design manages any potential material loss during plant operation by routing the material streams to the flare system for disposal.

Although the flare system is essential for the safe operation of the plant, proper design and optimization is necessary to control capital cost. Flaring design flow rates affect the size of the flare and determine the plot space requirements. [1] High flare loads increase flare stack and header size and require large plot space due to flare radiation limits which altogether can increase the overall project capital cost.

LNG plant design, train, and equipment configurations can significantly affect flare design loads. Addressing flare optimization at the design stage provides opportunities for controlling material loss during plant operation. Proper selection of design options for major LNG plant sections, guided by flare reduction strategies, can help in reducing flare design flow rate, size, and plot space requirements.

2. The Flare System

Disposal of gas and liquid streams in a process plant is achieved through flaring which involves controlled combustion of hydrocarbon streams. Flared hydrocarbons are routed to a remote location and are typically burned in an open flame during which stream components are ideally converted to carbon dioxide and water.

Proper design of the flare system creates an efficient mechanism so that complete combustion of hydrocarbon can be achieved. Without flaring, the alternate options of venting or draining of the process streams can create health, safety, and environmental problems. Local regulations may also make it mandatory to include a flare system for safe disposal of process streams from an operating plant.

A typical flare system (Figure 1) in a process plant consists of three primary components: the flare structure, a plant wide flare header piping system, and plot area for radiation zone. The process streams for disposal are sent to appropriate flares where the streams undergo combustion. Routing of the process streams to the flare is achieved through a plant wide piping system that also includes knockout drums for phase separation of any two phase streams. Liquid content to the flare must be avoided for safe operation.

At the flare tip, the continuous flare flame is maintained using a pilot system, and purge gas is used to prevent oxygen ingress into the flare system. In order to delineate safe work zone conditions, a sterile radiation zone is maintained around the flares.

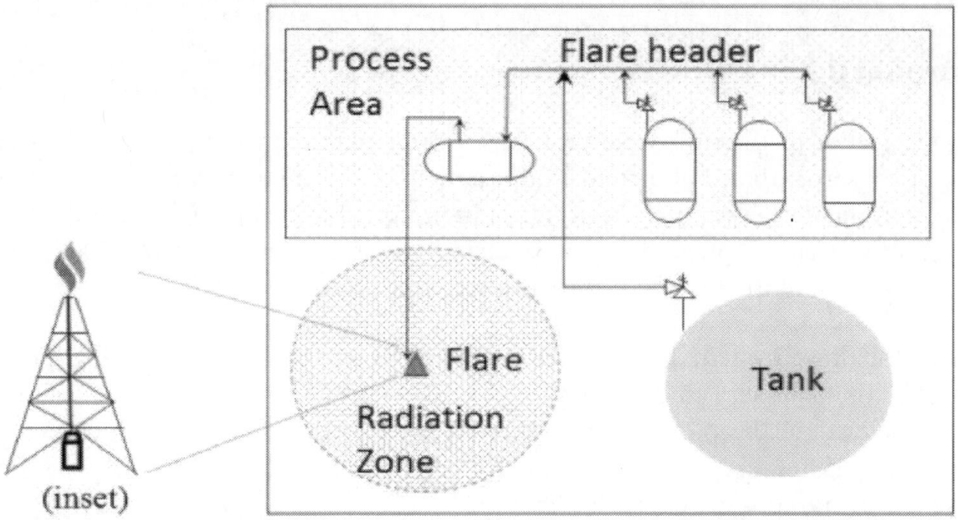

Figure 1. Simplified flare system sketch

2.1. Flare type:

The main component of the flare system is the flare structure that can be classified as elevated or ground flare based on the location of the open flame relative to the flare structure. Selection of the flare type is based on project and process requirements.

- Elevated flares: Elevated flares include a vertical stack with a flare tip that is located at the highest point of the stack. A process stream routed to the flare stack for disposal undergoes combustion at the top of the stack. An elevated combustion point helps in the dispersion of the flare products into the atmosphere.

  The support structure for the elevated flares can be self (Figure 2a), guy wire (Figure 2b), or derrick supported (Figure 2c) based on the height of the flare stack. High elevation of the flare also creates visual and noise impact from the flaring process. Efficient mixing of flaring stream with combustion air is important in achieving smokeless flare design.

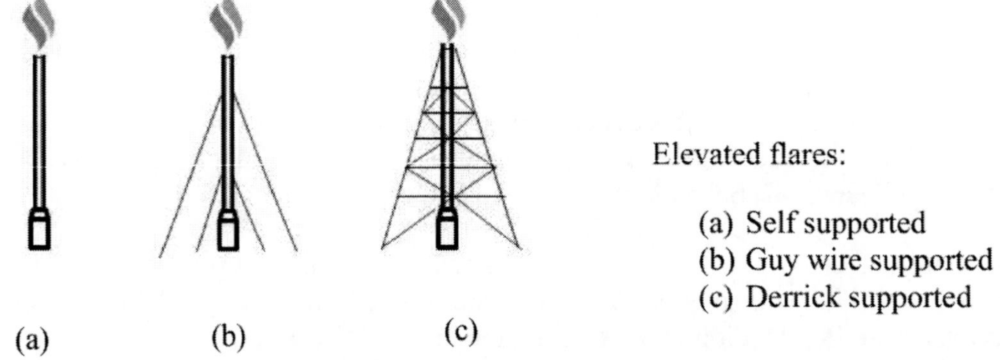

Figure 2. Elevated flares

While there are no regulatory restrictions for the stack height, Federal Aviation Administration (FAA) may require height limits for flare placements in the vicinity of an airport. Flare dispersion modeling during the design phase is essential to ensure that concentrations of toxic emissions do not exceed safe levels at any location where personnel can be present. Flare dispersion is affected by the type of flare tip. In the US, 40 CFR 60.18 (4)(iii) limits flare tip velocity to 400 ft/s with a flared gas lower heating value of at least 1000 Btu/SCF. [3]

- Ground Flares: Open multi-point ground flares (Figure 3a) or enclosed ground flares (Figure 3b) are the preferred option when project requirements prohibit visual and noise disturbances from the flaring process. Ground flares are designed smokeless but dispersion of flare combustion products is typically inefficient. Stream combustion in ground flares is achieved at grade level.

  In an enclosed ground flare the burners are located within a refractory lined physical structure that encloses the flame and contains thermal radiation. Open ground flares are not enclosed but typically are surrounded by a radiation fence which similarly blocks radiation.

  Open ground flares can be designed to handle full flare capacity similar to the elevated flares whereas enclosed ground flares are used for routine small scale flaring. Compared to the elevated flares, the ground flares are costlier and require larger plot area (open ground flares).

(a) Open ground flare (b) Enclosed ground flare

Figure 3. Ground flares [2]

2.2. Flare header piping system

Plant equipment are protected from overpressure scenarios using relief valves. The outlet flow from the relief valve is collected using a plant wide flare header system and transported to the flare for disposal. Process streams from intermittent plant operations such as start-up, shutdown, and emergency blowdown are also routed to the flare through the flare header piping to dispose of hydrocarbon inventory in a controlled manner to prevent or mitigate escalation of any unsafe event. Flare and flare header piping can be classified based on the operating conditions of the flaring stream:

Table 1. Flare and flare header classification:

• Based on flaring stream operating pressure:	
High Pressure (HP) Flare:	Sonic flare. Handles large high pressure flaring streams.
Low Pressure (LP) Flare:	Sub-sonic flare. Typically used for equipment with low design pressure specifications such as storage tanks.

• Based on flaring stream composition and operating temperature:	
Warm (Wet) Flare:	High temperature flaring streams with components that can potentially freeze at low temperatures
Cold (Dry) Flare:	Low temperature flaring streams with no freezable stream components

Separation of the flares in terms of pressure, temperature, and stream composition is necessary for safe and reliable flare operation. Depending on the operating conditions of the relieving stream, the flare header material can range from carbon steel to stainless steel.

Typically, for the Warm flare, carbon steel piping can be used to handle the flaring stream whereas the Cold flare will require stainless steel piping because of the possibility of very low temperatures that may exist in the flaring streams during relieving scenarios. High design flow flaring rate for the flare results in large flare header size and increases the cost of the system. Stainless steel metallurgy makes the Cold flare relatively more size and cost sensitive compared to the Warm flare.

2.3. Radiation zone

Flaring large quantities of hydrocarbon can result in high intensity thermal radiation and noise around the flare during plant operation. In order to maintain safe working conditions, the area around the flare is demarcated as sterile zone. The permissible levels for flare radiation are summarized in Table-3 below.

Table 3. Radiation and noise limits

• **Radiation: Permissible Radiation Level** [4]	
KW/m^2	
1.58	maximum radiant heat intensity at any location where personnel with appropriate clothing can be continuously exposed
4.73	maximum radiant heat intensity in areas where emergency actions lasting 2 min to 3 min can be required by personnel without shielding but appropriate clothing
6.31	maximum radiant heat intensity in areas where emergency actions lasting 30 seconds can be required by personnel without shielding but appropriate clothing
9.46	maximum radiant heat intensity at any location where urgent emergency action by personnel is required. When personnel enter or work in an area with the potential for radiant heat intensity greater than 6.31 KW/m^2, radiation shielding and/or special protective apparel (e.g. a fire approach suit) should be considered. Safety precaution- It is important to recognize that personnel with appropriate clothing cannot tolerate thermal radiation at 9.46 KW/m^2 for more than few seconds

• **Flare noise limits**		
85 dB (A)	max	continuous exposure
115 dB (A)	max	for short term exposure (up to 15 mins)
115 dB (A)	max	with hearing protection

Heat radiation calculations for flare typically take into consideration the wind velocity, relative humidity, and solar radiation at the flaring location to assess the heat radiation impact of flaring. The impact of flare radiation and flare front can be identified early in the project to determine locations for flare stack and plot area for radiation zone.

3. Flares in LNG Plant

In a typical LNG plant the feed gas to LNG conversion involves four primary process units that include feed gas pre-treatment, refrigeration, liquefaction, and product storage. The feed gas pre-treatment section removes water, acid gas, and mercury from the feed gas stream. The treatment process occurs at high pressure and warm temperatures. Refrigeration and liquefaction units operate at high to low pressure and temperature ranges. The LNG product storage unit operates at close to atmospheric pressure and very low temperature range.

The flare system in an LNG plant is comprised of different flares that are designed to handle streams originating from various units for safe disposal. The hydrocarbon flare streams mainly contain methane, ethane, propane and other heavy components from the natural gas streams or refrigerant streams:

- Warm (Wet) Flare: A high pressure carbon steel flare for warm and wet hydrocarbons, designed to handle streams containing heavy hydrocarbons, contaminated water streams and streams with any freezable components. These streams mainly originate from the pre-treatment units.

- Cold (Dry) Flare: A high pressure stainless steel flare for cold and dry hydrocarbon vapors. These light hydrocarbon streams originate from the refrigeration and liquefaction units of the plant.

- Low Pressure (LP) or Marine Flare: A low pressure stainless steel cold flare for emergency vapor (low pressure) releases from the LNG storage and loading system.

Apart from the Warm, Cold, and LP flares, the LNG plant may also include a Sour Gas Flare, and Thermal Oxidizer or Incinerator for flaring hydrocarbon streams containing a high percentage of acidic gases such as H_2S, CO_2, and volatile hydrocarbons (VOC) originating from pre-treatment units. Some facilities also include dedicated warm and cold liquid collection header systems to separately handle liquid disposal.

4. Flare Design Loads

The design flare loads in LNG plants are determined by identifying major relieving scenarios during process upset, intermittent or normal operations. The Warm and Cold flares are designed to handle the largest gas flows in the LNG plant while the LP or Marine flare is designed based on major flaring incidents from the tank or product storage area.

In the following sections, all discussions related to controlling cases for flare design are based on the Propane Pre-Cooled Mixed Refrigerant Process (C3MR) by Air Products and Chemicals Inc. (APCI).

4.1. Warm Flare Design Load:

The largest flow at ambient or warm temperatures (above 0 °C) in the LNG plant is the feed gas flow into the facility. The inlet pressure to the process units is typically maintained by single or multiple (parallel) pressure letdown valves. Any overpressure scenario in the feed gas line in the inlet or pre-treatment facilities can result in large feed gas flaring to the Warm flare. Another large flow that can determine the Warm flare capacity is the blowdown loads from the inlet and pre-treatment units of the LNG plant. The blowdown operation removes hydrocarbon inventory from the units and routes the hydrocarbon streams to the flare system.

4.2. Cold Flare Design Load:

The Cold flare design flow is typically based on refrigerant flows in the refrigeration and/or liquefaction units of the LNG plant. The primary refrigerants for the APCI process are Propane which is used for pre-cooling the feed gas and a Mixed Refrigerant (typically consists of nitrogen, methane, ethane, and propane) is used for liquefaction process. The Propane (PR) and Mixed Refrigerant (MR) flows are circulated using Propane and MR Compressors respectively. Process upset cases such as PR or MR compressor blocked outlet or local power failure may require routing PR or MR circuit flow to the Cold flare. Blowdown load from the cryogenic heat exchanger and emergency depressuring events are also sent to the Cold flare.

4.3. LP /Marine Flare Design Load:

Low Pressure or Marine flare handles any relieving streams from the LNG storage tanks or product loading operations. The capacity of the LP/Marine flare is typically set by the Boil-Off Gas (BOG) compressor blocked outlets case. The largest flow to the Boil-Off Gas compressor occurs during LNG product loading operations which may utilize multiple BOG compressors operating in parallel.

5. Flare System Optimization

Collectively, a significant quantity of gas is estimated to be flared annually from LNG export terminals. [5] Optimization of the flare system and minimization of flaring incidents can potentially reduce LNG plant's carbon foot print and overall plant cost. Flare design optimization and flare minimization will require consideration of design options that can minimize flaring incidents or reduce the amount of flare load during a flaring incident.

5.1. Warm Flare:

For the Warm flare, the largest flare load is typically from the feed gas pre-treatment section located upstream of the liquefaction unit. An overpressure scenario in which the feed gas forward flow is blocked in the pre-treatment section can result in routing all feed gas flow to the Warm flare through the activated relief valves.

In another scenario the failure of plant inlet control valve can significantly increase flow to the plant inlet. This control valve failure will activate relief valves downstream from the valve and route the relief valve outlet flow to the Warm flare (assuming that there is no additional heating provided and that the upstream pressure is not high enough to cause the downstream gas temperature to be too cold for the Warm flare metallurgy). The flow resulting from the control valve failure is a function of the pressure drop across the control valve. The higher the difference between the inlet and outlet pressures across the valve, the higher the potential flow to the flare subsequent to the control valve failure. High flare load will increase the flare height, the radiation zone, and associated costs.

5.1.1. <u>High Integrity Pressure Protection System (HIPPS) for the feed gas line</u>

An preventative safety system such as High Integrity Pressure Protection System (HIPPS) can be used to eliminate high Warm flare load from any blocked outlet overpressure scenario in the feed gas flow at plant inlet. For the pressure letdown control valve failure case, consideration of a HIPPS downstream of the plant inlet control valve will prevent a flaring incident by shutting down the forward flow to the plant. Implementation of a HIPPS at the plant inlet will trip any dedicated emergency shutdown valves at the plant inlet before reaching a scenario in which the operating pressure exceeds the design conditions. This scheme will significantly reduce the flow to the flare by eliminating full feed gas forward flow as a consideration for the Warm flare sizing.

HIPPS can provide a completely independent safety layer. Inclusion of redundant features such as additional sensors and instrument logic control increases the HIPPS reliability. ASME Sec VIII, Division 1, UG-140 accepts the instrumented protective device in consultation with and on the advice of the owners for certain services.

A HIPPS can provide sufficient reliability to replace a mechanical device such as a pressure relief valve. HIPPS is a particularly attractive option for revamps where limited capacity of an existing flare can prohibit any flare load increase from revamp changes. Implementation of HIPPS for such cases will reduce the occurrence of any flaring scenarios that can impact flare design loads.

HIPPS application requires frequent testing and maintenance to keep the device operative and available when needed. This can be a significant burden for the operator. The use of a HIPPS requires a hazard and risk analysis (per ISA S84.00.01/2/3) to determine the required risk reduction that must be achieved by the overpressure protection. This then sets the required safety integrity level for the instrumented protection. Any HIPPS based design should be thoroughly reviewed in HAZOP, and clearly understood by both the owner management and the plant operation personnel.

5.2. Cold Flare:

The design flare load for the Cold flare usually constitutes the highest flaring rate in LNG plants. Refrigerant compressor blocked outlet cases result in routing the refrigerant system flow to the Cold flare. The refrigerant system flow and inventory capacity affect the size and capacity of the Cold flare and the overall flare system.

5.2.1 Implementation of HIPPS for refrigerant compressor system

Similar to the HIPPS for feed gas inlet control valve, HIPPS can be considered for refrigerant compressor system. A two out of three pressure transmitter voting can be implemented on the refrigerant compressor outlet to trip the machine during a high pressure case. Using HIPPS, the overpressure scenario and high flare load from compressor blocked outlet can be eliminated for the Cold flare.

HIPPS application requires frequent testing and maintenance to keep the device operative and available when needed. This can be a significant burden for the operator. The use of a HIPPS requires a hazard and risk analysis (per ISA S84.00.01/2/3) to determine the required risk reduction that must be achieved by the overpressure protection. This then sets the required safety integrity level for the instrumented protection.. Any HIPPS based design should be thoroughly reviewed in HAZOP, and clearly understood by both the owner management and the plant operation personnel.

5.2.2 Partial condensation at propane condenser by natural convection.

Failure of the propane condenser results in overpressure in the propane refrigerant system. For air cooled LNG plant design, the resulting calculated flare load for the Cold flare sizing can be reduced by considering partial condensation at the propane condenser air coolers.

Power failure affects the propane condenser air cooler fans however partial condensation at elevated pressure will still occur due to natural convection cooling at the air coolers which are usually located high up on the top of pipe racks. Consideration of partial condensation will reduce the relieving load and the flaring requirement.

5.2.3 Recycle relief valve flow to compressor suction [6]

The relief valve downstream of the refrigerant compressor discharge sends the outlet flow to the Cold flare in case of overpressure scenarios in the compressor system. Recycling the relief valve outlet flow to suction of the compressor can significantly reduce the Cold flare load (Figure 4).

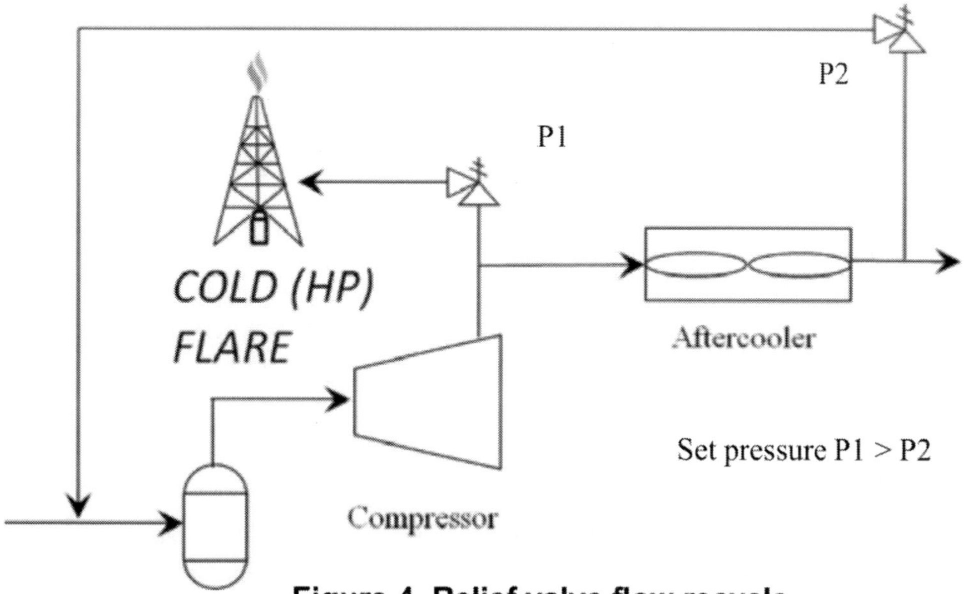

Figure 4. Relief valve flow recycle

The relief valve outlet flow recycle scheme eliminates the flare load from compressor blocked outlet case. The recycle route is parallel to the anti-surge valves that protect the compressors. This scheme can be implemented for both propane and mixed refrigerant compressors.

The proposed relief valve recycle scheme works based on differential pressure which makes it susceptible to unintended relief valve opening. Using pilot operated relief valves can potentially solve this problem. This scheme may require multiple relief valves within the compressor system. Long term operation and heat built up effects may have to be completely analyzed with inputs from compressor supplier and dynamic simulation analysis.

5.2.4 Split scheme for the flare load from the refrigerant compressors

The relief valve flow from refrigerant compressors blocked outlet or power failure cases is conventionally sent to the Cold flare. However the outlet flow from the relief valves may be routed to the warm flare if the stream temperature is at a higher temperature during the compressor system overpressure scenario. A high stream temperature can result in warmer letdown temperature downstream of the relief valves, and may not reach an unacceptable limit.

This scheme (Figure 5) will substantially reduce the size of the cold flare which is all stainless steel material and lower the cost by routing the relieving load to the warm flare which is typically based on carbon steel material.

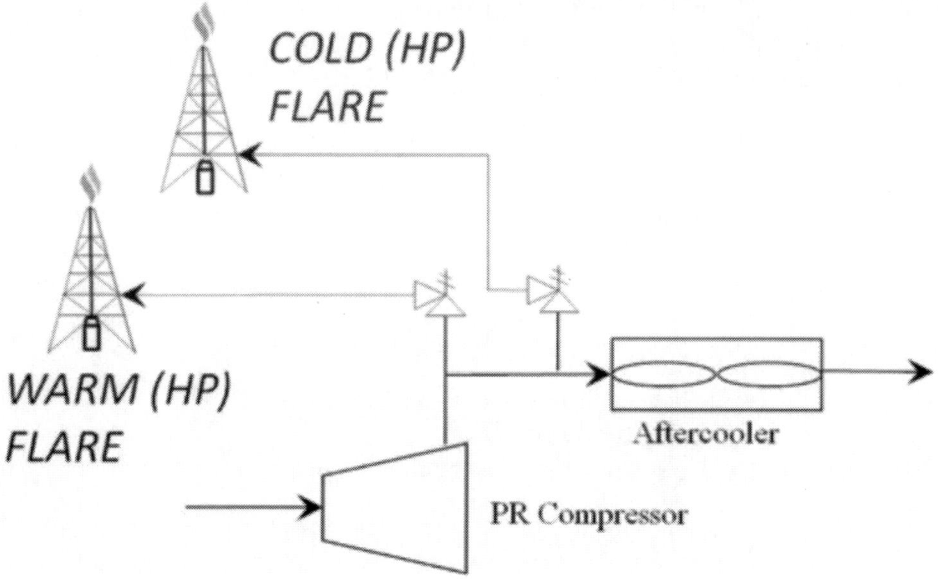

Figure 5: Flare load split scheme

If all of the flare stacks (Warm, Cold, and LP flares) are located at a single location, the heights of the flare stacks are usually the same height so that there is no flame impingement on any flare stack from an adjacent active stack. The choice to route the refrigerant flow to the Warm flare may not provide any benefit if it leads to increasing the Warm flare size beyond the Cold flare size and if other controlling scenarios for the Cold flare maintain the same initial Cold flare size.

The total radiation level from this scheme may stay the same because the total propane system load to flare remains the same based on the overpressure scenario for the compressor system.

5.2.5 Increasing the propane system design pressure

Increasing the propane system design pressure can potentially reduce the flare load to a level so that it no longer controls the Cold flare design. The propane refrigeration system is generally designed in carbon steel materials. The design pressure of the propane system can be increased to a level that is within flange limits of the piping material class. A higher set pressure for the relief valve will reduce flare flow rate from the compressor during overpressure scenarios. Higher relief valve set pressure may also reduce the frequency of overpressure scenarios for the compressor system.

This option will lead to higher design pressure for the air coolers, compressor casings, and other equipment in the propane compressor system. These design modifications can lead to the compressor surging at elevated pressure (if the anti-surge control system fails) which may damage the compressor. Many owners also view the propane compressor discharge relief valve as providing a back-up means of machinery protection in addition to its overpressure protection function.

5.2.6 Dynamic simulation for compressor overpressure scenarios

Dynamic simulation can provide realistic insight into the impact of overpressure scenario on the overall compressor system. Using information on inventory, driver power, compressor curves and other details, Dynamic simulation can predict realistic reliving rates during overpressure incidents. Based on dynamic simulations, the relief flow can be significantly lower compared to a conservative estimate of relieving all compressor refrigerant forward flow and system inventory capacity to the flare during overpressure scenarios. [7]

5.3. Flare system for multiple train design:

In multiple train design each train can have a common or independent pre-treatment, refrigeration, and liquefaction sections. For the independent train design, both Warm and Cold flare systems can be common for the overall plant with each flare sized for their single train controlling cases plus the respective start-up load for one additional train. For independent trains, lacking any common cause failure, double jeopardy considerations can eliminate multiple train failures and limit overall flare loads. Provision of common spare stack can be determined by the project philosophy and safety review of the flare design.

For a single train design with multiple parallel compressor strings (2x50%), the cold flare load may not decrease compared to a same capacity single compressor string (1x100%). In multiple parallel compressor string design, the blocked outlet for the compressor strings can occur at a location where compressor outlet flows from both the strings combine, sending the full combined 100% compressor outlet flow to the Cold flare.

6. Flaring Minimization

Various plant operations can result in flaring in an LNG plant. With increased emphasis on containing plant emissions and operating costs, design modifications can be considered to reduce flaring incidents.

Intermittent planned or unplanned operations such as start-up, shutdown, or plant trips can result in significant flaring as a percentage of feed gas [8]. A full pressure re-start/or partial depressurization for faster re-start of the liquefaction plant after a trip, based on available power, can reduce the amount of gas flared. Using aeroderivative gas turbine driven compressors for refrigeration service can also reduce flaring during start-up activities. Consideration of isolation and maintenance activities during the design phase can provide easy access and means of equipment isolation with minimum flaring.

Design options should also consider recycling or re-routing any gas stream from flare to either front-end of the plant for processing, or to the fuel gas system for consumption.

7. Conclusion

The flare system design in an LNG plant is affected by equipment and system configurations. With increased emphasis on containing plant emissions and costs, design modifications can be considered to reduce flaring incidents during normal or intermittent plant operations. An optimized flare system design will require design considerations for major LNG plant units including the feed gas inlet, pre-treatment, and refrigeration sections.

High flaring rates increase the plant cost by increasing the size of the flares, the flare headers and the required plot area for radiation zone. Flare system optimization for single or multiple train configurations reduces the flare loads by recycling, re-routing, or eliminating the flaring streams from various units of the LNG plant. Implementation of flare load optimization and minimization options during the design phase provides opportunities to limit the overall plant flaring rates, incidents, and related costs.

8. References

[1] L. B. Evans, W. M Vatavuk, D. K. Stone, S. K. Lynch, R. F. Pandullo, W. Koucky, Flares, Available at http://www3.epa.gov/ttncatc1/dir1/cs3-2ch1.pdf. Accessed on February 01, 2016.

[2] TxOGA, Flare Operation Discussion, Available at http://www.tceq.state.tx.us/ assets/public /implementation/air/rules/Flare/Resources_TXOGAFlareOperationDiscussion.pdf, Accessed on February 01, 2016

[3] Electronic code of federal regulations, General control device and work practice requirements, Available at http://www.ecfr.gov/cgi-bin/text-idx?SID=7ae903b94d41527119c 08be5a939da ed&mc= true&node=se40.7.60_118&rgn=div8, Accessed on February 01, 2016.

[4] API Standard 521. Pressure - relieving and Depressuring Systems. American Petroleum Institute, Sixth Edition, January 2014.

[5] C. D. Elvidge, M. Zhizhin, K. Baugh, F. Hsu, and T. Ghosh, Methods for Global Survey of Natural Gas Flaring from Visible Infrared Imaging Radiometer Suite Data, Energies, 9, (2016), 14, Available at http://www.mdpi.com/1996-1073/9/1/14, Accessed on February 01, 2016.

[6] C. E. Lucas, P. A. Brochu, W. C. Rooney, Gas compressor, WO 2005010375 A1, Available at https://patentscope.wipo.int/search/en/detail.jsf?docId=WO2005010375&redirectedID=true, Accessed on February 01, 2016.

[7] Y. N. Liu, C. E. Lucas, R. W. Bower, and J.C. Bronfenbrenner, Reducing LNG Costs by Better Capital Utilization, Available at http://www.airproducts.com/industries/Energy/LNG/ resource-center/whitepapers/reducing-lng-costs-better-capital-utilization.aspx, Accessed on February 01, 2016.

[8] Qatar Gas, Flare reduction programme, Available at https://www.qatargas.com/English /SafetyAndEnvironment/Documents/Qatargas%20Flaring_Fact%20Sheet_20%20Nov%202012. pdf, Accessed on February 01, 2016.

444973 Accurate Modelling of Acid Gas Cleaning Processes

Abstract:

The process of removing acid gas content, such as H_2S, CO_2, Mercaptans, COS, and CS_2, from natural gas has several business challenges, the most important of which is meeting safety and environmental regulations. H_2S is toxic and SO_2 emissions into the atmosphere can lead to acid rain once the product is consumed. In the presence of water, acid gases form acids that could corrode equipment and cause pipeline failures. Reducing the CO_2 content in the gas not only prevents damage to equipment, but also improves the heating value. Additional challenges include the increasing demand for the products, varying acid gas content in the feed, and performance degradation of the acid gas cleaning process over time.

Accurate modeling can offer valuable insight into understanding the intricacies and trade-offs involved in a process plant. Process simulators with strong acid gas cleaning modeling capabilities such as Aspen HYSYS, can accurately predict performance helping users to meet sales gas regulations, ensure product quality, and identify process bottlenecks and creep capacity. In addition to modeling acid gas cleaning processes, Organizations can also save time and money by optimizing the entire gas plant with a simulator capable of modeling acid gas cleaning, dehydration, NGL recovery, heat exchangers, and more.

This paper discusses Acid Gas Cleaning and its ability to model processes using chemical solvents, such as amines, and physical solvents, such as DEPG. While modeling acid gas cleaning processes, it is important to account for the kinetic reaction rates and the mass transfer resistance. The mass transfer resistance is the rate-limiting step at the vapor-liquid interface, and simulating this effect results in superior modeling accuracy. Another effect that must be captured in the model is the performance degradation from the presence of Heat Stable Salts.

The paper also details commercial case studies using Aspen HYSYS Acid Gas Cleaning, and the model validation work completed. Best practices will be shared on how companies have leveraged the power of this technology to improve their operations.

444997 Non-Traditional Model Predictive Control and Nonlinear Process Modeling for Return on Investment APC Projects

Abstract:

As Advanced Control technologies find broader and broader users in industrial automation and control, there are challenges on what to do when processes are sometimes not fully continuous.

For example gas processing plants in the field which face huge condensate feed surges known as pigging come from varying conditions in the gas reservoir or the gas and condensate collection system. In addition, there a major production lines that are not continuous but have start/stop operations for clean-in-place and to respond to feed availability changes. These production plants have to have well designed MPC improve process predictability and stability during these transitions.

Handling major process changes or transitions become critical to maximize the return on investment of the advanced control strategy. Success then becomes a sifnficant differentiator from traditional continuous control strategies. This paper addresses new advances in gas processing controls of fractionation, cryogenic processes and even production lines that periodically start and stop and highlights the challenges and methods in the industrial process control practice.

INHA UNIVERSITY

Strategic Approach for the Development of GTL-FPSO (Gas to Liquid-Floating Production Storage and Offloading) Process

Clean Energy Process Integration Lab.
Graduate School of Chemistry and Chemical Engineering
INHA University, South Korea

Sungwon Hwang*, Hyeju Song, Jeongeun Son, Soojin Kwon and Yongheon Cho

목차

1. Introduction of GTL Process

2. Research contents
   - Modeling of Reforming Process
   - Modeling of Fischer-Tropsch Process
   - Modeling of Upgrading Process
   - Heat Integration of GTL Process

3. Conclusions and Future Work

1. Introduction of GTL Process

1. Introduction of GTL Process

GTL(Gas-To-Liquid) Technology

- A technology to produce synthetic oil by converting natural gas
- Main sub-process units : Reforming unit to produce syngas, F-T(Fischer-Tropsch) unit to produce synthetic oil, and Upgrading unit for final treatment and high yields.
- Less contaminants than conventional crude (e.g. sulfur, aromatics, heavy metals, etc.)

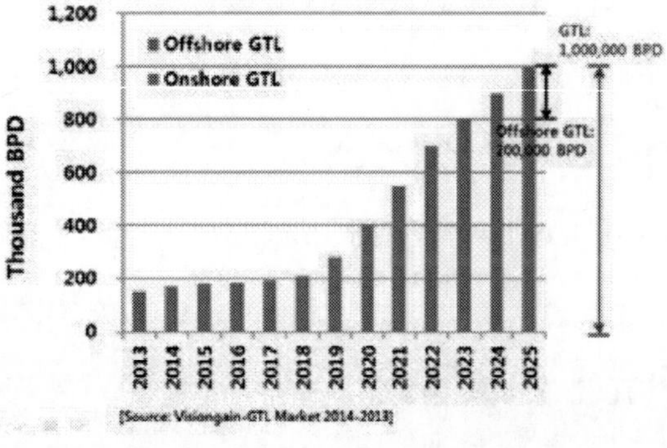

[Source: Visiongain-GTL Market 2014-2013]

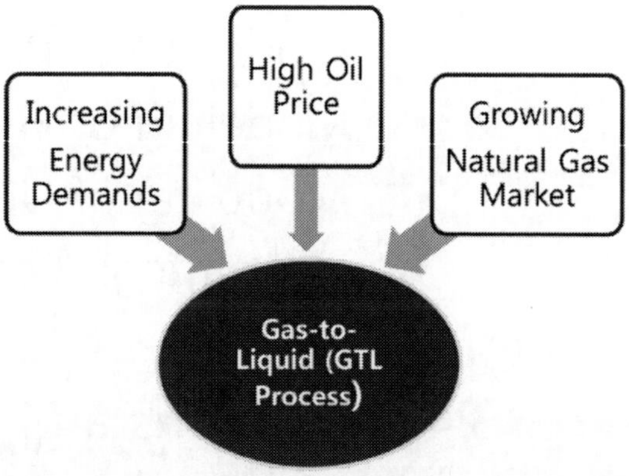

1. Introduction of GTL Process

Flow scheme of GTL(Gas to Liquid) Process

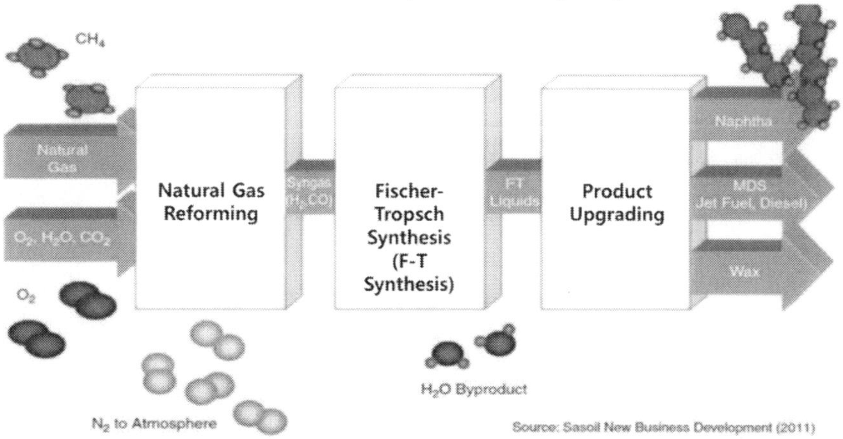

GTL-FPSO

- GTL Process (High valued technology) + FPSO (Flexible transportation)
- Can be relocated to other gas wells during its lifetime.

1. Introduction of GTL Process

- **Research Target**

  To develop "Conceptual Design Package of GTL-FPSO for the production of 20,000 BPD Gasoline and Distillates"

- **Research Contents**

  1) Modeling of three sub-process reaction systems
  2) Conceptual Design of Integrated GTL Process
  3) Heat Integration of the Process by using Pinch Technology

1. Introduction of GTL Process

- **Methodology and approach**

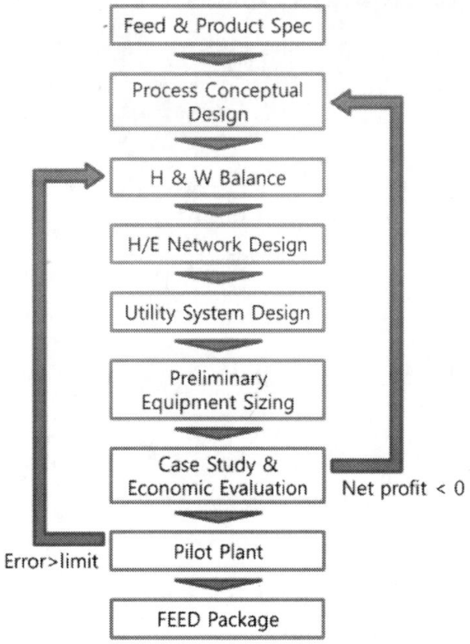

- **Offshore Process Selection Criteria**

INHA UNIVERSITY 7 Clean Energy Process Integration Lab

INHA UNIVERSITY

2.1. Modeling of Reforming Process

2.1. Modeling of Reforming Process

Modeling of Reforming Reaction System

① Reforming Reaction

: A technology to convert natural gas to Syngas (H_2, CO)

- Steam Reforming (with a feed of H_2O) √
- Dry Reforming (with a feed of CO_2) √
- Auto thermal Reforming (with a feed of O_2)

② Main Reactions

- SRM1 : $CH_4 + H_2O \leftrightarrow CO + 3H_2$ [$H_2/CO=3$]
- SRM2 : $CH_4 + 2H_2O \leftrightarrow CO_2 + 4H_2$
  (Steam Reforming of Methane)
- DRM : $CH_4 + CO_2 \leftrightarrow 2CO + 2H_2$ [$H_2/CO=1$]
  (Dry Reforming of Methane)
- WGS : $CO + H_2O \leftrightarrow CO_2 + H_2$
  (Water Gas Shift)

2.1. Modeling of Reforming Process

Modeling of Reforming Reaction System

③ Conceptual Design of a Reforming Reactor

- Constant heat input

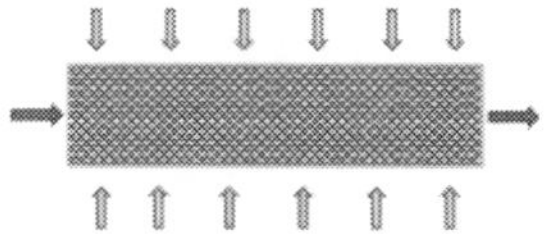

- Tubular reaction

- Control of heat supply

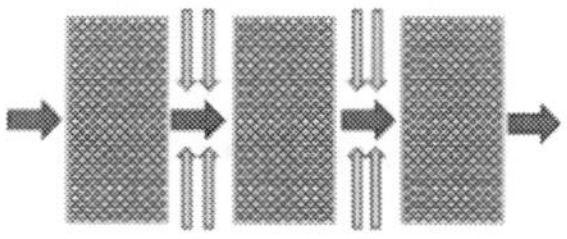

- Fixed bed with intermediate heating

- Variation of inert catalyst

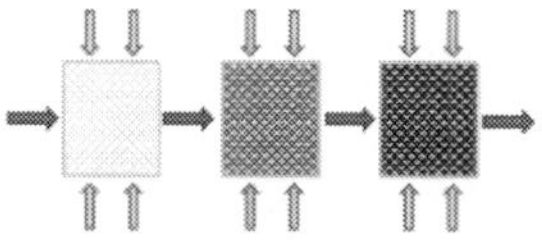

- Evenly distributed heat supply
- Sectional residence time distribution

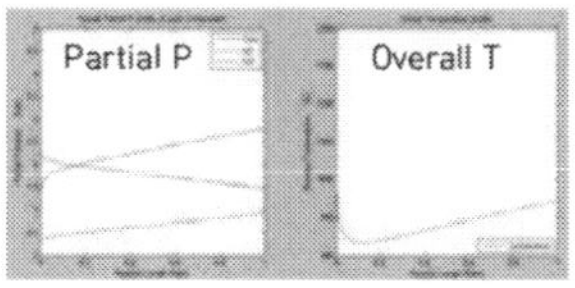

- Heat flux=93.7kW/m^2
- CH_4 conversion: 38.3%
- Minimum temperature : 918.0K

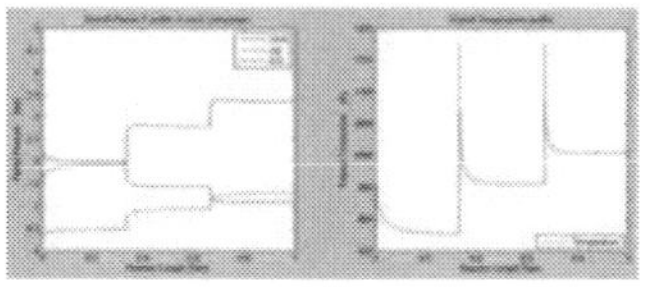

- CH_4 conversion: 51.2%
- Minimum temperature : 878.3K

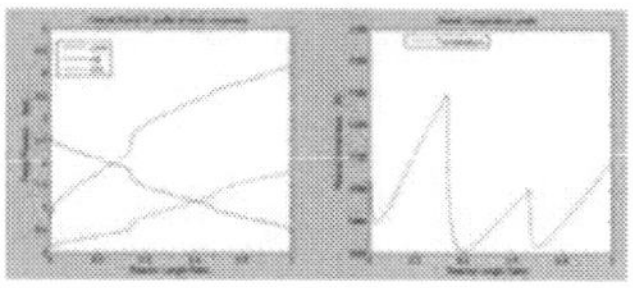

- Catalyst distribution : 0.2% / 9.8% / 90%
- CH_4 conversion: 72.4%
- Minimum temperature : 1042K

2.1. Modeling of Reforming Process

Modeling of Reforming Reaction System

④ Analysis of Commercialized Reformer Reactor and its Modeling

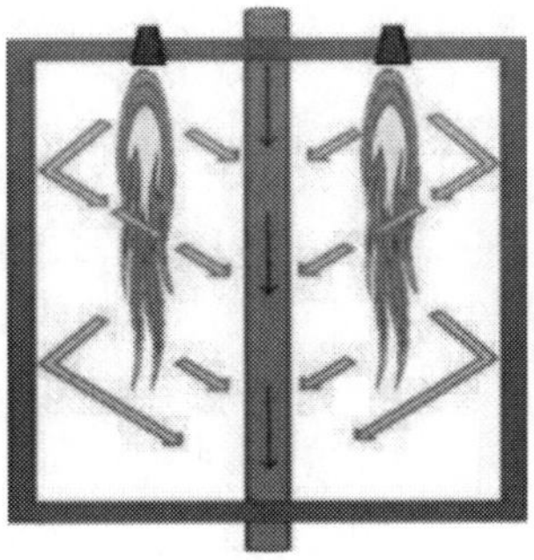

Top Fired

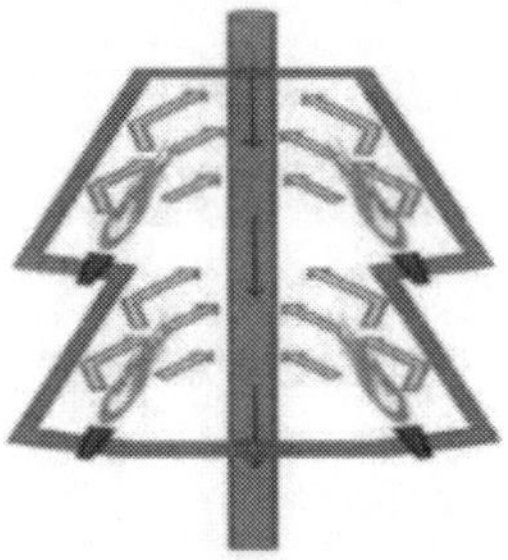

Foster Wheeler
Terrace Wall

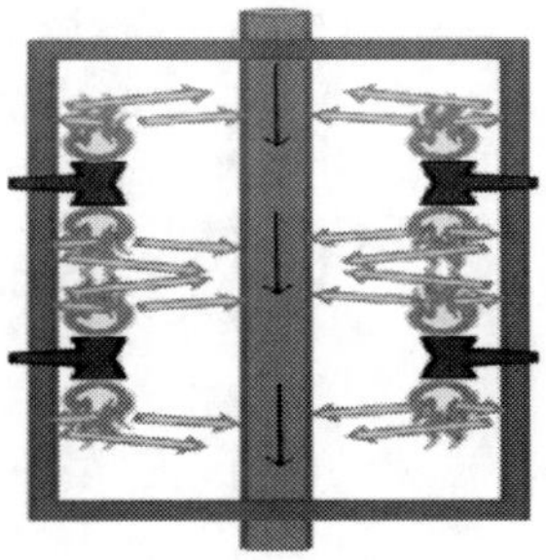

Side Fired

2.1. Modeling of Reforming Process

Modeling of Reforming Reaction System

⑤ Analysis of Fired Reformer Efficiency

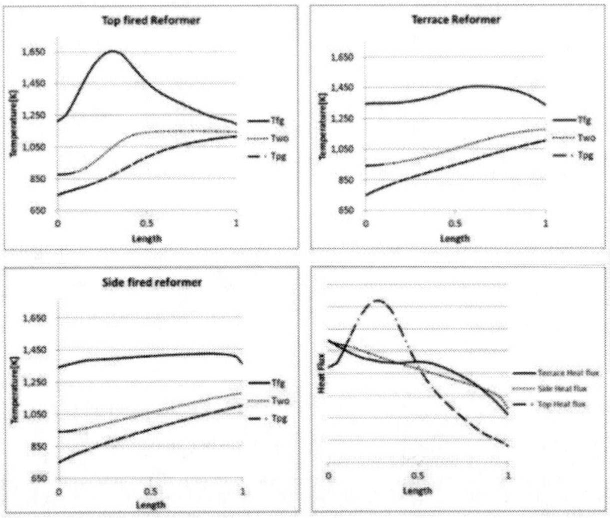

Conversion	Top fired	Foster Wheeler Terrace Wall	Side fired
CH4	78.18%	75.60%	74.72%
CO2	45.49%	42.40%	41.38%

2.1. Modeling of Reforming Process

Modeling of Reforming Reaction System

⑥ Development of Process Flowsheet

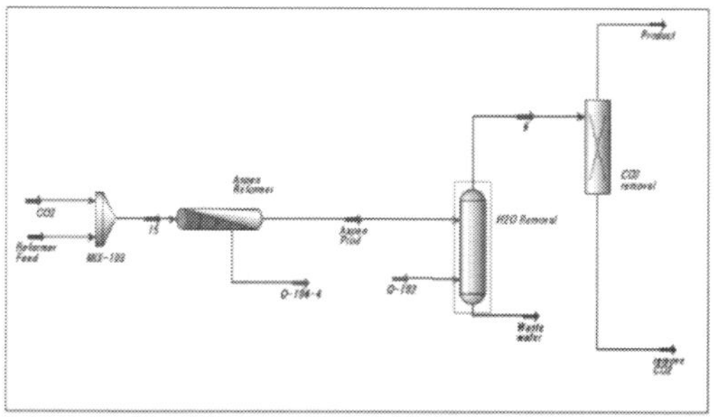

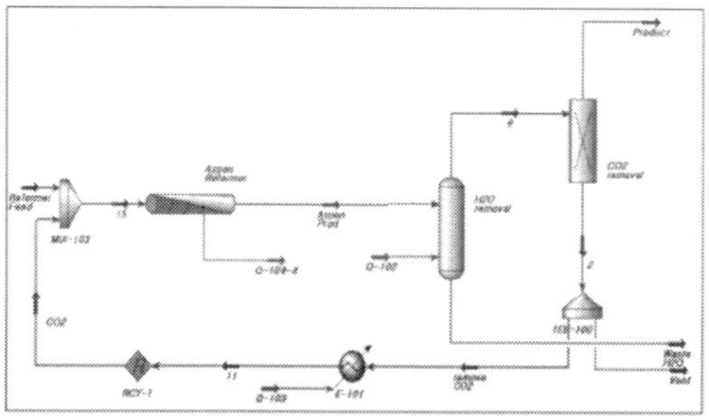

Case	Recycle	T (℃)	P (atm)	CH4 Conversion (%)	Product Eff. (%)	Energy (%)
A	Recycle	900	10	93.4	100	100
B		850	10	85.6	92	91
C		800	10	75.0	80	79

INHA UNIVERSITY

2.2. Modeling of Fischer-Tropsch Process

2.2. Modeling of Fischer-Tropsch Process

1) Introduction of F-T Synthesis Reaction Mechanism

① **Primary FTS Reaction**

a. Paraffin (55%) $(2n+1)H_2 + nCO \rightarrow C_nH_{2n+2} + nH_2O$
b. Olefin (35%) $2nH_2 + nCO \rightarrow C_nH_{2n} + nH_2O$
c. Alcohol (9%) $2nH_2 + nCO \rightarrow C_nH_{2n+2}O + (n-1)H_2O$
d. Acid (0.3%) $2(n-1)H_2 + nCO \rightarrow C_nH_{2n-1}OOH + (n-2)H_2O$
e. Aldehyde (0.2%)
f. Ketone (0.2%)
g. Ester (0.2%)

② **Secondary FTS Reaction**

– Re-adsorption of Olefin, Alcohol, Acid and Secondary Chain Growth

③ **Water Gas Shift Reaction** $CO + H_2O \rightarrow H_2 + CO_2$

2.2. Modeling of Fischer-Tropsch Process

1) Introduction of F-T Synthesis Reaction Mechanism

- **FTS** : alkyl mechanism with CH_2 insertion (Example)

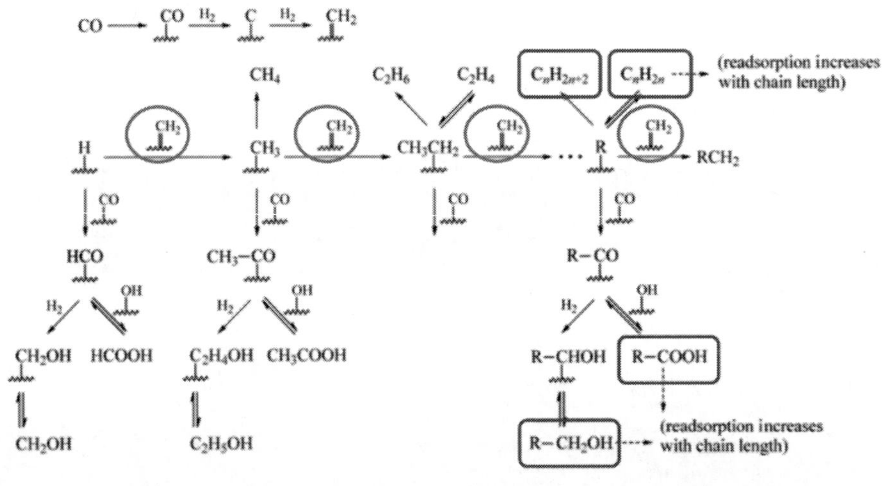

2.2. Modeling of Fischer-Tropsch Process

2) Procedure of the F-T Modeling Process

F-T Kinetic Model → Kinetic Parameter Estimation → Kinetic Parameter Evaluation → Result Analysis

↓

Optimized F-T Synthesis

2.2. Modeling of Fischer-Tropsch Process

3) Estimation of Product Distribution

ASF Law
(Anderson-Schulz-Flory)

- Conventional Prod. Est. Method
- Independent of the number of carbon atoms in the product molecule
- Chain growth factor(α) = Constant

Relatively lower accuracy

LHHW method
(Langmuir-Hinshelwood-Hougen-Watson)

- Latest Prod. Est. Method
- Detailed Adsorption Mechanism
- Chain growth factor(α_n) = f(n)

Relatively Higher accuracy

2.2. Modeling of Fischer-Tropsch Process

4) Modeling of Reaction Kinetics (Elementary Reaction Steps)

1	$CO + s = CO\text{-}s$	
2	$CO\text{-}s + H\text{-}s = C\text{-}s + OH\text{-}s$	
3	$C\text{-}s + H_2 = CH_2\text{-}s$	
4	$H_2 + 2s = 2H\text{-}s$	
5(*n*)	$CH_2\text{-}s + H\text{-}s \rightarrow CH_3\text{-}s + s$	RDS
	$CH_2\text{-}s + C_nH_{2n+1}\text{-}s \rightarrow C_nH_{2n+1}CH_2\text{-}s + s$	RDS
6	$HO\text{-}s + H\text{-}s = H_2O + 2s$	
7	$H\text{-}s + CO\text{-}s = HCO\text{-}s + s$	
	$C_nH_{2n+1}\text{-}s + CO\text{-}s = C_nH_{2n+1}CO\text{-}s + s$	
8	$HCO\text{-}s + H_2 = HCHOH\text{-}s$	
	$C_nH_{2n+1}CO\text{-}s + H_2 = C_nH_{2n+1}CHOH\text{-}s$	
9(*n*)	$HCHOH\text{-}s + H\text{-}s = CH_3OH + 2s$	RDS
	$C_nH_{2n+1}CHOH\text{-}s + H\text{-}s = C_nH_{2n+1}CH_2OH + 2s$	
10(*n*)	$HCO\text{-}s + OH\text{-}s = HCOOH + 2s$	RDS
	$C_nH_{2n+1}CO\text{-}s + OH\text{-}s = C_nH_{2n+1}COOH + 2s$	
11(*n*)	$CH_3\text{-}s + H\text{-}s \rightarrow CH_4 + 2s$	RDS
	$C_nH_{2n+1}\text{-}s + H\text{-}s \rightarrow C_nH_{2n+2} + 2s$	
12(*n*)	$C_nH_{2n+1}\text{-}s = C_{n+1}H_{2n} + H\text{-}s$	RDS

RDS: rate determining step.

2.2. Modeling of Fischer-Tropsch Process

4) Modeling of Reaction Kinetics (Expression of Kinetics)

$$R_{CH_4} = k_{11,1}[CH_3 - s][H - s] = k_{11,1}\alpha_{T,1}K_2 p_{H_2}[s]^2$$

$$R_{C_nH_{2n+2}} = k_{11}[C_nH_{2n+1} - s][H - s] = k_{11}K_4 p_{H_2}[s]^2 \prod_{i=1}^{n} \alpha_{T,i}$$

$$R_{C_2H_4} = k_{12,2}[C_2H_5 - s] = k_{12,2}\sqrt{K_4 p_{H_2}}[s] \prod_{i=1}^{2} \alpha_{T,i}$$

$$R_{C_nH_{2n}} = k_{12}e^{-C_0 n}[C_nH_{2n+1} - s] = k_{12}e^{-C_0 n}\sqrt{K_4 p_{H_2}}[s] \prod_{i=1}^{n} \alpha_{T,i}$$

$$R_{CH_3OH} = k_{9,1}K_1K_4K_7K_8 p_{CO} {p_{H_2}}^2 [s]^2$$

$$R_{C_nH_{2n+1}OH} = k_9 e^{-C_{AL}n} K_1K_4K_7K_8 p_{co} {p_{H_2}}^2 \prod_{i=1}^{n-1} \alpha_{T,i}[s]^2$$

$$R_{C_nH_{2n-1}OOH} = k_{10}e^{-C_{AC}n} K_1K_7 p_{co} p_{H_2O} \prod_{i=1}^{n-1} \alpha_{T,i}[s]^2/K_6$$

$$R_{CO_2} = \frac{k_{W2}K_{W1}K_{W3}p_{CO}p_{H_2O} - k_{-W2}p_{H_2}p_{CO_2}}{\sqrt{K_{W3}p_{H_2}} + p_{H_2} + K_{W1}p_{CO}p_{H_2O}/p_{H_2}}$$

$$\alpha_{T,1} = \frac{k_5K_1K_2K_3K_4K_6 {p_{H_2}}^2 p_{CO}}{k_5K_1K_2K_3K_4K_6 {p_{H_2}}^2 p_{CO} + (k_{9,1}K_1K_7K_8 p_{H_2} p_{CO} + k_{11,1})\sqrt{K_4 p_{H_2}}\, p_{H_2O}}$$

$$\alpha_{T,n} = \frac{k_5K_1K_2K_3{K_4}^{1.5}{K_6}^2 {p_{H_2}}^{2.5} p_{CO}}{k_5K_1K_2K_3{K_4}^{1.5}{K_6}^2 {p_{H_2}}^{2.5} p_{CO} + \left(k_9 e^{-C_{AL}n}K_1K_7K_8 p_{H_2}p_{CO} + k_{11}\right)K_4K_6 p_{H_2}p_{H_2O} + k_{10}e^{-C_{AC}n}K_1K_7 p_{CO}{p_{H_2O}}^2[s] + k_{12}e^{-C_0 n}K_6\sqrt{K_4 p_{H_2}}\, p_{H_2O}}$$

$$[s] = 1 \Bigg/ \left[1 + K_1 p_{CO} + \sqrt{K_4 p_{H_2}} + K_1K_7 P_{CO}\sqrt{K_4 p_{H_2}} + K_1K_7K_8 p_{CO} p_{H_2}\sqrt{K_4 p_{H_2}} + \frac{K_1K_2K_4K_6 p_{H_2} p_{CO}}{p_{H_2O}} + \frac{K_1K_2K_3K_4K_6 {p_{H_2}}^2 p_{CO}}{p_{H_2O}} + \frac{p_{H_2O}}{K_6\sqrt{K_4 p_{H_2}}} + \sum_{i=1}^{n}\prod_{i=1}^{n} \alpha_{T,i}\left(1 + K_1K_7K_8 p_{CO} p_{H_2}\sqrt{K_4 p_{H_2}} + K_1K_7 p_{CO}\sqrt{K_4 p_{H_2}}\right)\sqrt{K_4 p_{H_2}}\right]$$

2.2. Modeling of Fischer-Tropsch Process

5) Fitting of Reaction Kinetic Parameters

- **Objective Function**

$$\text{Minimize,}\quad \text{Error} = \sum_{i=1}^{N_{exp}} \sum_{j=1}^{NRS} w_j \left(\frac{m_{i\,j\,exp} - mi_{j\,cal}}{m_{i\,j\,exp}} \right)^2$$

- **Optimization Algorithm**

Genetic Algorithm(GA), Levenberg-Marquardt (LM) Algorithm

2.2. Modeling of Fischer-Tropsch Process

5) Fitting of Reaction Kinetic Parameters

- **Range of operating condition :**
  Temperature: 533-573K
  Pressure: 1.00-2.65MPa
  Ratio of H_2/CO in the feed: 0.66-2.00
  Space velocity: 0.66-2.7*10-3Nm^3·kg^{-1}·s^{-1}

- **22 세트의 실험 데이터를 이용**

	1	2	3	4	5	6	7	8	9	10	11
T(K)	533	533	533	543	543	543	543	543	553	553	553
P(bar)	15.8	20.5	25.1	16.5	16.3	11	26	22	16.7	21.5	26.1
H2/CO ratio	1	1.51	2.05	0.67	1.01	0.99	1.5	1.99	1.01	0.68	1
Space Velocity (Nm3·kg-1·s-1)	6.7E-04	8.3E-04	1.0E-03	6.6E-04	9.9E-04	6.7E-04	1.3E-03	1.3E-03	9.9E-04	1.7E-03	1.3E-03

	12	13	14	15	16	17	18	19	20	21	22
T(K)	553	553	563	563	563	563	563	573	573	573	573
P(bar)	12	16.5	16.7	16.8	20.3	25.7	11.7	17.5	19.8	20.9	25.5
H2/CO ratio	1.49	2.02	1	1.5	1	0.68	2.02	1	1	1.5	1.9
Space Velocity (Nm3·kg-1·s-1)	6.6E-04	1.0E-03	1.3E-03	1.3E-03	1.7E-03	1.7E-03	1.0E-03	1.7E-03	2.0E-03	2.3E-03	2.7E-03

2.2. Modeling of Fischer-Tropsch Process

6) List of the Identified Kinetic Parameters

Parameter	Value		Parameter	Value	
$k_{5,0}$	1.382×10^{6}	$mol\cdot kg^{-1}\cdot s^{-1}$	$E_{12,2}$	9.705×10^{4}	$kJ\cdot mol^{-1}$
E_5	8.341×10^{4}	$kJ\cdot mol^{-1}$	K_1	1.323×10^{1}	MPa^{-1}
$k_{9,1,0}$	5.508×10^{7}	$mol\cdot kg^{-1}\cdot s^{-1}$	K_2	1.397	-
$E_{9,1}$	1.075×10^{5}	$kJ\cdot mol^{-1}$	K_3	8.475×10^{-1}	MPa^{-1}
$k_{9,0}$	5.672×10^{7}	$mol\cdot kg^{-1}\cdot s^{-1}$	K_4	8.118	MPa^{-1}
E_9	9.122×10^{4}	$kJ\cdot mol^{-1}$	K_6	9.600×10^{-2}	MPa
$k_{10,0}$	5.561×10^{8}	$mol\cdot kg^{-1}\cdot s^{-1}$	K_7	7.752×10^{-1}	
E_{10}	9.790×10^{4}	$kJ\cdot mol^{-1}$	K_8	1.037	MPa^{-1}
$k_{11,1,0}$	4.041×10^{6}	$mol\cdot kg^{-1}\cdot s^{-1}$	C_O	2.136×10^{-1}	-
$E_{11,1}$	8.785×10^{4}	$kJ\cdot mol^{-1}$	C_{AL}	4.416×10^{-1}	-
$k_{11,0}$	7.400×10^{6}	$mol\cdot kg^{-1}\cdot s^{-1}$	C_{AC}	4.518×10^{-1}	-
E_{11}	9.220×10^{4}	$kJ\cdot mol^{-1}$	$k_{w2,0}$	7.547×10^{5}	$mol\cdot kg^{-1}\cdot s^{-1}$
$k_{12,0}$	9.351×10^{5}	$mol\cdot kg^{-1}\cdot s^{-1}$	E_v	7.858×10^{4}	$kJ\cdot mol^{-1}$
E_{12}	8.662×10^{4}	$kJ\cdot mol^{-1}$	K_{W1}	3.286×10^{1}	MPa^{-2}
$k_{12,2,0}$	1.138×10^{6}	$mol\cdot kg^{-1}\cdot s^{-1}$	K_{W2}	3.303×10^{1}	MPa

2.2. Modeling of Fischer-Tropsch Process

7) Experimental data vs. Modeling results (I)

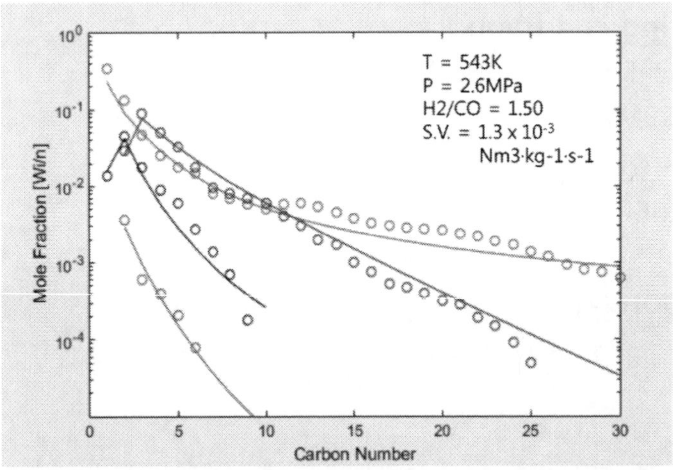

(1) Paraffins
(2) Olefins
(3) Alcohols
(4) Acids

Lines-predictions

symbols-experimental data

MARR Error = 29.2% (Increased compared previous research results, 32.8%)

※ Teng Botao, 2007

2.2. Modeling of Fischer-Tropsch Process

7) Experimental data vs. Modeling results (II)

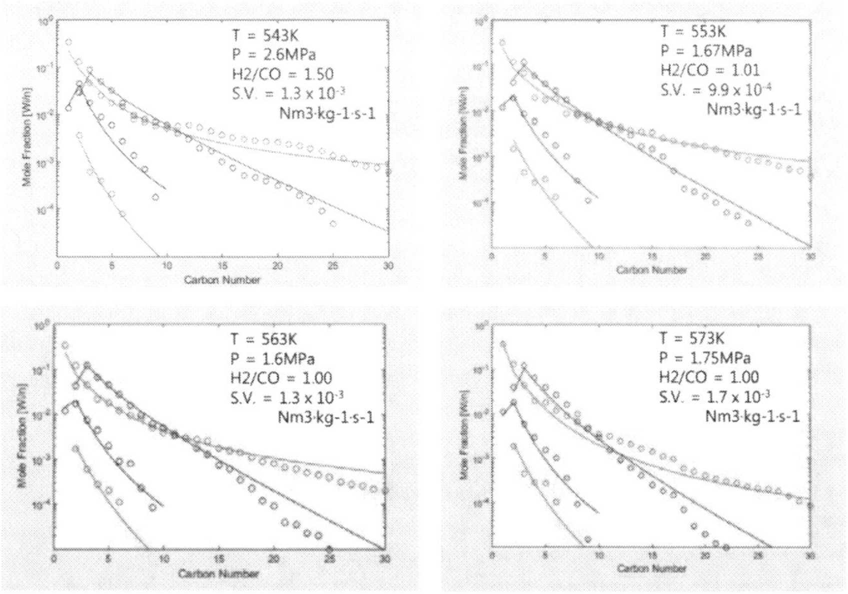

(1) Paraffins
(2) Olefins
(3) Alcohols
(4) Acids

Lines-predictions

symbols-experimental data

2.2. Modeling of Fischer-Tropsch Process

8) Temperature vs. Production Distribution

Paraffins

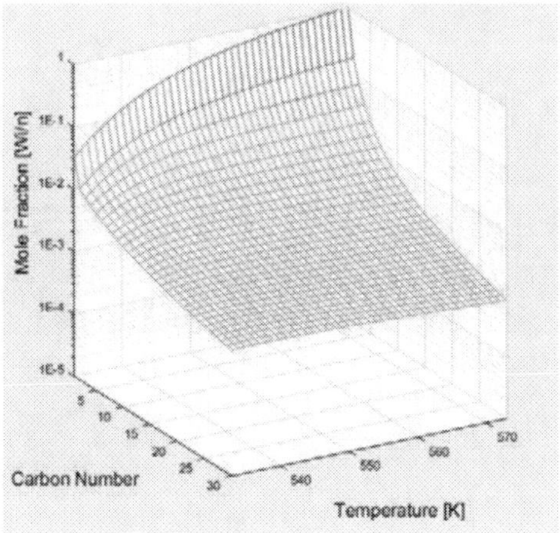

Olefins

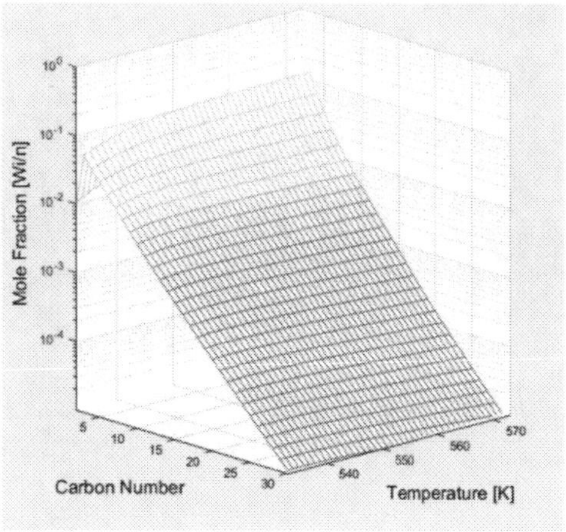

(P = 1MPa, H_2/CO ratio =1, SV = 1.3 x $10^{-3}Nm^3 \cdot kg^{-1} \cdot s^{-1}$)

2.2. Modeling of Fischer-Tropsch Process

8) Pressure vs. Production Distribution

Paraffins

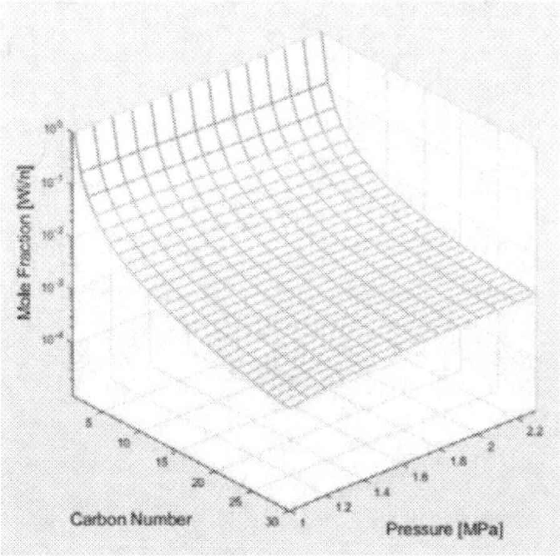

Olefins

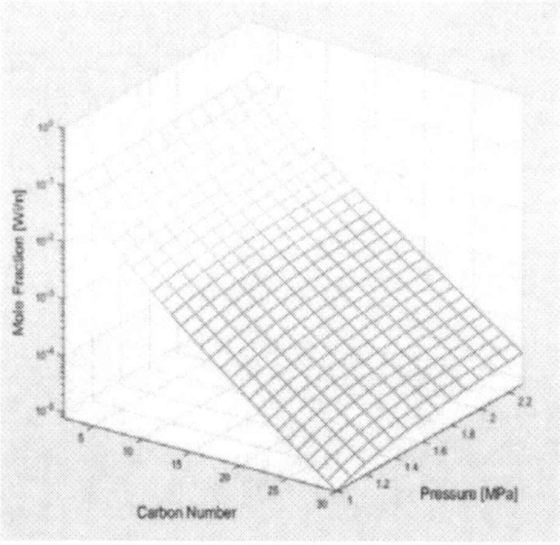

(T = 553K, H_2/CO ratio =1, SV = 1.3 x 10^{-3} $Nm^3 \cdot kg^{-1} \cdot s^{-1}$)

INHA UNIVERSITY

2.3. Modeling of Upgrading Process

2.3. Modeling of Upgrading Process

1) Reaction Mechanism of Upgrading Process (Example)

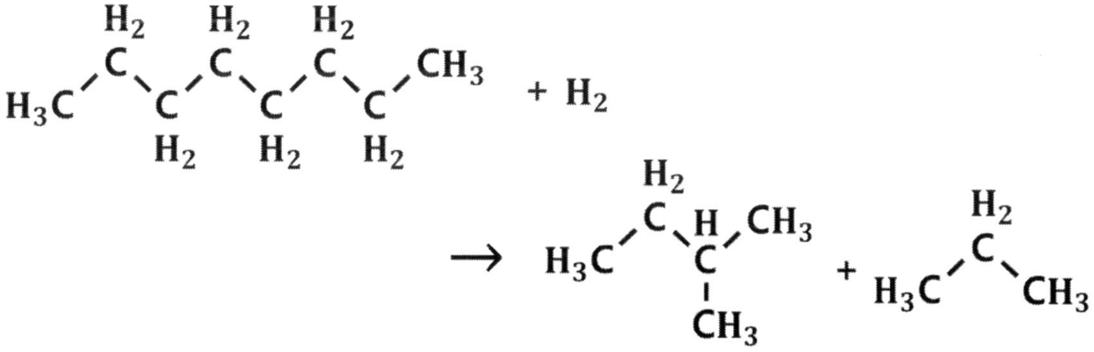

- Feed: Product of Fischer-Tropsch Process
- Main Composition: Normal & Iso Paraffins (Carbon number : 1 – 34), H_2

2.3. Modeling of Upgrading Process

1) Reaction Mechanism of Upgrading Process

Assumptions:

- One type of iso-paraffin exists per each hydrocarbon
- Hydrocarbons with carbon number 7 to 34 are involved with cracking reaction.
- N-butane does not have isomerization reaction.

(Martens JA, Jacobs PA, Weitkamp J. Attempts to rationalize the distribution of hydrocracked products)

2.3. Modeling of Upgrading Process

1) Reaction Mechanism of Upgrading Process

① Isomerization & cracking reaction kinetics

$$r_{isom}(i) = \frac{k_{isom(i)} \times (P_n(i) - P_{iso}(i)/K_{eq}(i))}{P_{H_2} \times [1 + \sum_{i=1}^{n} P_n(i) \times K_{L,n}(i) + \sum_{i=4}^{n} P_{iso}(i) \times K_{L,iso}(i)]}$$

$$r_{ck}(i) = \frac{k_{cr(i)} \times P_{iso}(i)}{P_{H_2} \times [1 + \sum_{i=1}^{n} P_n(i) \times K_{L,n}(i) + \sum_{i=4}^{n} P_{iso}(i) \times K_{L,iso}(i)]}$$

② Mass balance for product formation

$$r_{prod,n}(3) = \frac{2}{2(7-6)} r_{ck}(7) + \frac{2}{2(8-6)} r_{ck}(8) + \frac{2}{2(9-6)} r_{ck}(9) + \cdots + \frac{2}{2(34-6)} r_{ck}(34)$$

$$r_{prod,n}(i) = (1-\gamma) \times \left[\frac{2}{((i+4)-6)} r_{ck}(i+4) + \frac{2}{((i+5)-6)} r_{ck}(i+5) + \cdots + \frac{2}{(34-6)} r_{ck}(34) \right]$$

$$r_{prod,iso}(i) = \gamma \times \left[\frac{2}{((i+4)-6)} r_{ck}(i+4) + \cdots + \frac{2}{(34-6)} r_{ck}(34) \right] + \left(\frac{2}{2((i+3)-6)} r_{ck}(i+3) \right)$$

(Chiara Gambaro, hyrocracking of Fischer-Tropsch waxes:kinetic modeling via LHHW approach)

2.3. Modeling of Upgrading Process

2) Feed distribution

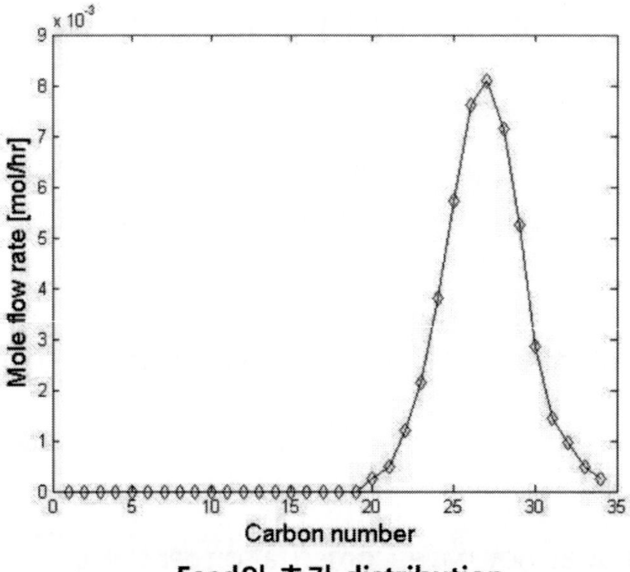

Feed의 초기 distribution

2.3. Modeling of Upgrading Process

3) Experimental condition

Operating condition vs. wax conversion and C10-17 mol fraction

No.			A						B		C		D	
			1	2	3	4	5	6	1	2	1	2	1	2
Operating conditions		T	320	340	359	361	380	400	361	361	361	361	361	361
		P	40	40	40	40	40	40	50	60	40	40	40	40
		Wax	13.03	13.03	13.03	13.03	13.03	13.03	13.03	13.03	8.68	21.71	13.03	13.03
		H2	0.134	0.134	0.134	0.134	0.134	0.134	0.134	0.134	0.134	0.134	0.08	0.18
		totcat	1	1	1	1	1	1	1	1	1	1	1	1
Experiment	Experiment	conversion	50	88	90	90	95	98	85		99	52	62	98
		C1017	26	41	42	42	44	43	82		52	28	36	52
	Calculated	conversion	42	70	88	88	99	97	81		98	76	82	98
		C1017	24	35	39	39	42	45	38		41	38	36	42
WABT		conversion	24.32	70.32	98.54	99.17	100	836.79	97.83	95.95	99.56	202.95	99.22	98
		C1017	16.2	43.46	52.17	50.82	26.41	989.69	53.52	54.51	48.91	100.44	-160.83	53.34
ODE		conversion	21.44	48.99	82.12	84.83	98.52	99.97	82.54	80.3	91.3	74.25	87.72	83.64
		C1017	8.4	28.26	49.31	50.81	54.7	42.86	49.6	48.36	53.75	44.87	52.25	50.19

2.3. Modeling of Upgrading Process

4) Analysis of Modeling Results

Parity plots for residue (C20+) conversion

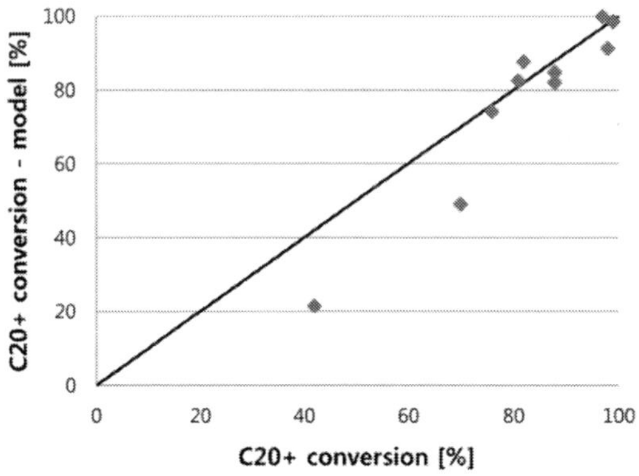

Parity plots for middle distillates fraction (C10-17)

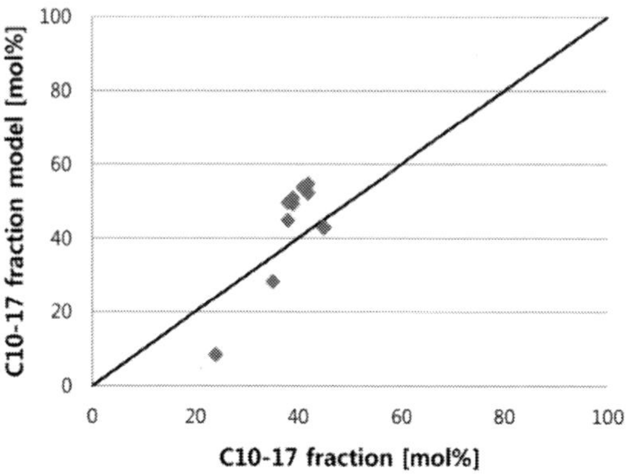

Parity plots for residue(C20+) conversion and middle distillates fraction(C10-17)

2.3. Modeling of Upgrading Process

5) Distribution of Final Products

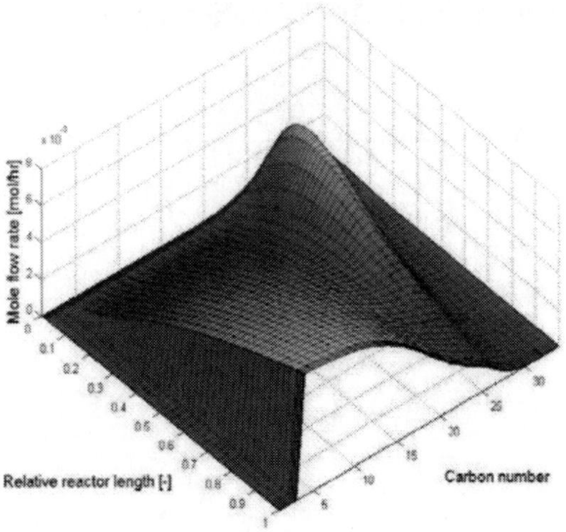

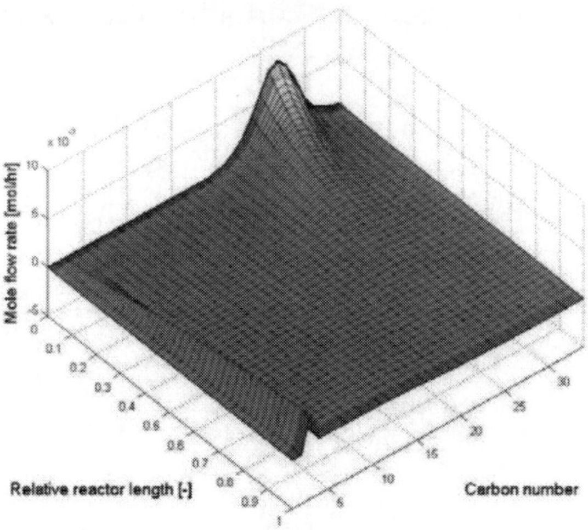

2.3. Modeling of Upgrading Process

6) Temperature vs. Conversion

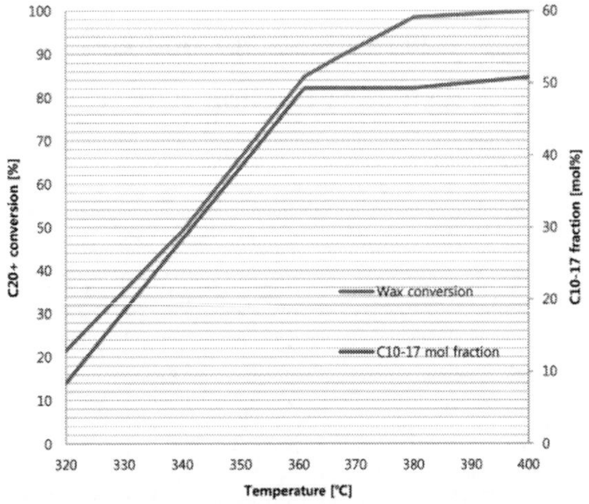

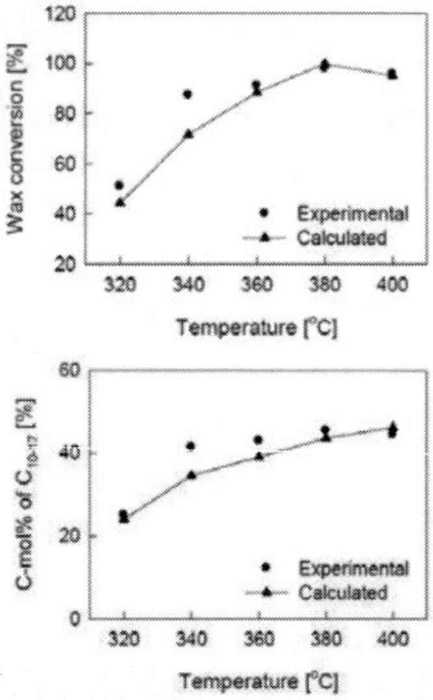

Experimental data and modeling results for the impact of various temperature on 'C20+ conversion' and 'C10-17 mol fraction'

2.3. Modeling of Upgrading Process

6) Wax Flowrate vs. Conversion

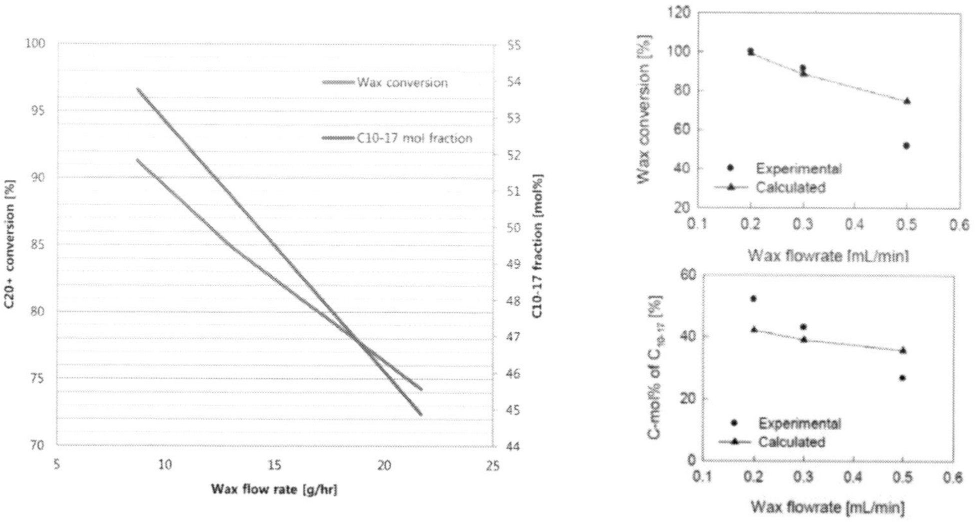

Experimental data and modeling results for the impact of various feed flowrates on 'C20+ conversion' and 'C10-17 mol fraction'

INHA UNIVERSITY

2.4. Heat Integration of GTL Process

2.4. Heat Integration of GTL Process

- GTL Process Flowsheet

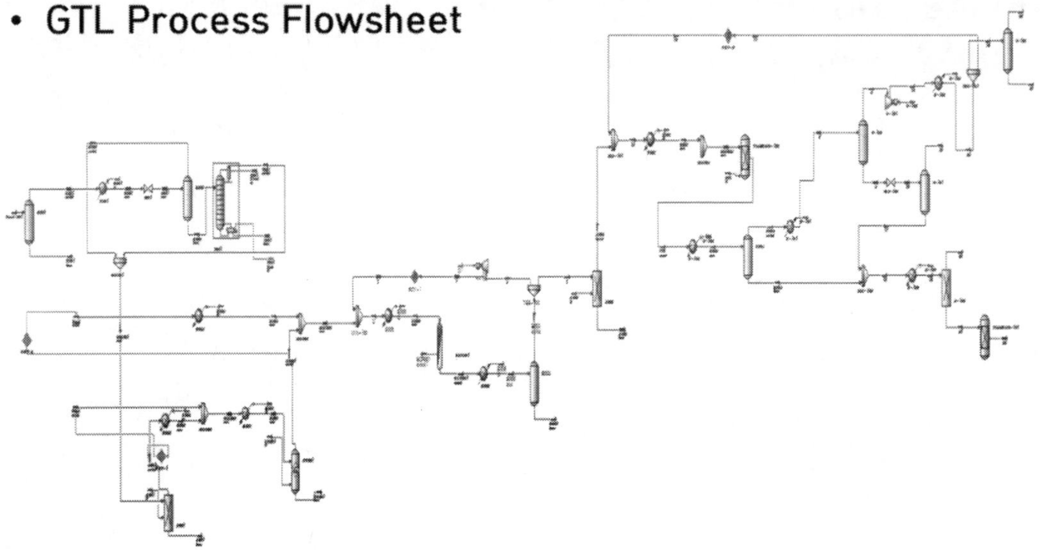

2.4. Heat Integration of GTL Process

High Energy Consumption

Requires an Analysis for Process Energy Consumption

E.g. Pinch analysis

2.4. Heat Integration of GTL Process

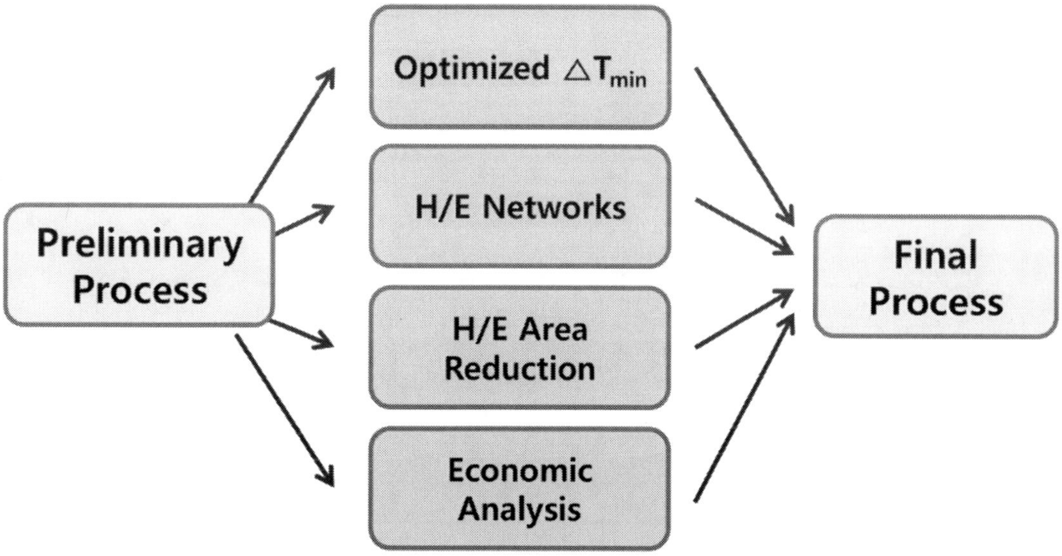

2.4. Heat Integration of GTL Process

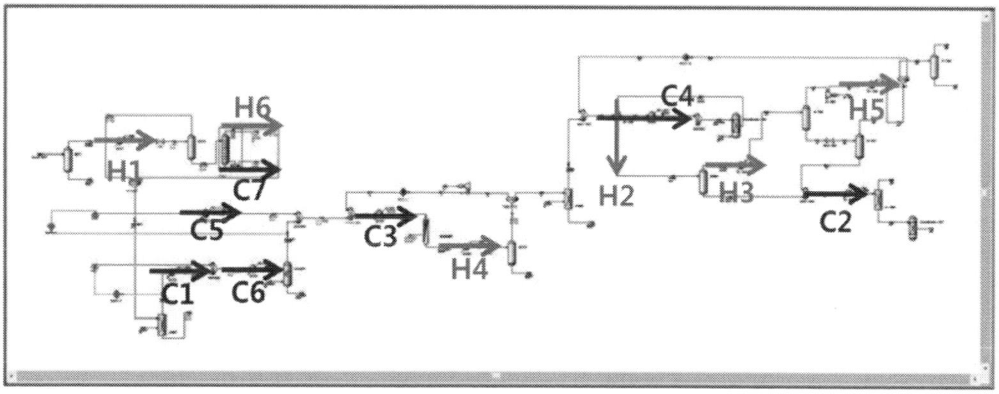

Process flow diagram of GTL process

2.4. Heat Integration of GTL Process

Extracted stream data for GTL process

stream name	Initial T [℃]	Target T [℃]	Enthalpy [kJ/s]
H1	60	25	4763
H2	230	188	25437
H3	188	30	232568
H4	900	30	686454
H5	41	28	4271
H6	55	31	11
C1	11	360	50169
C2	35	200	181506
C3	458	900	284188
C4	25	230	122768
C5	25	550	30541
C6	283	600	118665
C7	249	261	327

2.4. Heat Integration of GTL Process

Minimum Utility Consumption

- Confirmation of Thermodynamic Minimum Utility Consumption

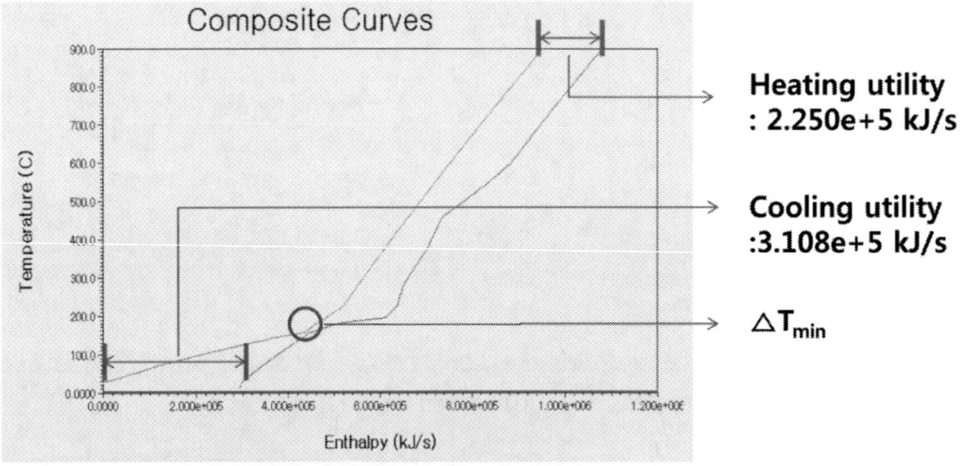

Example of Composite Curve

2.4. Heat Integration of GTL Process

Grid Diagram for Heat Exchanger Networks

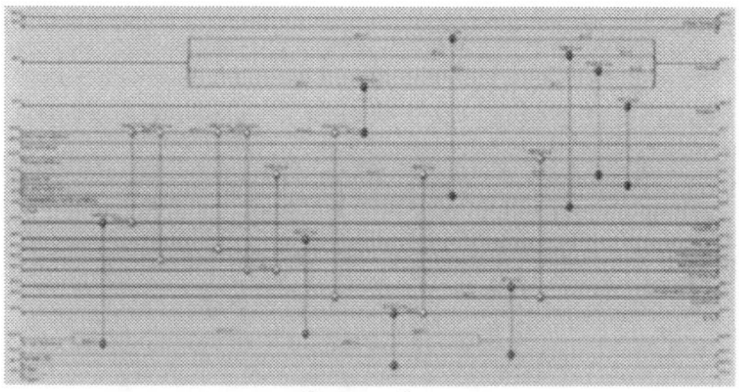

Case 1

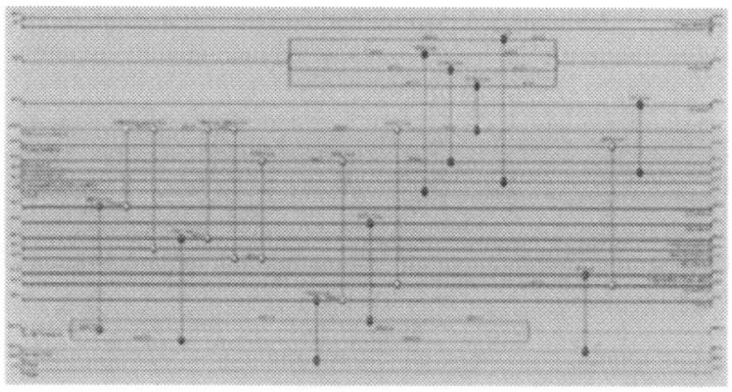

Case 2

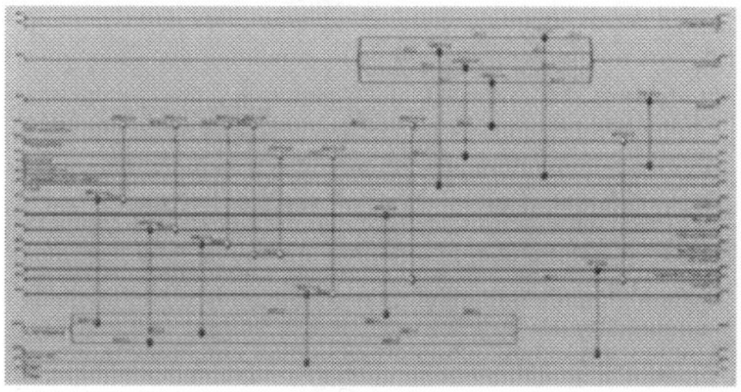

Case 3

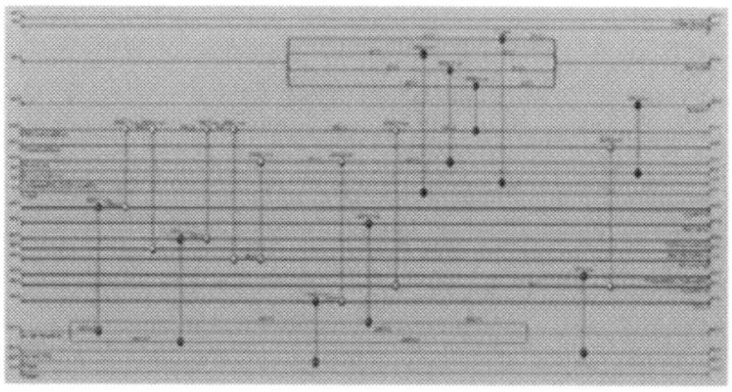

Case 4

2.4. Heat Integration of GTL Process

Analysis of Various Case Results

		Initial	CASE1	CASE2	CASE3	CASE4	CASE5	CASE6	CASE7
Utility	Heating [kJ/h]	217.5	102.8	97.2	95.1	88.4	80.9	88.4	88.4
	Cooling [kJ/h]	263.5	126.6	121.0	119.0	112.3	104.7	112.3	112.3
Total Area [^5m^2]		0.56	1.2	1.3	1.3	1.8	2.9	1.8	1.8
Cost	Heating [$/s]	4.124	2.508	2.328	2.302	2.122	1.879	2.122	2.122
	Cooling [$/s]	0.164	0.109	0.105	0.103	0.098	0.092	0.098	0.098
	Operating [$/s]	4.29	2.617	2.433	2.405	2.220	1.972	2.220	2.220
	Capital [^7$]	5.37	5.081	5.480	5.527	6.525	8.906	6.432	6.427
	Total [$/s]	4.836	3.136	2.993	2.970	2.886	2.881	2.877	2.876

2.5. Heat Integration of GTL Process

Achievement of High Energy Efficiency for GTL Process through Pinch Technology

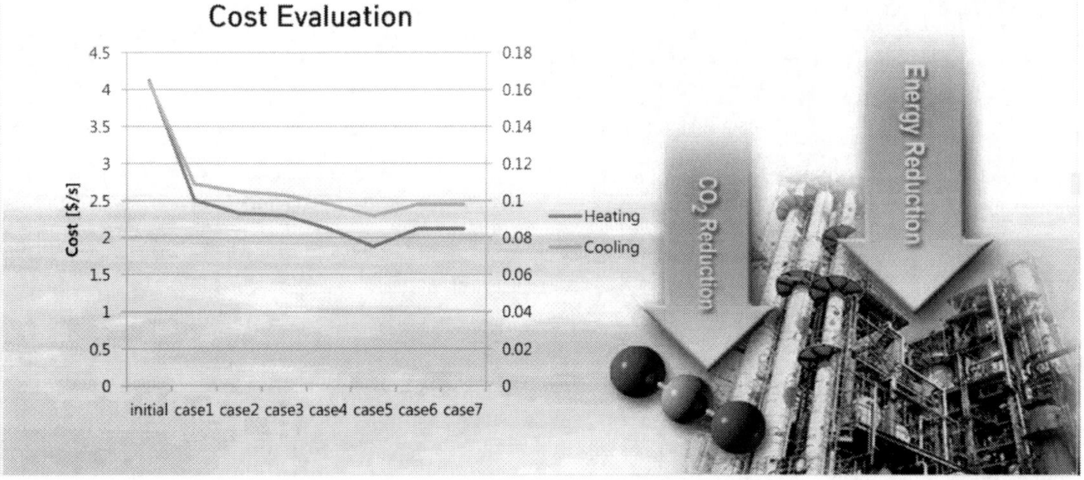

INHA UNIVERSITY

3. Conclusions and Future Work

3. Conclusions and Future Work

Conclusions

1) Identification of Reaction Mechanisms and Development of Reaction Kinetics for Reforming, Fischer-Tropsch, Upgrading Processes
2) Reactor Modeling
3) Development of Overall GTL Process
4) Pinch Analysis of GTL Process to achieve High Energy Efficiency

3. Conclusions and Future Work

Future Works

1) Continuous Development of Reaction and Separation System for each sub-process
2) Modeling and Optimization of Utility System (e.g. Electricity, steam ,etc.)
3) Economic Analysis of Various GTL Process Configurations

Q & A

- This work was supported by Korea Institute of Science and Technology and funded by Ministry of Trade, Industry and Energy (Project No. 20142010102790)

- This work was funded by Ministry of Trade through the Project of "Development of Integrated Interactive Model for Subsea and Topside System to Evaluate the Process Design of Offshore Platform" (Project No. 10060099)

Using a Large Data Bank to Develop a Robust and Simple Model to Estimate the Saturation Pressure of Oil Reservoirs

By

Hassan Alkandari[1], Adel Elsharkawy[1], Adel Malallah, Osamah Alomair

Petroleum Engineering Department, College of Engineering & Petroleum
Kuwait University
Po Box 5969 Safat 13060-Kuwait

Abstract

Measurements of saturation pressure are crucial for all hydrocarbon reservoir fluids. Below the crude oil saturation pressure gas reaches a critical saturation; then, a two phase flow occurs and results in decreasing oil production and recovery. To maximize oil production and recovery, the reservoir pressure has to be maintained closer to the original saturation pressure. This pressure is normally measured using bottom-hole samples or surface recombined samples of oil and gas. Occasionally, the samples become unavailable and the pressure needs to be estimated using computational methods.

In this study, a large data bank of 231 crude oil compositions and saturation pressure measurements including literature data, unpublished data, and newly measured data were used to develop two empirical models to predict the saturation pressure of a variety of crude oils if the oil sample is unavailable or the experimental measurements are unreliable. The first proposed model utilizes the extended compositions of hydrocarbons up to the heptane plus fraction in addition to non-hydrocarbons. The second model uses the lumped compositions of light, intermediate, and heavy components in addition to non-hydrocarbon components as the input variables. The lumped model has the advantage of using fewer input parameters while maintaining the thermodynamics. The accuracy and validity of both models to calculate the saturation pressure for volatile oils, black oils, and heavy oils are presented using several compositional data. The models performance is also compared to the Peng-Robinson equation-of-state (PR-EOS), and the Soave-Redlich-Kwong equation-of-state (SRK-EOS) as well as all published methods that use compositions as input variables. The comparison indicates that the proposed models are much simpler and more accurate than the other computational methods. The proposed models treat the heptane-plus fraction as a single component, thus eliminating the splitting and characterizing the plus fractions and the binary interaction parameters needed for the EOS's calculations.

Key words: Saturation pressure (Ps), Bubble-point pressure (BPP), Equation of state (EOS), Correlations, and PVT Data

[1] Corresponding Authors: Hassan Alkandari , Hassan@hkandari.com, Adel Elsharkawy, adel.elsharkawy@ku.edu.kw

Introduction:

Reservoir fluid properties are an essential source of information for all calculations on hydrocarbon reserves. Accurate estimation of the pressure-volume-temperature (PVT) measurement is crucial. Of all the PVT properties, saturation pressure (Ps) is considered the most important parameter. It is well known that gas starts to flow once the reservoir pressure drops below the bubble point pressure and the gas saturation reaches a critical value. To minimize the gas flow and maximize the oil recovery, it is very important to maintain the reservoir pressure close to the original bubble point pressure.

Although the PVT properties are experimentally measured in laboratories and provide reliable results, this process is expensive, time consuming, and sometimes unavailable; therefore, measured field data (separator gas oil ratio, separator pressure, stock-tank oil specific gravity, and reservoir temperature) or reservoir fluid compositions are used to estimate the bubble point pressure.

There are two methods in the petroleum industry to calculate the saturation pressure of crude oil when experimental is unavailable; either using production data or compositional data.

Methods Based on Production Data (Composition is unavailable)

Numerous attempts have been made since the 1940s to develop correlations to estimate the saturation pressure from production data, such as producing gas oil ratio, oil gravity, gas gravity, reservoir pressure and temperature. Table 1 presents summary of all the methods that use production data to estimate the saturation pressure.

Methods based on Compositional data

Many authors have reported that the application of any correlation in other regions results in serious errors. These conclusions arise from the fact that regional crude oils represent a physical-chemical trend; therefore, all attempts to develop a general correlation using field measured parameters to estimate the saturation pressure from a large data bank have failed. For this reason, Elsharkawy (2003) made the first attempt to correlate the saturation pressure to the reservoir fluid composition instead of using the production data. In recent years, there have been several attempts to estimate the saturation pressure of oil reservoirs. These methods are reported in Table 1.

Elsharkawy (2003) presented a correlation of the bubble point pressure for the oil reservoir using the same input data needed for the equation-of-state (EOS) calculations, i.e., the compositional data and the reservoir temperature. Elsharkawy's model is based on the saturation pressure measurements of 60 crude oil samples from the Middle East and 75 collected from the literature. He demonstrates that the calculation of the saturation pressure depends on the type of EOS, the number of pseudo components, the method used to characterize the pseudo-components and the binary interaction number. Elsharkawy compared the performance of his model to the SRK-EOS and PR-EOS for black oils, volatile oils, heavy oils, and synthetic oils. He indicated that his model eliminates splitting the heavy plus fraction, characterizing the pseudo-components and inclusion of the binary interaction parameters needed for the EOS calculations.

Later, Tale et al. (2009) made the second attempt to model the saturation pressure of crude oil using 132 data samples published by Elsharkawy (2003). Tale's model, which is based on the alternating conditional expectation (ACE), has a set of 14 equations with 42 constants. The average absolute relative error (AARE) reported by Tale et al. indicates that the model did not perform well with the selected data. The ACE model has an AARE of 12.3% compared with AARE of 6.7% by Elsharkawy (2003) and 8.1% AARE for the SRK-EOS.

AlQuraishi (2009) attempted to model the saturation pressure using Elsharkawy's (2003) data. The model was developed using linear genetic programming. He correlated the saturation pressure to the formation temperature, methane content, and heptanes plus molecular weight. Thus, AlQuraishi's model neglected the effect of the intermediate and heptane plus content on the saturation pressure. The error analysis reported by AlQuraishi indicates that his model has a 5.8% error compared to 7.7% by Elsharkawy, 9.9% using the SRK-EOS, and 10.1% using the PR-EOS.

Bandyopadhyay and Sharma (2011) presented a model to predict the bubble point pressure, using temperature and fluid composition. They used 129 experimental and literature data originally published by Elsharkawy (2003) to develop their model. The model introduced a temperature interaction parameter that depends on compositions. The proposed model has a set of 24 coefficients and the estimated bubble point pressure is calculated as the sum of the series calculations. They reported an average absolute error (AAR) of 7.83% using their model compared to 8.3% using Elsharkawy's simple model (2003), 10.15% using SRK-EOS, and 10.58% using the PR-EOS.

Khamehchi et al. (2011) used an intelligent proxy simulator (IPS), which is based on an adaptive neural fuzzy inference system (ANFIS) and a structured artificial neural network (ANN) to predict the bubble point pressure and asphaltene onset pressure. The model uses a reservoir fluid composition and temperature to predict the Ps. It is not clear how the Ps is calculated from the proposed method, and they did not present the accuracy of the proposed method compared to the other published methods.

Farasat et al. (2013) presented a new mathematical approach to calculate the crude oil saturation pressure as a function of the temperature and the reservoir fluid composition. The model developed by Farasat et al used 130 experimental data, which were originally published by Elsharkawy (2003). Farasat used a support vector machine (SVM) to model the saturation pressure. They did not clearly demonstrate how to calculate the saturation pressure from the SVM model. Moreover, they did not indicate the overall average absolute deviation (AAD) for their model compared to other published models and both the PR-EOS and SRK-EOS for the 130 data.

Ahmadi et al. (2014) also presented an ANN technique called gene expression programming (GEP) to determine the bubble point pressure of the oil samples. Similar to Farasat et al. (2013), Ahmadi's model used most of the data previously published by Elsharkawy (2003) to construct a non-linear equation using reservoir fluid compositions and temperatures as input parameters. The proposed model is highly complicated and consists of 8 equations that have several constants compared to the one simple equation proposed by Elsharkawy (2003). They compared the performance of their proposed model to the SRK-EOS and PR-EOS; however, their paper does not mention how the heptanes plus fraction were treated and characterized for the EOS calculations. They reported an AAD of 4% using the proposed model compared to 8% using Elsharkawy's model (2003), 10.9% by PR-EOS, 10.5% by SRK-EOS, and 15% by Bandyopadhyay and Sharma (2011).

Gholami et al. (2014) estimated the bubble point pressure using a support vector regression SVR and a supervised learning algorithm SLT based on the statistical learning theory, and then the ACE was used. The model used reservoir fluid compositions and temperature to estimate the bubble point pressure. Gholami et al. used the same data originally published by Elsharkawy (2003). They presented three attempts based on genetic programming to correlate the bubble point pressure: SVR, ACE, and power-law committee machine (PLCM). The ACE-based method resulted in a set of three equations with a matrix of [8 by14] constants. Gholami et al. compared the accuracy of their model to the SRK-EOS and PR-EOS predictions reported by Elsharkawy (2003). Based on their AARD analysis, the PLCM model has the lowest errors followed by the SVR model, Elsharkawy's model, ACE model, PR-EOS, and SRK-EOS.

Lately, Jarrahian et al (2015) presented two empirical models to estimate the saturation pressure of black oils. The first model is based on production information. The second model is based on compositional information mostly from Elsharkawy (2003) and Heidaryan and Moghadasi (2011). They have not presented comparison of the accuracy of their model to Farasat (2013) model, Ahmadi (2014) model, and Gholami (2014) model.

This paper is concerned with the estimation of saturation pressure using compositional data. Therefore, the objective is to develop a robust yet simple model to estimate the saturation pressure for various types of reservoir oils using a larger data bank than that used by other published studies. The accuracy of the proposed models is compared to previously published models as well as SRK-EOS and PR-EOS.

Data bank

This study uses a large data set of 231 crude oil compositions and saturation pressure measurements. The data set consists of 130 data sets originally reported by Elsharkawy (2003); in addition to others collected from literature (Jacopy, 1958; Vogel, 1980; Coats, 1986; Droham, 1980; Dansh, 1992; Moharam, 1996; Rosenegger, 1999); 94 new data sets that have not been utilized before; as well as 7 newly measured samples at the Petroleum Fluid Research Center (PFRC) in Kuwait University. These data represents a wide range of compositions, such as volatile oils, black oils, and heavy oils. The data cover a pressure range of 537 to 5065 psia and a temperature range of 58 to 319 °F, crude oil compositions from C_1 to C_{7+}, non-hydrocarbons, such as N_2, H_2S, and CO_2, and the molecular weight and the specific gravity of the heptane plus fraction.

Data screening and classification

The compositions and the saturation pressure measurements in the data bank are subjected to a meticulous screening process to check the validity of the data. Our screening criteria uses the following steps: the sum of the molar compositions of all the components must add up to one, otherwise, the sample is eliminated. Samples with a missing molecular weight or specific gravity of the plus fraction are also eliminated.

After filtering, the data were reduced to 214 and divided into 179 measurements used to develop the new models, and 35 data sets are used for the blind test of the robustness of the proposed models. Statistical analysis of the input variables of the data set used in training and testing the extended and the lumped models are illustrated in Tables 2.

Several of the reported compositional data have extended analysis up to (C_{20+}). The following techniques are used for lumping or regrouping the extended analysis into single pseudo-fractions (C_{7+}). The mole fraction (X_i) of the individual components of the extended analysis is equal to the plus fraction (Ahmad, 2011).

$$X_{C_{7+}} = \sum_{i=7}^{n} X_i \qquad i=7, 8 \ldots n \qquad (1)$$

The average molecular weight (Mw) of the individual component of the extended analysis is equal to the molecular weight of the plus fraction.

$$Mw_{C_{7+}} = \sum_{i=7}^{n} \frac{X_i Mw_i}{X_{C_{7+}}} \qquad i=7, 8 \ldots n \qquad (2)$$

The density of the total plus-fraction can be calculated as follows:

$$\gamma_{C_{7+}} = \frac{X_{C_{7+}} Mw_{C_{7+}}}{\sum_{i=7}^{n} \frac{X_i Mw_i}{\gamma_i}} \qquad i=7, 8 \ldots n \qquad (3)$$

It is important to note the significance of the density of plus fraction because its density determines its paraffinic naphthenic aromatic (PNA) content.

Equation-of-state calculations

The EOS calculation of the saturation pressure is compared with the newly developed models as well as the other methods. The two EOS calculations considered in this paper are the ones most considered and widely used in the petroleum industry to evaluate the volumetric and phase behavior of reservoir fluid properties: the Soave-Redlich-Kwong EOS (SRK-EOS) and the Peng-Robinson EOS (PR-EOS). It should be noted that the accuracy of any EOS calculation of the saturation pressure largely depends on the number of pseudo-components and the splitting schemes used to divide the heptane plus fraction; as well as the characterization and calculation of the critical properties of the pseudo-components and the binary interaction parameters. Therefore, the two EOS calculations of the saturation pressure are not the same, even though the number of pseudo-components, characterization of the pseudo-components, and binary interaction parameters were the same.

Model development

As previously stated, the objective of this work is to provide the petroleum reservoir engineer with a simple yet accurate tool to estimate the saturation pressure for a variety of crude oils using compositional data when experimental is unavailable. Two models are presented in this study to calculate the saturation pressure of the crude oil samples. The first model uses all input parameters, such as the molar compositions of the hydrocarbon and non-hydrocarbon components similar to the EOS models. This model is referred to as the extended model. The second model uses fewer input parameters in which some of the hydrocarbon components have been lumped into a single group; this is designated as the lumped model. Both models are simpler than all published correlations, physically correct, and have an acceptable accuracy compared with the well-known EOS.

Figure 1 shows a correlation matrix of all the data available under consideration. This figure also indicates the relative importance of each of the input variables into the first proposed model. It agrees with most published correlations and the EOS that the methane content, heptane plus content and temperatures have the greatest impact on the saturation pressure followed by the intermediate content. The data understudy indicates that the non-hydrocarbon has the smallest impact. This result is simply a reflection of the molar contribution of each component comprising the crude oil composition.

Model I - Extended Model

The extended model uses 13 variables comprising the molar compositions of the pure components, heptane plus properties, and reservoir temperatures, which are similar to the input parameters for the EOS. Figure 2 shows the correlation coefficient graph of all the input parameters in the extended model. Thus, the elimination of the intermediate effect has a negative impact on the accuracy of the predictive model. Our proposed extended model has the following form:

$$Ps = 124.72N_2 + 17.57CO_2 + 22.55H_2S + 64.22C_1 - 9.80C_2 - 52.49C_3 + 6.16C_4 - 19.22C_5 - 23.63C_6$$

$$- 21.43C_{7+} + 435.31\gamma_{C7+} + 1.14Mw_{C7+} + 4.29T \qquad (4)$$

Where Ps is the saturation pressure in psia, T is the temperature in °F, and the hydrocarbon and non-hydrocarbon compositions are expressed in mole.

Model II - Lumped Model

The second model is much simpler and uses only 7 -variables (the lumped model). Several lumping schemes are adapted in the petroleum industry for phase behavior modeling and reservoir simulation of the petroleum reservoir fluids (Ahmed, 2011). Hydrocarbon components are grouped into a pseudo fraction based on their molecular weight (Whitson, 1980), the similarity of their physical-chemical properties (Lee, 1979), their carbon number, their true boiling point or their number of families based on the chemical structure (Becker et al., 2015). (Emera and Samara, 2005) proposed lumping the hydrocarbon components into volatile (C_1 & N_2), intermediate (CO_2, H_2S, C_2 to C_4) and heavy categories (C_{5+}) to calculate the minimum miscibility pressure. (Duan et al., 2013) proposed lumping several components into a single carbon number (SCN) with predetermined values for properties, such as critical pressure, temperature, and eccentric factor. Others have proposed optimal lumping schemes based on the objectives (Lin et al., 2008). In lumped model, $C_{1+}N_2$, C_2, CO_2, and H_2S, the intermediate (C_3 to C_6), heptanes plus (C_{7+}) and its properties, and temperature are the input variable. The correlation matrixes of the lumped components are provided in Figure 3. Even though the non-hydrocarbon has the least effect their contribution is important, especially the H_2S in the case of sour crudes. Both models presented by AlQuraishi (2009) and Bandyopadhyay and Sharma (2011) have eliminated the effect of H_2S.

The proposed model has the following form:

$$Ps = 82.115*(C_1+N2) - 11.635*C_2 + 39.158*CO_2 + 38.244*H_2S - 1.217*(C_3—C_6) - 890.701*(C_{7+}/MW_{7+}*\gamma_{7+}) + 4.217*T - 1042 \quad (5)$$

Where Ps is the saturation pressure in psia; and T is the temperature in °F.

Result and discussion

Model accuracy

The accuracy of the two proposed models used to calculate the saturation pressures of the previously described crude oil samples is shown in Table 3. This table shows reports the average relative error, average absolute relative error standard deviation and the correlation coefficients for all the methods considered in this study. Table 3 also reports the accuracy of the PR-EOS, SRK-EOS, Elsharkawy's model, AlQuraishi's model, and Bandoypandyay and Sharma's model and Jarrahian model. The methods proposed by Gholami et al., Ahmadi et al., and Jarrahian et al are included in Table 3, even though they produce a unreasonable estimate of the saturation pressures, as explained in the appendix.

This table indicates that the extended and the lumped models have an (AARE) in the order of 8% and 8.5%, respectively. It should be noted that the complications in the Bandoypandyay and Sharma model did not improve the accuracy of the saturation pressure, because it had a large error. It is also noted that Jarrahian et al model produces an average error of 35%. Compared with the EOS, the lumped model has a few constants and eliminates the splitting, characterizing the pseudo-components and binary interaction parameters necessary for the EOS calculations. The lumped model as well as the extended model are very simple and require one-step calculations compared with all the other methods reported in Table 3.

A cross-plot of the calculated saturation pressure using the extended model based on the training data versus the measured data is illustrated in Figure 4A. This figure indicates that most of the data falls on the unit slope line, and there is no over-fit. This trend is supported by the high correlation of regression of the model (97.6%) presented in Table 3. Figure 4B shows a cross-plot of the testing data (28 from literature and 7 new measured in PFRC at Kuwait University). The extended model indicates an excellent agreement with all of the 28 data sets chosen at random, with the exception of a few data points at higher pressures. Additionally, the model has a perfect match for the 7 newly measured samples.

Figure 5A shows the performance of the lumped training-model for capturing the measured saturation pressure for most of the 179 samples. A cross-plot of the calculated saturation pressure using the lumped model based on the training data versus measured data is depicted in Figure 5B. This figure indicates that most of the data scatters falls to the unit slope line. The lumped model also has a high coefficient of regression of 97.4%, which is indicated in Table 3.

Models Validity

The accuracy and validity of the models for estimating the saturation pressure of various types of crude oil are reported in Table 4. This table illustrates the capability of the newly proposed models to predict the saturation pressure for volatile oils, black oils, and heavy oils. Table 4; provides the compositions, properties of the heptane plus fraction, temperature, and measured saturation pressures of the samples. Additionally, the calculated saturation pressure using the PR-EOS, SRK-EOS, Elsharkawy's model (2003), AlQuraishi's model (2009), and Bandoypandhyay and Sharma's model (2011), Jarrahian et al (2015); as well as the extended and lumped models are provided in Table 4. It is clear from this table that the proposed models have a wide range of applications regardless of crude oil types. We have also included information regarding the 7 newly measured data; at the PFRC, compositional, and saturation pressure. At the bottom of Table 4, we have provided the unpublished data of the compositions; and the saturation pressures at low temperatures. It is well known that the saturation pressure of a crude sample is measured at low temperatures in addition to the reservoir temperature as a method to check the quality of the sample.

Figure 6A reports the error of all the methods considered in this paper to predict the saturation pressure for volatile oils (heptane plus content below 30%). Both the extended and the lumped models have an AARE of nearly 4% followed by Bandoypandhyay and Sharma's model (2011); and Elsharkawy's model (2003); the SRK-EOS has nearly twice the AARE of the newly proposed model. AlQuraishi's model (2009) has an AARE of 12% because the model is based on methane and neglects the effect of heavy and intermediate components, whose contribution is highly significant in the reservoir and production performance of volatile crudes. Jarrhaine et al (2015) model is not suitable for volatile oils. It shows a 35% error in estimated saturation pressure.

The performance of the extended and lumped models compared to the various methods for the black oil samples ($35< C_{7+} <55\%$) is provided in Figure 6B. The newly proposed models have an AARE close to 10%; followed by Bandoypandyay and Sharma's model 14%, the SRK-EOS and PR-EOS 16%, and Elsharkawy 17%. AlQuraishi's model has an AARE of 24%. Jarrhain et al (2015) model shows a 33% error in estimated saturation pressure even though the authors claim that the model is suitable for black oils.

Figure 6C shows the error level of all the methods as well as the proposed models for the newly measured crude oil samples at Kuwait University using the PFRC. Again, Jarrhain et al (2015) model show a 40% error in estimated saturation pressure.

The performance of all published models at low temperature is illustrated in Figure 6D. The low-temperature calculation indicates that Bandoypandhyay and Sharma's model (2011) has accuracy problems and is not valid at low temperatures because the authors proposed a temperature-composition related coefficient. Also, Jarrhain et al (2015) model is not suitable at low temperature. It shows a 40% error in estimated saturation pressure.

The ability of the proposed models to capture the thermodynamic is thoroughly checked against its independent variables. The validity of the model is compared to the experimentally measured data as well as the predictions by other methods used to capture the physics of increasing the saturation pressure as a function of temperature is illustrated in Figure 7. Two main points are observed here; first, the model matches the measured saturation pressures much better than the other methods. Second, it follows the physical trend predicted by the EOS's. Figure 8 shows the model performance for capturing the effect of changing the saturation pressure as a function of the molecular weight of the heptane plus fraction compared with the SRK-EOS and PR-EOS. The extended model follows the trend of the PR-EOS in increasing the saturation pressure. Elsharkawy's model (2003) was unable to predict the correct trend as predicted by the EOS's. The model sensitivity in capturing the change in the saturation pressure as a function of increasing the methane content is shown in Figure 9. The proposed model as well predicts the correct trend of increasing the saturation pressure when increasing the methane content.

Conclusion

In this paper, two models are presented to estimate the saturation pressure of a variety of crude oils when measures are unreliable or samples are unavailable. The two models are developed using compositional data and saturation pressure measurements from a large data bank comprising literature data, unpublished data, and newly measured and reported data of 231 crude oil samples. The input parameters in the first model are the extended hydrocarbon and non-hydrocarbon molar percentages and reservoir temperatures. The second model uses lumped hydrocarbon fractions of intermediate, and heptane plus fractions in addition to its properties; and the reservoir temperature as input parameters. The accuracy of the two models has been tested and compared to that of the SRK-EOS and the PR-EOS; as well as published correlations that use compositional data as the input parameters. The comparison indicates that the two models are much more accurate than all the published methods. The validity of both models has also been tested using various types of crude oils samples and compared to experimentally measured saturation pressures and estimates using the EOS as well as the published correlations. The proposed models are physically correct and much simpler than all the published methods. The models do not require tedious computations and eliminate splitting of the plus fraction, characterization of the pseudo-components; as well as the binary interaction parameters that are required for the EOS calculations.

Acknowledgement

The authors thank the General Facility Lab at Kuwait University for their cooperation and conducting the compositional and saturation pressure measurements of the 7 new samples described in this paper using Petroleum Fluid Research Center (PFRC) project no. GE 01107.

References

Agarwal, R.K., Li, Y.K., Nghiem, L., 1990. A regression technique with dynamic parameter selection for phase – behavior matching. SPE Res. Eng. February, 115 – 120.

Ahmadi, M. A., Zendehboudi, S, James, L., Elkamel, A., Dusseault, M., Chatsiz, I., Lohi, A., 2014, New Tools to determine Bubble Point Pressure of Crude Oils: Experimental and Modeling Study, *Journal of Petroleum Science and Engineering, v.123, 204-216.*

Ahmed, T., 2011, Splitting and lumping schemes of the plus fraction, working guide to vapour-liquid phase equilibria calculations, Chapter 9, Hydrocarbon Phase Behaviour, Gulf Publishing Co. 111-126.

Al-Gathe, A., Abd-el Fattah, Kh., El-Mtwally, K., 2014, New artificial intelligent approach for bubble point pressure, Int. J innovation and Sci. Research, Vol.4, No.2, July, 121-135.

Al-Marhoun, M.A. 1985. Pressure-Volume-Temperature Correlations for Saudi Crude Oils. Presented at the SPE Middle East Oil Technical Conference and Exhibition, Bahrain, 11–14 March. SPE-13718.

Almehaideb, R.A. 1992. Improved PVT Correlations for UAE Crude Oils. Presented at the Middle East Oil Show and Conference, Bahrain, 15-18 March. SPE-37691.

Al-Najjar, H.S., Al-Soof, N.B.A., and Al-Khalisy, K.M. 1988. Correlations for Bubble-Point Pressures, Gas Oil Ratios and Formation Volume Factors for Iraqi Crude Oils. *Journal of Petroleum Research* (June 1988): 13.

AlQuraishi, A. A., 2009, Determination of crude oil saturation pressure using linear genetic programing, Energy & Fuel, 23, 884-887.

Al-Shammasi, A.A. 2001. A Review of Bubblepoint Pressure and Oil Formation Volume Factor Correlations. *SPE Res. Eval. & Eng.* 4 (2): 146-160. SPE71302.

Asgarpour, S., McLauchlin, L., Wong, D. et al. 1989. Pressure-Volume-Temperature Correlations for Western Canadian Gases and Oils. *J Can Pet Tech.* 28 (4): 103.

Asoodeh, M., Bagheripour, P.,2012, Estimation of bubble point pressure from *PVT* data using a power-law committee with intelligent systems, *J. Pet. Sci. and Eng., Vol. 90–91, July 2012, Pages 1-11.*

Bandyopadhyay, P., Sharma, A., 2011, Development of a new semi analytical model for prediction of bubble point pressure of crude oils, *Journal of Petroleum Science and Engineering, Volume 78, Issues 3–4, September 2011, Pages 719-731.*

Becker, P. J., Celse, B., Guillaume, D., Dulot, H., Costa, V., 2015, Hydrotreatment modelling for variety of VGD feedstocks: a continuous lumping approach, Fuel, 139, 133-143.

Coats, K.H., and Smart, G.T., 1986, Application of regression base EOS PVT Program to laboratory data, SPERE, May 277-299.

Dansh, A. Xu, D. H., Todd, A. C., 1992 A grouping method to optimize reservoir description for compositional simulation of gas injection process, SPERE, Aug. 343-348.

De Ghetto, G. and Villa, M. 1994. Reliability Analysis on PVT Correlations. SPE-28904, Presented at the European Petroleum Conference, London, United Kingdom, 25-27 October.

Dindoruk, B. and Christman, P.G. 2001. PVT Properties and Viscosity Correlations for Gulf of Mexico Oils. Presented at the SPE Annual Technical Conference and Exhibition, New Orleans, 30 September-3 October. SPE-71633

Dokla, M.E. and Osman, M.E. 1992. Correlation of PVT Properties for UAE (United Arab Emirates) Crudes. *SPE Form Eval.* (March 1992): 41.

Droham, J.K., 1988, Enhancing the evaluation of PVT data, OSEA paper presented the 7th Offshore South East Asia, Feb. 2-5, Singapore.

Duan, J., Wang, W., Liu, H., Cong, J., 2013, modelling the characterization of the plus fraction by using continuous distribution function, Fluid Phase Equilibria, 345, 1-10.

Elam, F.M. 1957. *Prediction of Bubble Point Pressures and Formation Volume Factors from Field Data.* MS thesis, University of Texas at Austin, Austin, Texas.

Elmabrouk, S., Zekri, A., Shirif E., 2014, The prediction of bubble point pressure and bubble point oil formation volume factor in absence of PVT analysis, PET. Sci. & Tech., 32, 1168-1174.

Elsharkawy, A. M., 2003, An empirical model for estimating the saturation pressures of crude oils, Journal of Petroleum Science and Engineering 38 (2003) 55-77.

Elsharkawy, A.M. and Alikhan, A.A. 1997. Correlations for predicting solution gas/oil ratio, oil formation volume factor, and undersaturated oil compressibility. *J. Pet. Sci. Eng.* 17 (3–4): 291-302.

Emera, M., Samara, H., 2005, Use of genetic algorithm to estimate CO2-oil minimum miscibility pressure- a key parameter in design of CO2 miscible flood, J. Pet. Sci. Eng., Vol. 46, issue 1-2, 37-52.

Farasat, A., Shokrollahi, A., Arabloo, M., Gharagheizi, F., Mohammadi, A., 2013, toward an intelligent approach for determination of saturation pressure of crude oil, Fuel Processing Technology 115 (2013) 201-214.

Frashad, F., LeBlanc, J.L., Garber, J.D. et al. 1996. Empirical PVT Correlations for Colombian Crude Oils. SPE-36105 Presented at the SPE Latin American and Caribbean Petroleum Engineering Conference, Port of Spain, Trinidad and Tobago, 23–26 April. -MS.

Gharbi, R.B.C., Elsharkawy, A.M., 1999, neural network model for estimating the PVT properties of Meddle East crude oils, SPE Reservoir Eval. Eng. 2 (3), 255-265.

Gholami, A., Asoodeh, M., Bagheripour, P., 2014 How committee machine with SVR and ACE estimates bubble point pressure of crudes, *Fluid Phase Equilibria, Volume 382, 25 November 2014, Pages 139-149.*

Glasø, Ø. 1980, Generalized Pressure-Volume-Temperature Correlations. *J Pet Tech.* 32 (5): 785-795.

Heidaryan, E., Moghadasi, J., 2011, Crude saturation pressure through imperial correlation, Thermodynamics 2011, Athens, 31 August- 3 September 2011.

Hasan, M., 1993, PVT Correlations for Indonesian Crude Oils. *Proc.* 22nd Annual Convention of the Indonesian Petroleum Association, Indonesia, Vol.1, IPA93-2.1-070.

Jarrhian, A., Moghadsi, J., Heidaryan, E., 2015, Empirical estimating of black oil bubblepoint (saturation) pressure, J. Petroleum Science and engineering, (126), 69-77.

Jacopy R.H., Berry V.J.,1958, A method for predicting the pressure maintenance for reservoir producing volatile oil, AIME 213, 59-65.

Kartoatmodjo, T., Schmidt, Z. 1994, Large Data Bank Improves Crude Physical Property Correlations. *Oil Gas J.* 92 (27): 51–55.

Kartoatmodjo, T.R.S. and Schmidt, Z. 1991, *New Correlations for Crude Oil Physical Properties*, SPE 23556.

Khairy, M., El-Tayeb, S., Hamdallah, M., 1998. PVT Correlations Developed for Egyptian Crudes. *Oil Gas J.* 96 (18): 114.

Khamekhcki, E., Behvandi, R., Rashid, F., 2011, Prediction of bubble point pressure & asphaltene onset pressure during CO2 injection using ANN & ANFIS models, Pet. Sci. & Tech. Vol. 1 No.2, 35-45.

Labedi, R.M. 1982, *PVT Correlations of the African Crudes*, PhD thesis, CSM.

Labedi, R.M. 1990, Use of Production Data to Estimate the Saturation Pressure, Solution GOR, and Chemical Composition of Reservoir Fluids, SPE-21164-MS. Presented at the SPE Latin America Petroleum Engineering Conference, Rio de Janeiro, Brazil, 14-19 October,

Lasater, J.A. 1958, Bubble Point Pressure Correlations. *J Pet. Tech.*10 (5): 65–67. SPE-957-G.

Levitan, L.L. and Murtha, M. 1999. New Correlations Estimate Pb, FVF. *Oil Gas J.* 97 (10): 70.

Lee, S., 1979, Experimental and theoretical studies on the fluid properties required for simulation of thermal process, Paper SPE 8393.

Li, Y.K., Nghiem, L.X., Siu, A., 1985. Phase behavior computations for reservoir fluids: effect of pseudo-components on the phase diagrams and simulation results. J. Can. Pet. Technol. Nov. – Dec., 29 – 36

Lin, B., Leibovici, C. F., Jorgensen, S. B., 2008, optimal component lumping: problem formulation and solution techniques, Computer & Chem. Eng., 32, 1167-2272.

Macary, S.M. and El-Batanoney, M.H. 1992, Derivation of PVT Correlations for the Gulf of Suez Crude Oils. *Proc. of the* 11th EGPC Pet. Exp. & Prod. Conference, Vol. 1, 374, Cairo, Egypt,

Moharam, H.M., 1995. Prediction of viscosity of heavy petroleum fractions and crude oil using a corresponding states method. Ind. Eng. Chem. Res. 34, 4140 – 4144.

Moradi, B., Malekzadeh, E., Amani, M., Boukadi, F.H., Kharat, R., 2101, Bubblepoint pressure imperial correlation, SPE 132756.

Obomanu, D.A. and Okpobiri, G.A. 1987, Correlating the PVT Properties of Nigerian Crudes. *J. Energy Res. Tech.* 109 (4): 214-217.

Okoduwa, I. G., Ikiensikimama, S.S., 2010, Bubble point pressure correlation for Niger Delta Crude Oils, SPE 136968.

Omar, M.I. and Todd, A.C. 1993. Development of New Modified Black Oil Correlations for Malaysian Crudes. SPE-25338 Presented at the SPE Asia Pacific Oil and Gas Conference, Singapore, 8-10 February.

Owolabi, O.O. 1984. *Reservoir Fluid Properties of Alaskan Crudes*. MS thesis, University of Alaska, Fairbanks, Alaska (May 1984).

Pedersen, K.S., Thomassen, P., Fredenslund, Aa., 1988a. Characterization of gas condensate mixtures. AIChE Spring National Meeting, March 6 – 10, New Orleans, LA.

Peng, D. y., Robinson. D. B., 1976. A new two-constant equation of state , Ind. Eng. Chem. Fund. 15, 59-63.

Petrosky, G.E. Jr. 1990. *PVT Correlations for Gulf of Mexico Crude Oils*. MS thesis, University of Southwestern Louisiana, Lafayette, Louisiana.

Petrosky, G.E. Jr. and Farshad, F. 1998. Pressure-Volume-Temperature Correlations for Gulf of Mexico Crude Oils. *SPE Res Eval & Eng.*1 (5): 416-420. SPE-51395.

Riemens, W.G., Schutle, A.M., Jong, L.N.J., 1988. Birba field PVT variations along the hydrocarbon column and confirmatory field tests. J. Pet. Technol. Jan., 83 – 88

Rosengger, L., Wu., R.1990, Integrated PVT data characterization- lessons from four case histories, J. Can. Per. Tech., 38, (13), (paper 97-05)

Shojaei, M., Bahrami, F., Barati, P., Riahi, S., 2014, Adaptive neuro-fuzzy approach for reservoir oil bubble point pressure estimation, *J. of Natural Gas Sci. and Eng., Volume 20, September 2014, Pages 214-220.*

Soave, G. 1972. Equilibrium constants from a modified Redlich-Kwong equation of state, Chem. Eng. Sci. 27, 1197-1203.

Standing, M.B. 1947. A Pressure-Volume-Temperature Correlation for Mixtures of California Oils and Gases. *API Drilling and Production Practice* (1947): 275-287.

Tale, M., Alquraishi, A., Al Mushaigeh, E. 2009, New model for crude oil saturation pressure prediction using alternation conditional expectation, Offshore Mediterranean Conf. & Exhibit. Italy 25-27, 2009.

Talebi, R., Ghiasi, M. Talebi, H., Mohammadyian, M., Zendehboudi,S., Arabloo, M., Bahadori, A., 2014, Application of soft computing approaches for modelling saturation pressure of reservoir oil, *J. Nat. Gas Sci. and Eng., Vol. 20, Sep. 2014, 8-15.*

Vazquez, M., Beggs, H. D. 1980, Correlations for Fluid Physical Property Prediction. *J Pet. Tech.* 32 (6): 968-970. Vazquez, M.E. 1976. *Correlations for Fluid Physical Property Prediction,* MS thesis, University of Tulsa.

Velarde, J., Blasingame, T.A., and McCain Jr., W.D. 1997. Correlation of Black Oil Properties at Pressures below Bubble Point Pressure - A New Approach. Presented at the Annual Technical Meeting of CIM, Calgary, Alberta, 8–11 June. PETSOC 97-93.

Vogel, J.R., Yarborough, L. 1980, the effect of nitrogen on phase behavior and phase properties of reservoir fluids, SPE 8815.

Whitson, C. H., 1980, Characterizing hydrocarbon plus fraction, SPEJ, 24 (4), 683-694.

Williams, C.A., Zana, E.N., 1980. Use of the Peng-Robinson equation of state to predict hydrocarbon phase behavior and miscibility for fluid displacement. SPE 8817, SPE/DOE on Enhanced Oil Recovery, April 20–23, Tulsa, OK.

Appendix

In this appendix, sample calculations are presented to show the deficiency of some of the published methods.

Input parameters:

N_2	CO_2	H_2S	C_1	C_2	C_3	C_4	C_5	C_6	C_{7+}	GR.C7+	MWC7+	T, °F	Ps, psia
0.65	0.02	0	45.02	12.45	8.93	6.03	3.02	1.44	22.44	0.81	184	140	3002

Ahmadi et al (2014)

Ahmadi et al. proposed an 8 steps calculation (copied from their paper) to estimate the BPP as follows:

$$BPP = A \times Mw_{C7+} - \frac{B}{C_{7+}} + C \times Vol./Inter. + D \times SG_{C7+} \frac{E}{SG_{C7+}} + \frac{F}{Mw_{C7+}} + T \times 505.967 + \frac{2.45789 \times 10^6}{T} + \frac{G}{Vol./Inter.} \quad (5)$$

$$A = 1488.46 - SG_{C7+} \times 1087.64 - \frac{601.512}{SG_{C7+}} + 1670.58 + T \times 0.40771 + \frac{6345.73}{T} + \frac{8.72863}{Vol./Inter.} \quad (6)$$

$$B = \frac{4132.53}{Vol./Inter.} + \frac{6.1977 \times 10^7}{Mw_{C7+}} + T \times 3181.39 + \frac{1.06393 \times 10^8}{T} \quad (7)$$

$$C = \frac{20964.7}{C_{7+}} - \frac{1.76162 \times 10^6}{Mw_{C7+}} - T \times 19.1963 - Mw_{C7+} \times 21.5427 + \frac{12851.5}{SG_{C7+}} - \frac{712613}{T} - 3477.49 + SG_{C7+} \times 9532.73 \quad (8)$$

$$D = 1.05345 \times 10^{+06} - \frac{1.50748 \times 10^8}{Mw_{C7+}} - T \times 676.048 - \frac{1.20252 \times 10^7}{T} - \frac{12178.2}{Vol./Inter.} \quad (9)$$

$$E = 641357 - \frac{1.02891 \times 10^8}{Mw_{C7+}} - T \times 301.034 - \frac{1.91446 \times 10^6}{T} - \frac{3873.62}{Vol./Inter.} \quad (10)$$

$$F = 2.51471 \times 10^{+08} + T \times 25332.5 + \frac{3.47893 \times 10^8}{T} - \frac{244573}{Vol./Inter.} \quad (11)$$

$$G = 20125.6 - T \times 23.9271 - \frac{703193}{T} \quad (12)$$

Ahmadi et al. define the Vol. /Inter. as the mole percentage ratio of the volatile (C_1& N_2) to the intermediate components (C_2 to C_6, CO_2 and H_2S). Application of the above equations from their paper using the input parameters results in the following:

Vol./Inter.	A	B	C	D	E	F	G	BPP
1.43211	1643.9	1545062	-271.127	45122.71	23642.47	2.57E+08	11753	1,794,123

The calculated saturation pressure (BBP) from Eq.5 is 1,794,123 psia which is unreasonable. This indicates a deficiency in the model.

Gholami et al (2014)

Gholami et al. presented 3 steps calculation (copied from their paper) to estimate the BPP as follows:

$$Tr(M) = \sum_{i=0}^{5} B_i M^i \quad (15)$$

Application of Eq. 15 resulted in the following

N_2	0.00584
CO_2	0.060917
H_2S	0.050636
C_1	-0.47658
C_2	-0.41547
C_3	-0.45083
C_4	-0.02003
C_5	0.068769
C_6	0.273921
C_{7+}	1.814523
SG.C7+	-0.0197
MWC7+	37.03525
T., F	-0.11139

Application of Eq. 16

$$Tr(P_b) = Tr(N_2) + Tr(CO_2) + Tr(H_2S) + Tr(C_1) + Tr(C_2) + Tr(C_3) + Tr(C_4) + Tr(C_5) + Tr(C_6) + Tr(C_{7+}) + Tr(SG_{C7+}) + Tr(MW_{C7+}) + Tr(T) \quad (16)$$

Resulted that Tr(Pb) = 37.8159

$$P_b = (57.08403877741 \times (Tr(P_b))^2) + (1166.9102304199 \times (Tr(P_b))) + 2207.3573204358 \quad (17)$$

Application of Eq. 17 resulted in saturation pressure (P_b) = 127,967 psia, which is unreasonable, compared to the measured one is 3002 psia. This indicates that there is a deficiency in the model.

Jarrahian et al (2015)

Jarrahian et al. proposed the following Eq. (copied from their paper) to estimate the saturation pressure:

$$P_{sat} = A_0 \exp\Big(A_1 z_{N_2} + A_2 z_{CO_2} + A_3 z_{H_2S} + A_4 z_{C_1}^{A_5} + A_6 \sum_2^5 z_{C_i} + A_7 z_{C_{7+}}^{A_8} \gamma_{C_{7+}}^{A_9} Mw_{C_{7+}}^{A_{10}} + A_{11}(T - 459.67)\Big) \quad (2)$$

Application of Eq. 2 resulted in saturation pressure (Ps) = 1812 psia, compared to the measured one is 3002 psia. This indicates that model has inaccuracy problem. This has been the pattern in all the crude oil samples shown in Table 4 and figures 6A through 6D.

Table 1- Summary of attempts to calculate the saturation pressure of crude oils

Method based on production information	Method based on compositional data
Standing, 1947; Elam, 1957; Lasater, 1958; Vasquez, 1976; Vasquez and Beggs, 1980; GlasØ, 1980; Labedi, 1982&1992; Owolabi, 1984; Al-Marhoun, 1985; Obomanu and Okpobiri, 1987; Asgarpour et al., 1989; Al-Najar et al., 1988; Dokla and Osman, 1990; Petrosky, 1990; Kartoatmodjo and Schmidt, 1991&1994; Macary and El-Batanoney, 1992; Omar and Todd, 1993; Hasan, 1993; De Ghetto and Villa, 1994; Farashad et al., 1996; Almehaidaeb, 1997; Elsharkawy, 1997; Velarde et al., 1997; Kairy et al., 1998; Petrosky and Farshad, 1988; Al-shammasi, 2001; Gharbi & Elsharkawy 1999 Levitan and Murtha, 1999; Dindrouk and Christman, 2001 Moradi et al., 2010 Okoduwa and Ikiensikimama 2010 Asoodeh and Bagheripour, 2012 Shojaei et al., 2014 Talebi et al., 2014 Elmabrouk et al., 2014 Kanu and Ikiensikimama, 2014 Jarrahian et al 2015	Elsharkawy , 2003 Tale et al., 2009 AlQuraishi, 2009 Bandyopadhyay and Sharma, 2011 Khamehchi et al., 2011 Farasat et al., 2013 Ahmadi et al., 2014 Gholami et al. 2014 Jarrahian et al 2015

Table 2- Description of the data used to develop and test the extended model

Input	Training model (179 Data)			Testing (28 Data)			Testing (7 new experimental)		
	Min	Ave	Max	Min	Ave	Max	Min	Ave	Max
N_2	0.00	0.35	1.67	0.00	0.27	1.67	0.10	0.26	0.35
CO_2	0.00	1.27	11.37	0.00	1.17	11.37	0.14	0.41	0.83
H_2S	0.00	0.15	3.68	0.00	0.15	3.22	0.00	0.00	0.00
C_1	13.16	33.86	74.18	19.50	34.79	73.36	20.64	28.25	42.79
C_2	3.36	8.07	13.71	4.64	8.26	11.63	6.69	7.88	10.82
C_3	0.89	6.59	11.87	2.46	6.76	12.03	6.05	6.59	7.10
C_4	0.95	4.46	8.4	1.66	4.41	6.58	4.03	5.05	5.93
C_5	0.40	3.21	5.95	1.00	3.11	5.23	3.36	4.33	5.52
C_6	0.00	3.14	6.37	0.99	3.15	5.46	2.88	3.74	4.33
C_{7+}	10.72	38.89	57.73	11.18	37.88	51.22	27.61	43.49	51.31
$SG.C_{7+}$	0.74	0.87	0.959	0.70	0.86	0.93	0.85	0.89	0.92
MWC_{7+}	134	240	368.9	105	223	292	216	262	298
T, °F	58	156	319	58	165	261	121	157	230
Ps, psia	537	2198	5065	1025	2307	5000	1227	1779	3344

Table 3: Error analysis of the proposed models in comparison with all the investigated methods

Method	ARE	AARE	SD	R^2%
PR-EOS	6.24%	10.19%	1028	97.9
SRK-EOS	-3.12%	10.36%	1024	96.7
Elsharkawy (2003)	0.85%	8.57%	1015	96.4
Alquraishi (2009)	-6.06%	10.16%	1088	96.7
Bandyopadhyay (2011)	8.05%	12.28%	1054	96.8
Farasat (2013)	NA	NA	NA	NA
Gholami et al. (2014)	?	?	?	?
Ahmadi et al. (2014)	?	?	?	?
Jarrahian et al (2015)	35.52	35.56	983	95.2
New Extended Model	-0.37%	8.20%	1030	97.6
New Lumped Model	-0.50%	8.46%	1036	97.4

ARE = Average Relative Error

AARE = Average Absolute Relative Error

SD = Standard Deviation

R^2 = Correlation Coefficient

NA = Calculation was not possible, because the weight and bias were not provided.

? = The method produces unreasonable results as indicated in the appendix.

Table 4- Comparing model accuracy for calculations of saturation pressures for various crude oils

Volatile-Oil-Samples (C7+<30)

Fluid Source	index	N2	CO2	H2S	C1	C2	C3	C4	C5	C6	C7+	GR.C7+	MWC7+	T, °F	Ps, psi	SRK	PR	Elsh	Alq	Bandy	Jarr	Extd	Lump
Rosenegger	7	0.03	8.39	0	47.4	10.3	6.12	4.2	2.89	2.05	18.6	0.83	180	295	4000	3933	3879	4460	4420	4165	3317	4139	4178
Vogel-80	13	0.65	0.02	0	45	12.5	8.93	6.03	3.02	1.44	22.4	0.81	184	140	3002	2624	2548	2742	2895	2643	1812	3019	2996
Dansh-92	18	0	0	0	46.8	8.77	7.44	4.01	2.56	4.02	26.4	0.77	158	212	2941	2550	2594	3248	3617	3269	1944	3244	3374
	31	0.44	0.38	0	49.1	7.6	6.13	3.84	2.52	2	28	0.84	231	199	3739	3522	3507	3418	3951	3508	2309	3625	3644
Middle East	45	0.03	0.97	0	41.6	10.8	7.52	4.5	3	2.43	29.1	0.85	208	241	3069	3109	2979	3080	3513	3138	1962	3127	3140
Coats-86	51	0.55	1.03	0	36.5	9.93	8.85	6	3.78	3.56	30.4	0.84	200	234	2746	2537	2472	2714	2945	2643	1682	2777	2720

Black-Oil -Samples (35<C7+<55)

Fluid Source	index	N2	CO2	H2S	C1	C2	C3	C4	C5	C6	C7+	GR.C7+	MWC7+	T, °F	Ps, psi	SRK	PR	Elsh	Alq	Bandy	Jarr	Extd	Lump
Coats-86	73	1.64	0.08	0	28.4	7.16	10.5	8.4	3.82	4.05	36	0.84	252	131	1694	1674	1618	1615	1738	1644	1084	1713	1713
Middle East	88	0.08	1.37	0.69	36	8.67	5.94	2.97	1.5	3.1	39.7	0.9	274	168	2505	2710	2318	2247	2608	2321	1476	2441	2445
Middle East	114	0.29	0.48	0	28.4	8.29	7.38	5.06	3.42	4.41	42.3	0.88	252	133	1632	1672	1477	1544	1748	1517	984	1587	1598
Middle East	122	0	1.25	0	33.4	9.24	6.07	2.4	1.62	2.78	43.3	0.85	252	230	2110	2313	2278	2424	2737	2438	1410	2388	2412
Middle East	134	0	0.17	0	28.6	8.63	7.43	3.81	2.61	4.11	44.7	0.86	249	134	1548	1551	1429	1477	1764	1433	937	1503	1567
Middle East	146	0.44	0.83	0	27.8	7.54	6.9	4.61	3.1	3.11	45.7	0.9	272	133	1565	1799	1505	1558	1730	1528	1002	1602	1590
moharam-96	149	0.21	0.34	0	20	7.93	8	6.6	5.87	5.08	45.9	0.86	230	235	900	1353	1290	1531	1514	1225	773	1303	1294
Middle East	162	0.35	0.47	0	26.5	7.71	6.05	4.03	4.08	4.22	46.6	0.89	253	135	1500	1613	1380	1427	1631	1391	923	1478	1456
Middle East	176	0.12	0.51	0	28.8	7.91	6.46	3.17	2.07	3.25	47.7	0.88	250	134	1615	1665	1476	1574	1783	1502	981	1571	1614
Middle East	194	0.21	0.15	0	27.8	7.68	6.19	3.41	1.81	2.84	49.9	0.86	228	134	1590	1532	1377	1545	1680	1426	923	1466	1493
Rosenegger	207	0.37	0.02	0	17.3	6.11	8.23	5.47	3.76	3.96	54.8	0.85	242	156	537	938	888	914	1064	836	558	661	731

Newly-Measured Kuwait University Laboratory																						
Fluid Source	index	N2	CO2	H2S	C1	C2	C3	C4	C5	C6	C7+	GR.C7+	MWC7+	T, °F	Ps, psi	SRK	PR	Elsh	Alq	Bandy	Jarr	Extd	Lump
KU-1	209	0.33	0.22	0	25.6	6.87	6.39	5.61	4.68	4.33	46	0.879	222	133	1595	1360	1530	1442	1514	1394	860	1357	1337
KU-2	210	0.35	0.47	0	26.5	7.71	6.05	4.03	4.08	4.22	46.6	0.893	253	135	1500	1400	1620	1427	1631	1391	924	1478	1456
KU-3	211	0.27	0.74	0	23.1	7.2	6.45	5.93	4.66	4.16	47.5	0.906	290	138	1283	1140	1360	1284	1448	1296	800	1287	1219
KU-4	212	0.22	0.26	0	20.6	6.69	6.65	5.11	5.52	3.6	51.3	0.919	298	171	1227	1120	1300	1188	1415	1170	742	1176	1131
KU-5	213	0.19	0.14	0	27.8	8.37	6.88	5.28	4.33	3.85	43.2	0.89	278	121	1524	1300	1560	1402	1659	1409	928	1534	1489
KU-6	214	0.33	0.24	0	31.3	7.48	6.58	4.86	3.7	3.17	42.3	0.9	274	171	1978	1800	2200	1975	2242	1985	1224	2060	2026
KU-7	215	0.1	0.83	0	42.8	10.8	7.1	4.51	3.36	2.88	27.6	0.85	216	230	3344	3070	3400	3083	3568	3139	2003	3213	3200

Low-Temperature Oil Samples (59°F<T<71°F)																						
Fluid Source	index	N2	CO2	H2S	C1	C2	C3	C4	C5	C6	C7+	GR.C7+	MWC7+	T, °F	Ps, psi	SRK	PR	Elsh	Alq	Bandy	Jarr	Extd	Lump
Middle East	101	0	1.12	0.21	27	10.8	9.23	4.97	2.62	3.33	40.8	0.88	249	71	1247	1090	1033	1040	1235	564	808	1148	1205
Middle East	117	0.36	0.17	0	27.2	8.93	8.6	6.07	3.71	2.55	42.4	0.88	271	68	1130	1028	1245	1094	1248	766	820	1239	1228
Middle East	135	0	0.17	0	28.6	8.63	7.43	3.81	2.61	4.11	44.7	0.86	249	70	1270	1055	1255	1130	1311	783	826	1229	1297
Middle East	138	0.14	0.54	0	29.4	7.96	6.19	3.96	3.25	3.84	44.7	0.9	290	59	1265	1065	1350	1191	1301	880	894	1408	1390
Middle East	147	0.44	0.83	0	27.8	7.54	6.9	4.61	3.1	3.11	45.7	0.9	272	67	1320	1125	1400	1201	1268	921	879	1319	1312
Middle East	173	0.06	0.81	0	31.3	7.42	5.68	2.82	1.63	2.64	47.6	0.89	270	68	1460	1215	1535	1441	1463	1026	975	1548	1576
Middle East	193	0.16	0.3	0	24.7	7.51	6.92	4.76	2.65	3.43	49.6	0.88	239	71	1050	975	1156	1008	1107	701	714	958	986

Elsh: Elsharkawy (2003)
Alq: AlQuraishi (2009)
Band: Bandyopadhyay and Sharma (2011)
Jarr: Jarrahian et. al (2015)
Extd: Extended model
Lump: Lumped model

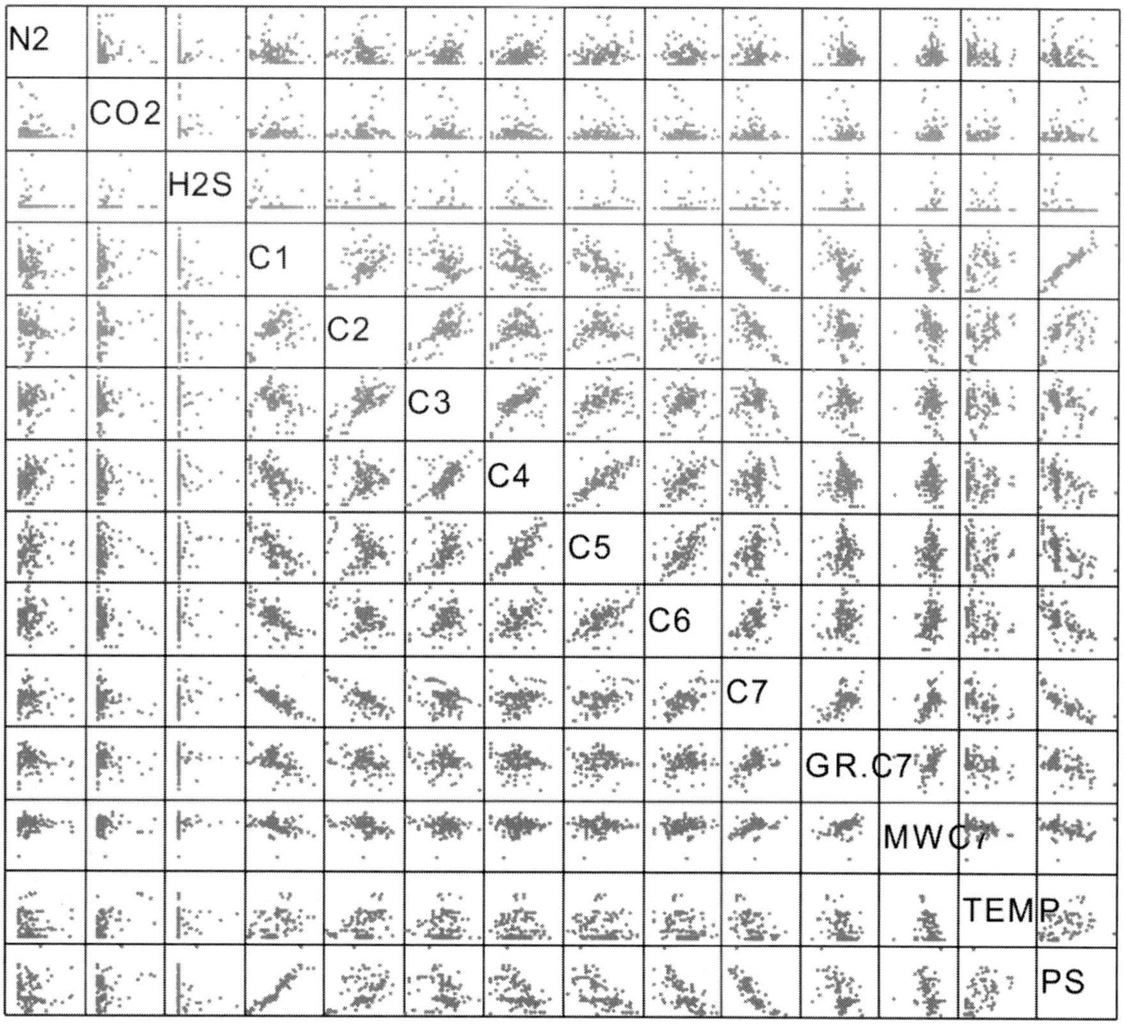

Figure 1: Correlation matrix for all the variables in the oil data bank

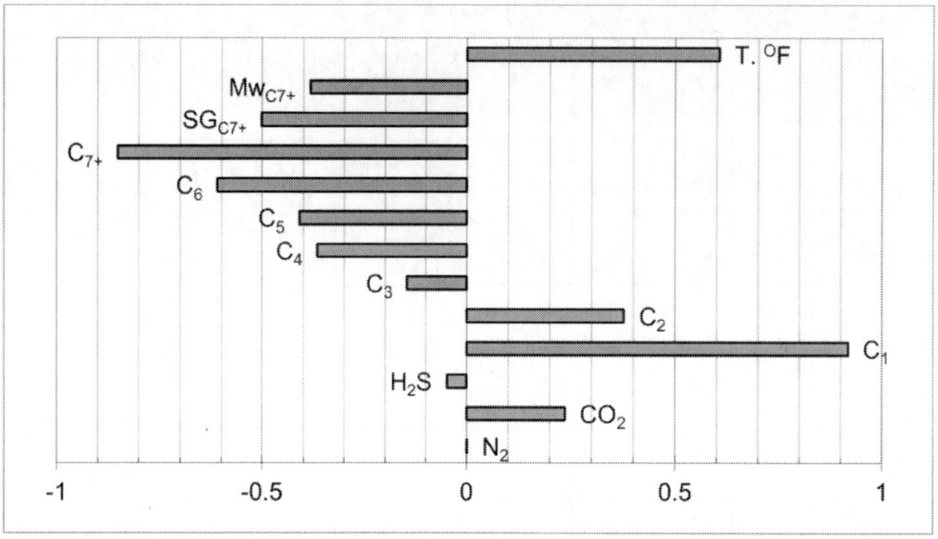

Figure 2: Correlation coefficient for all the variables in the training model (179 data)

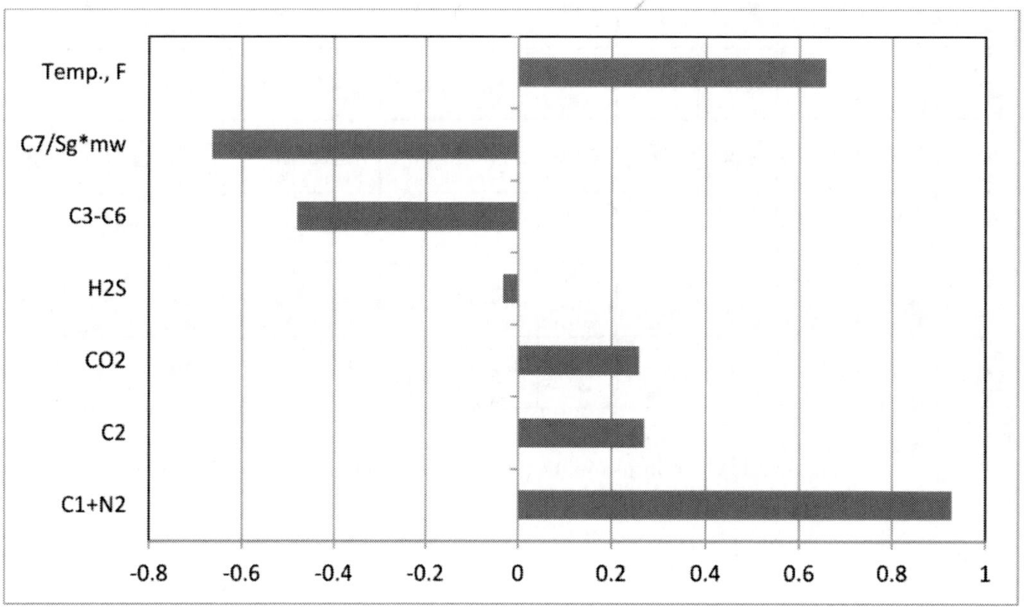

Figure 3: Correlation matrix for the lumped model (179 data)

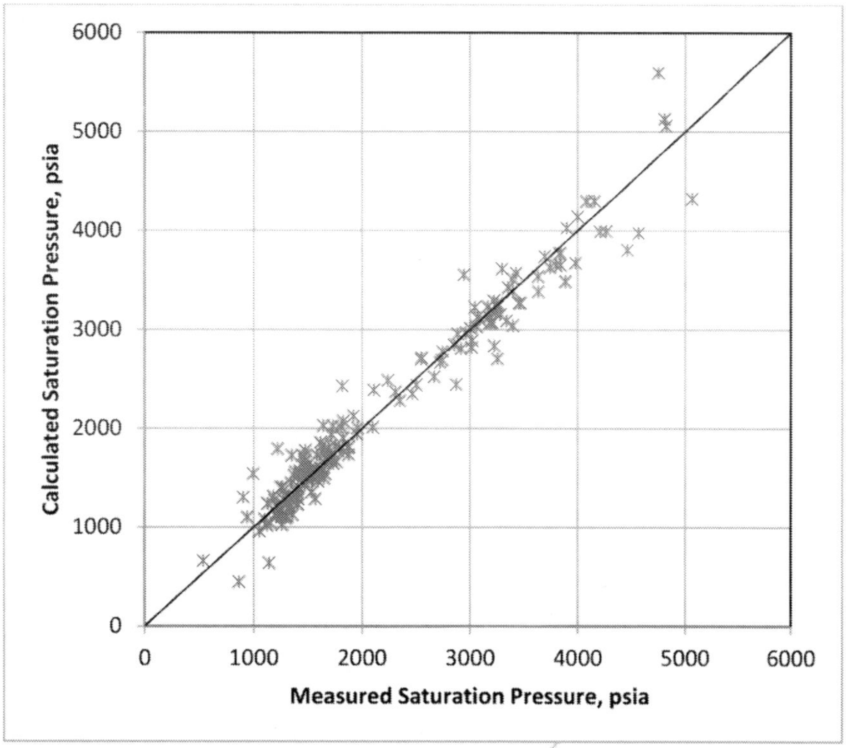

Figure 4A: Cross plot of the extended model using 179 training data

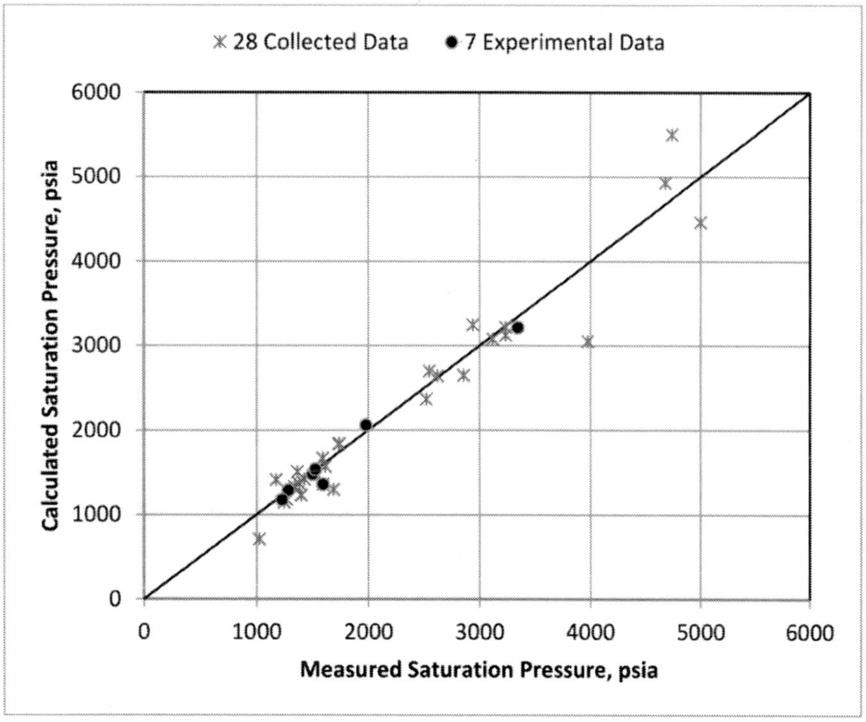

Figure 4B: Cross-plot of the extended model using 35-testing data

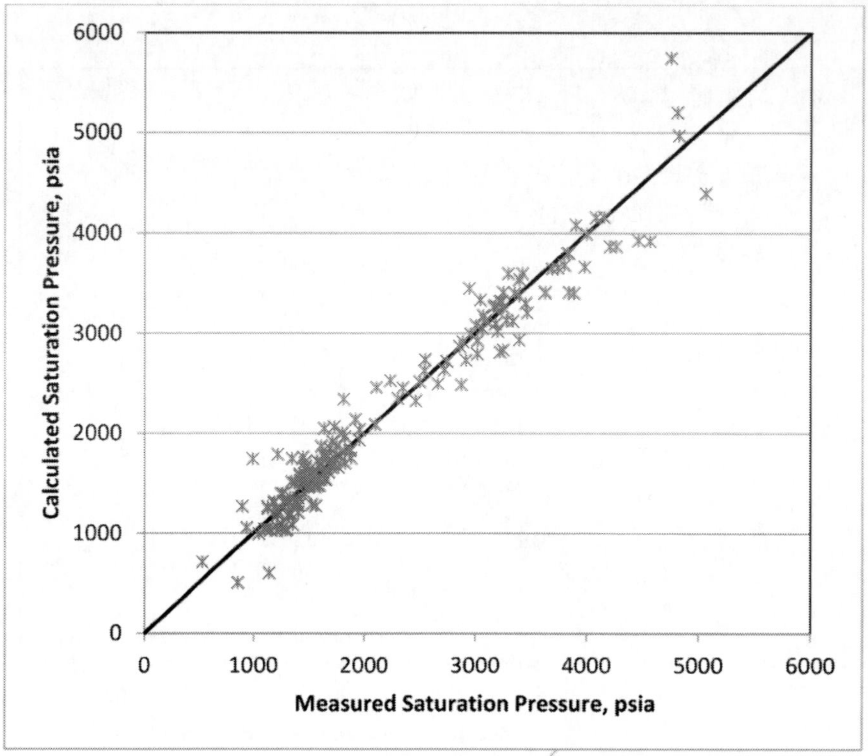

Figure 5A: Cross-plot of the lumped model using 179 training data

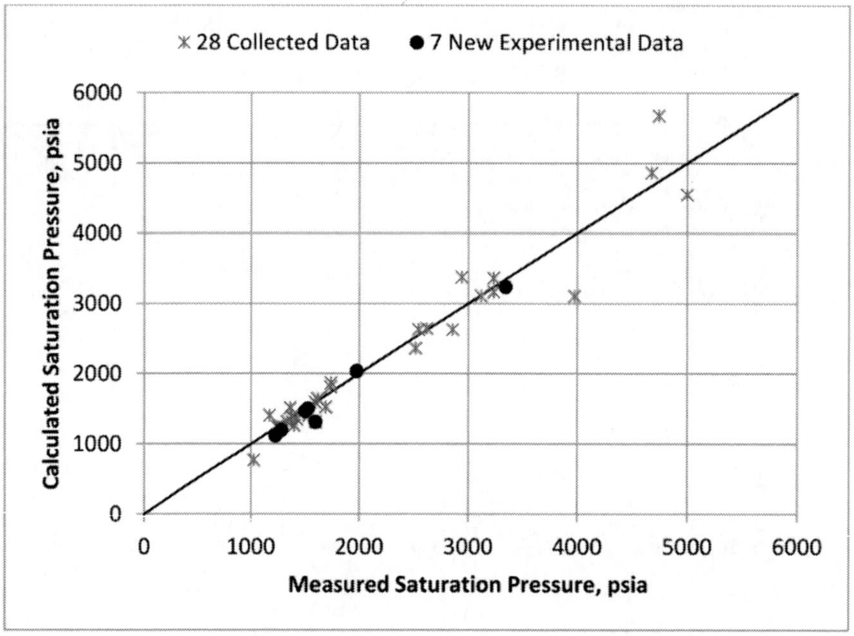

Figure 5B: Cross-plot of testing data for the lumped model

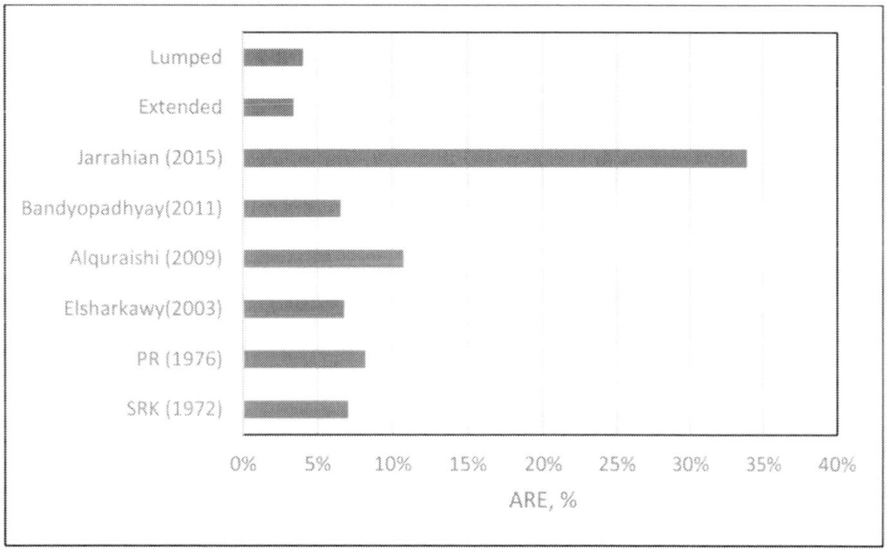

Figure 6A: Comparison of all the models' accuracy for the volatile oil samples

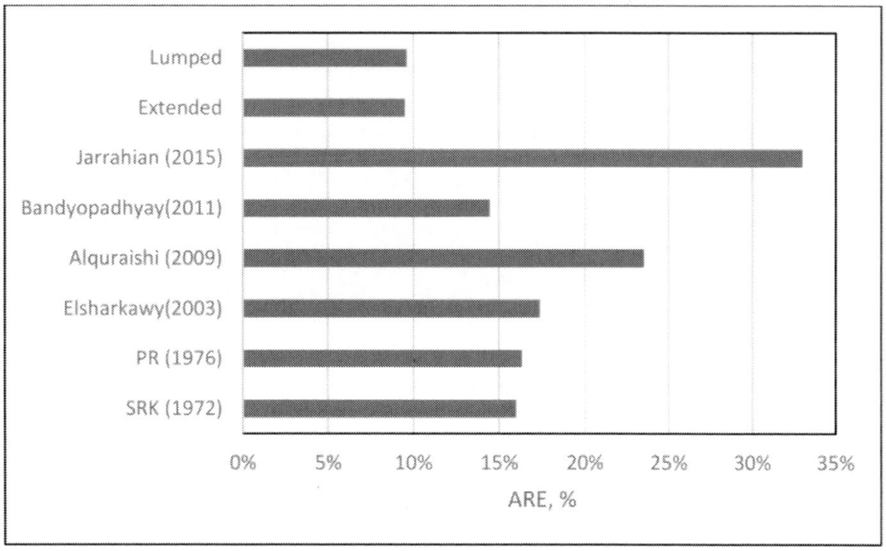

Figure 6B: Comparison of all the models' accuracy for the black oil samples

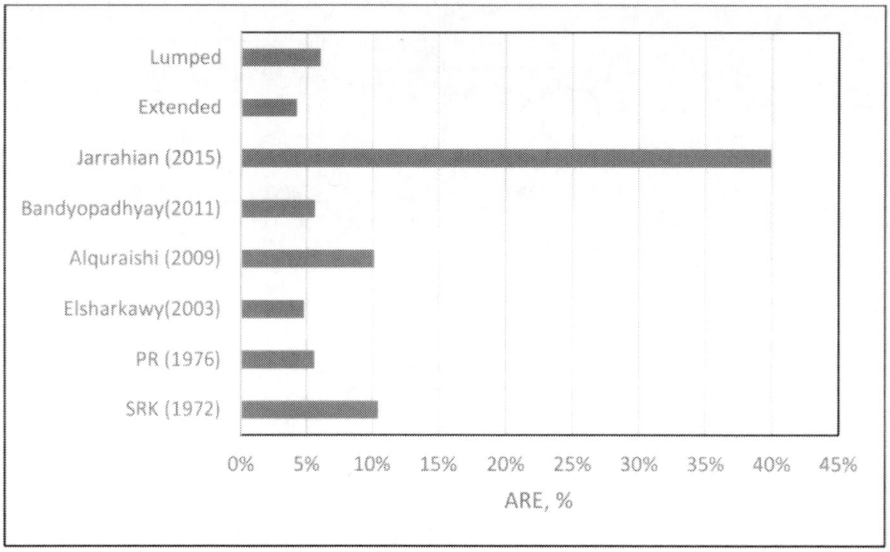

Figure 6C: Comparison of all the models' accuracy for the newly measured samples at KU

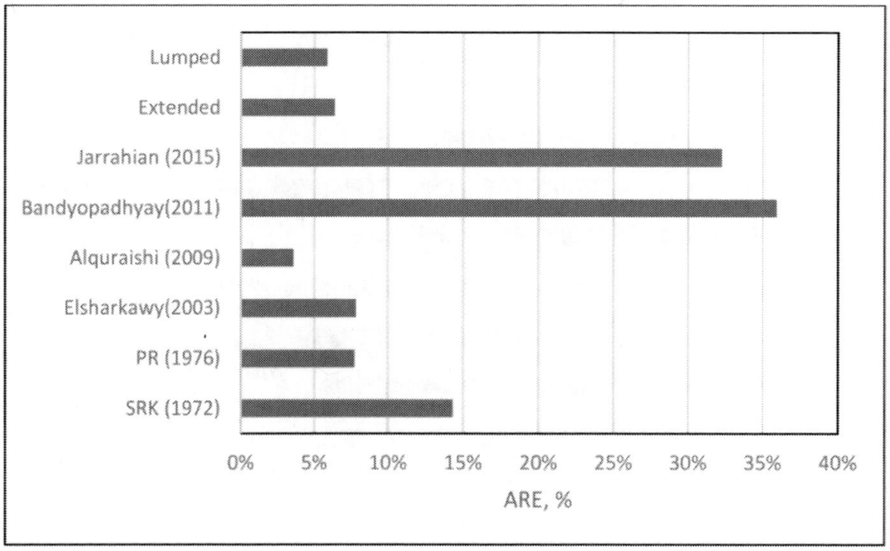

Figure 6D: Comparison of all the models' accuracy for low-temperature oil samples

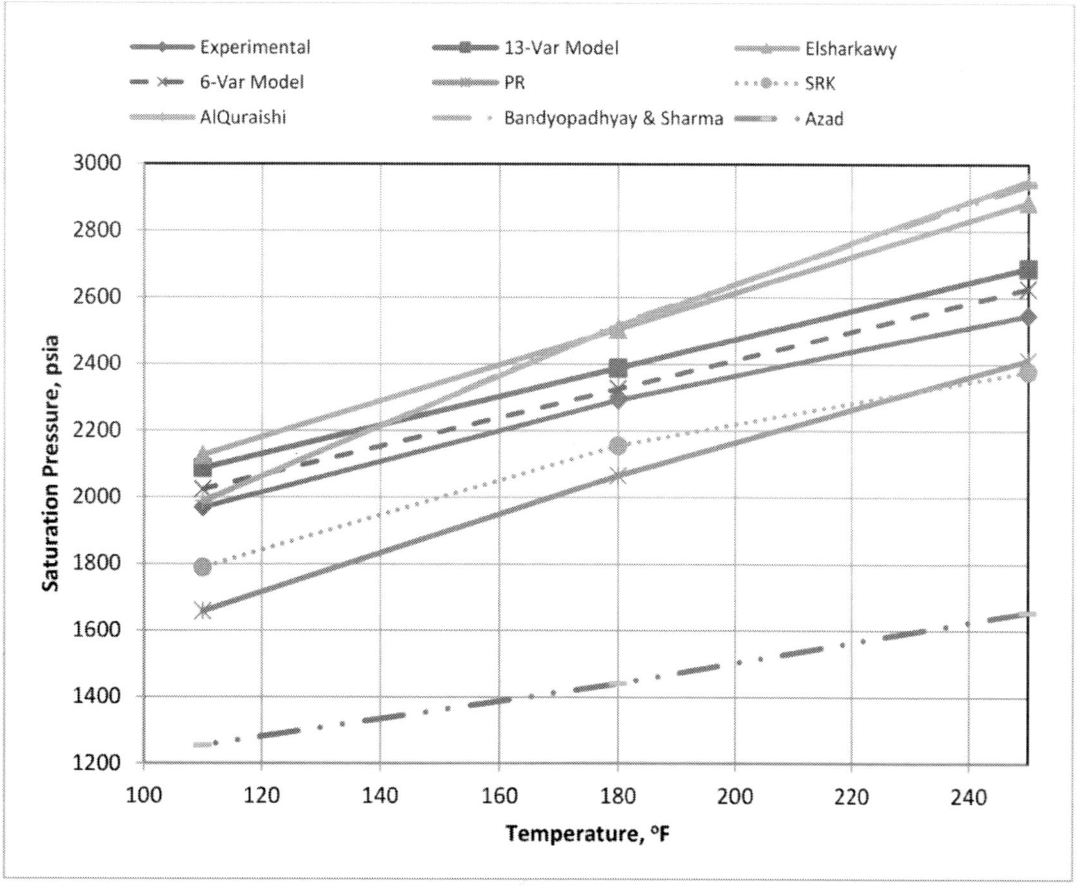

Figure 7: Changes in saturation pressure as a function of temperature for the different models (Sample #70)

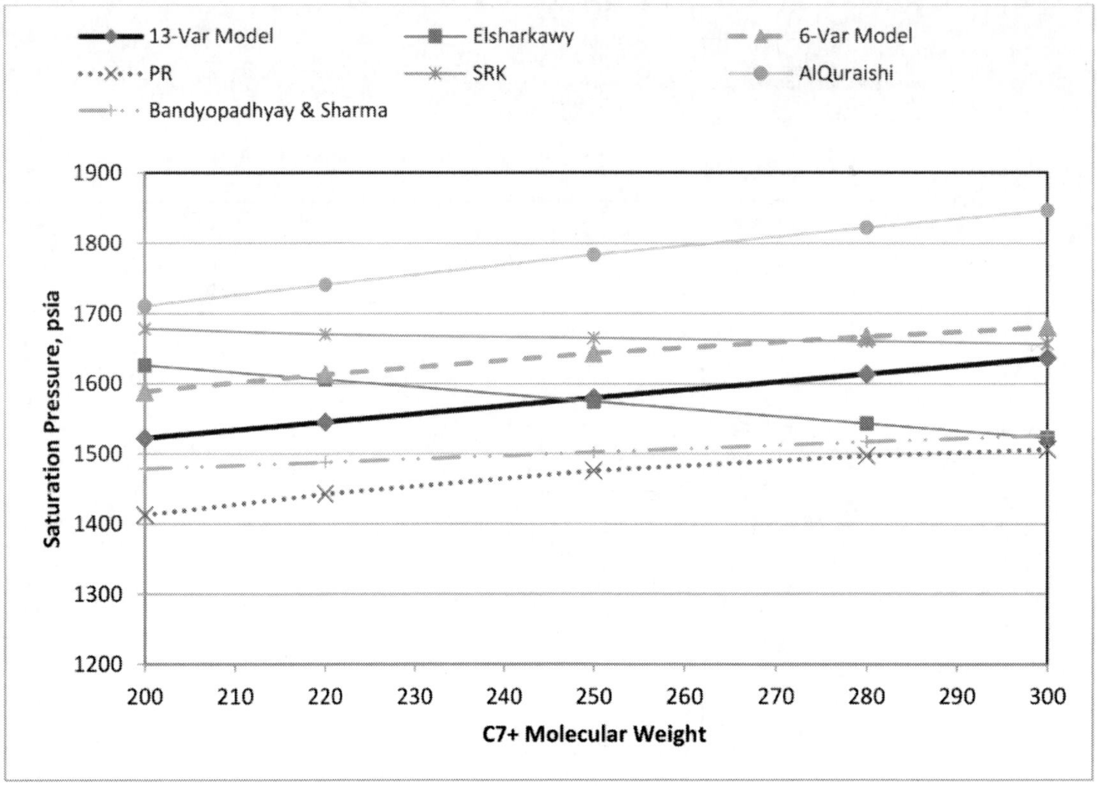

Figure 8: Comparing model performances: Changes in BPP as a function of C_{7+} molecular weight at 134°F (Sample #176)

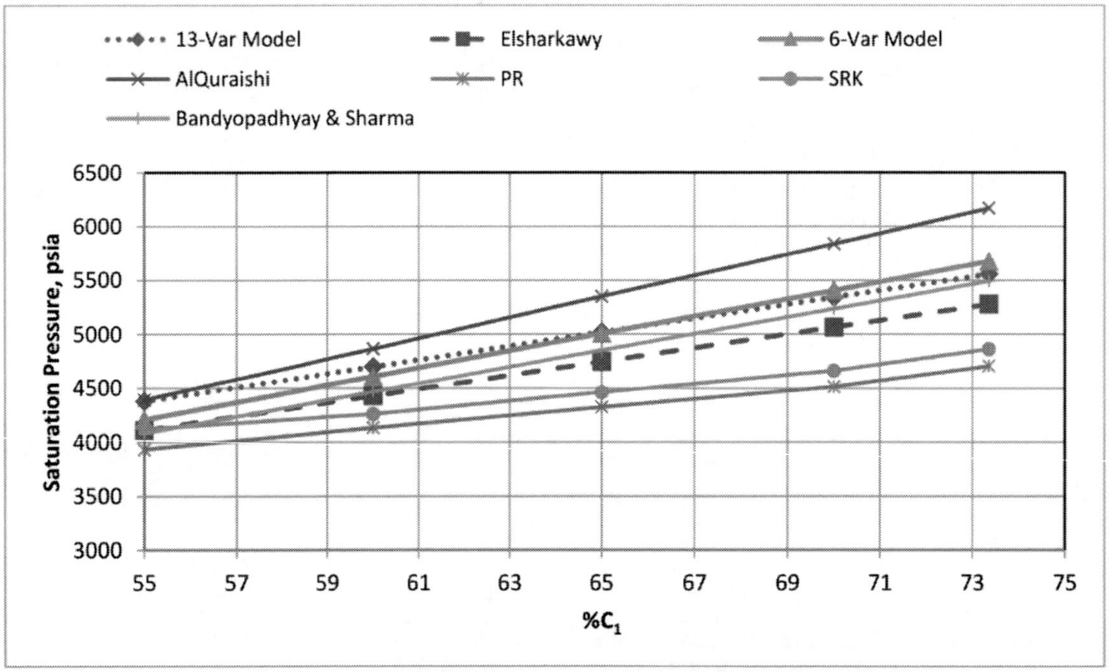

Figure 9: Sensitivity of saturation pressure to methane content at 212°F (sample #2)

439811 Purification of Ibuprofen Enantiomers By a New Technique Combining Distillation and Crystallization

Abstract:

Stripping crystallization (SC) is a new separation technique which combines distillation and crystallization. Basically, SC is operated at a triple-point condition, in which the liquid is simultaneously vaporized and crystallized due to the three-phase equilibrium. In essence, the process is continued until the liquid phase is completely eliminated and only the pure solid crystals remain in the feed. Thus, crystal washing is not required since only pure solid crystals remain in the feed and no impurities are adhered on the crystal surfaces at the end of SC. In the previous research, SC has been successfully applied to separate p-xylene from the mixed xylenes by Shiau et al. (2005, 2008, 2013).

Ibuprofen, also called 2-(isobutylphenyl)-propionic acid, is a non-steroidal anti-inflammatory drug widely used to treat headaches and minor pains. Although it is known that S-ibuprofen is responsible for the anti-inflammatory effects while R-ibuprofen might cause some long-term side effects, ibuprofen is still sold as a racemic mixture in the market due to the complexity involved in the separation of enantiomers. A thermodynamic model is developed in this work to simulate the three-phase equilibrium conditions during the SC operation. The experiments show that seeding with ultrasound mixing or magnetic stirring can be efficiently employed to purify S-ibuprofen from ibuprofen enantiomers. The experimental results, including the final enantiomeric purity and recovery ratio of S-ibuprofen, are consistent with the simulation results predicted by the model.

References:

1. Shiau, L. D.*, Wen, C. C. and Lin, B. S. "Separation and purification of p-xylene from the mixture of m-xylene and p-xylene by distillative freezing", *Ind. Eng. Chem. Res*, **44(7)**, p2258-2265, 2005.

2. Shiau, L. D.*, Wen, C. C. and Lin, B. S. "Separation of p-xylene from the multicomponent xylene system by stripping crystallization", *AIChE Journal*, **54(1)**, p337-342, 2008.

3. Shiau, L. D.*; Liu, K. F. "Investigations into the effects of the cooling rate on stripping crystallization", *Ind. Eng. Chem. Res*, **52**, p1716-1722, 2013.

441535 Hydrogen Disposal Flare System Hazard Identification and Mitigation Analysis

Abstract:

A hazard assessment on a hydrogen disposal flare system was conducted in preparation for a hydrogen turbopump test at the Air Force Research Laboratory (AFRL), Aerospace Systems Directorate at Edwards AFB, CA. The test operations required the disposal of large quantities of liquid and gaseous hydrogen through a commercial flare system to prevent the formation of an explosive environment. As part of the assessment, the primary steps in the test operation process were evaluated for the formation of a potential hazard by looking at each associated failure mode and the resulting affect on personnel and facilities. Evaluation of each operational step was conducted in a Process Failure Modes and Effects Analysis (PFMEA) that identified the probability and severity of the identified effects and associated hazards. If the probability and severity were unacceptable, mitigating steps were applied until the overall hazard risks were reduced to acceptable levels. The recommendations of this assessment were incorporated into both the flare system design and the operational procedures. This paper outlines the results of the PFMEA which include identification of the effects/hazards, available risk-mitigating techniques, and implementation of these techniques used for the mitigation of the identified hazards. Based upon the findings significant design modification to the vender's commercial equipment was needed to comply with recent changes in the code. Most of the identified hydrogen flare system hazards/effects were categorized into four groups: internal explosions, obstructions, external explosions, and maintenance degradation. Many corrective actions were implemented to mitigate both the probability and severity of the identified hazards. In addition, this paper discusses the installation of oxygen monitoring systems on the flare system, determination of appropriate system purge requirements, installation of purge system status alarms, incorporation of a triple redundant flare ignition system with status indicators, and a number of other hazard risk mitigation techniques. This comprehensive description of the hazards and mitigation of a hydrogen disposal system is a plant is vital part to safe operation of both hydrogen commercial and aerospace facilities. <align="center">

442068 Enhanced Furfural Production By an Innovative Bottom Dividing Wall Column with a Decanter Configuration

Abstract:

Enhanced Furfural Production by an Innovative Bottom Dividing Wall Column with a Decanter Configuration

Nguyen Van Duc Long, Le Cao Nhien, Le Quang Minh and Moonyong Lee[†]

School of Chemical Engineering, Yeungnam University, Gyeongsan 712-749, South Korea

Furfural, which are used to derive other industrial chemicals, has been identified as one of the major bio-based platform chemicals that can compete with petroleum-based chemicals. However, the current commercial furfural process has low yield and requires intensive energy. This paper reports the simulation, detailed design and optimization of an industrial process recovering furfural from lignocellulosic biomass, a renewable and inexpensive feedstock. The aqueous feed mixture from the acidic hydrolysis of biomass was fed to the separation process to recover furfural as the main product, along with methanol as valuable byproducts. The purification process was intensified by applying an innovative bottom dividing wall column with a decanter configuration (BDWC-D). The response surface methodology (RSM), which allows the interactions between variables to be identified and quantified, was used to optimize the BDWC-D structure. The predictions by the RSM showed satisfactory agreement with the rigorous simulation results. The heat pump technology and heat integration were then employed to improve the purification process. The results showed that the proposed sequence could save the overall energy requirements and total annual cost significantly as compared to the conventional sequence. This sequence can be employed for both grass-root and retrofit designs.

Keywords: Lignocellulosic Biomass; Furfural; Energy Efficiency Improvement; Dividing Wall Column; Response Surface Methodology

Acknowledgments

The authors wish to acknowledge the financial support from the R&D Convergence Program of NST (National Research Council of Science & Technology) of Republic of Korea and KITECH (Korea Institute of Industrial Technology) (ES150001).

442557 Flow Field Simulation and Distribution Performance Analysis of a Gas Pipe Distributor in Fccu

Abstract:
A fluidized catalytic cracking (FCC) is an important petroleum refinery process, which, aided by highly selective zeolite catalysts, cracks large-molecule low-value heavy oil feedstock into small-molecule highvalue

products. When spend catalyst is fed into the regenerator for reactivating, it is fluidized by the gas distributor at the bottom of regenerator. The gas distributor is designed to uniformly distribute gas to maintain the fluidized state of the fluidized bed and to provide the reaction gas to guarantee the burning coke deposited on the spend catalyst. Therefore, the gas distribution performance of the gas distributor influences the quality of initial gas distribution directly. So the pipe distributor used in a industrial FCC unit, which is composed by main pipes and a number of branch pipes with many nozzle tubes, was investigated by means of flow field calculation. The gas distribution performance of the pipe distributor was analyzed based on the simulated flow field. The simulation result shows that there are a obvious nununiform

gas distribution from the different nozzle gas flow rate. The static pressure increases gradually along the flow direction in the nozzle tube while the axial flow rate decreases, which leads to the nozzle gas flow rate increasing gradually. Furthermore, an obvious bias flow occurs at the inlet of the nozzle tube, which influences the gas flowing into the nozzle. The pressure drops performance of nozzle and pressure varying in the branch pipe are analyzed comprehensively. Due to the nonuniform

gas distribution by the gas pipe distributor in the practical usage, the problems such as overlarge temperature difference among the fluidized bed layers in the regenerator have often happened, which will not only reduce burning coke efficiency and but also decrease activity of the regenerated catalyst. Meanwhile, the erosion of the distributor also occurs, and in consequence the function of the distributor fails, which even causes the nonscheduled shutdown of the FCC unit and leads to tremendous economic loss. The support is provided for optimizations of both structural parameters and operation parameters of the gas pipe distributor and improvement of the gas distribution performance. Furthermore, improvement suggestions were provided based on the above reasoning. The results are expected to be useful for the distributor design.

442925 Identifying Gas Liquid Contactor Performances for Optimized Natural Gas Desacidification Amines Units - from Process Calculations to Experimental Strategy

Abstract:

Amine units are widely used for gas sweetening. The absorber performances of these reactive absorption processes are typically mass transfer limited. The proper selection of gas liquid contactor is thus critical to optimize the size of towers, the solvent flowrate and the duty of the reboiler. This work proposes an approach to identify the most suitable packings for a given application.

The case-study selected here to illustrate the approach deals specifically with a CO_2removal application where the sour gas is at 7 % vol, and the sweet gas specification is 2% vol. The gas flowrate is fixed at 0,18 MMSCFD with a solvent flowrate fixed at the minimum of the thermodynamic constraints (via equilibrium curves). An in-house rate-based simulator with accessible forms of mass transfer correlations, based on a double film model, is used to calculate the absorber height and diameter which are needed to reach the sweet gas specifications. A reference structured packing is compared with standard valve trays.

The calculations show that the bed height required with the first contactor is considerably higher than with the second one while the column diameter is reduced. In this case the use of standard valve trays is more attractive for cost optimization. Interestingly, because mass transfer is accelerated, the absorption rate with the structured packing is found to be thermodynamically limited due to a more severe temperature increase which reduces the driving force in the bottom section of the absorber.

The study determined that the optimum design for the investigated case is a compromise between a full trayed tower, that generally imposes a larger diameter, and a packed tower where the higher mass transfer forces result in a temperature rise in the bottom of the tower which in turn imposes a height increase and a much higher solvent flowrate. . To determine the mass transfer characteristics of a gas-liquid contactor providing this optimum, a sensitivity analysis has been conducted. Mass transfer parameters, i.e., liquid and gas side mass transfer coefficients and interfacial area, were varied by a factor 2 relative to the ones obtained with the reference structured packing. Calculations show that one should use a contactor that reduces liquid phase resistance, specifically, the mass transfer conductance $k_L a_e$ should be increased by at least 50%. This target could be reduced to +30% if one could simultaneously reduce the gas phase conductance $k_G a_e$ by 20%. This result can be explained by the link between mass and heat transfer (Chilton-Colburn analogy) whereby reducing $k_G a_e$decreases the temperature level which in turn thermodynamically increases the mass transfer driving force.

The next step will be to review the literature more extensively to make a short list of pre-selected random and structured packings that best meet the targets. Experimental tests will then be run on IFPEN facilities to measure flooding limits and mass transfer parameters in a small 150 mm diameter column. This step is required to confirm packing's behavior, considering the disparity of the values reported by different authors in the literature. The last step is to consolidate the existing correlations for the most promising packing, and mitigate differences between various authors by performing characterization tests on larger diameter test columns. The final correlations will be implemented in our in-house rate based simulator for industrial use.

The present approach can be applied to determine the required mass transfer characteristics for optimizing other gas treatment applications. However, the conclusions reached here as regards to mass transfer targets and the problematic temperature levels, are valid only to the present case -study, and are not directly applicable to other gas sweetening applications (selectivity, LNG, ...). For example, in a case where H_2S and CO_2 are treated simultaneously the mass transfer limitations could be very different.

443365 Technical Reconstruction of Double Effect Distillation for Methanol Process

Abstract:

Abstract: Methanol is one of important basis raw materials of chemical industry and also an excellent environment-friendly fuel, which make it as a significant commodity in the global market. Distillation is an essential unit operation for purified production of methanol. Product quality, production cost, environmental impact and energy consumption are largely related to distillation process.

Our research effort, under this circumstance, has been devoted since 2002 to energetically optimize 4-column methanol distillation system that is widely adopted in methanol units from coal, with reduced cycling water requirement. This patented process technology saw its industrial application in Chinese largest single unit of 1800Kt/a of Baofeng Group located at North China, with the lowest steam consumption of 0.9/1 t(steam)/t(methanol) compared with normally 25% higher for both simulative case study and on-site operation. Methanol product is in accordance with O-M-232G and the recycling water is reduced significantly from 67 t(CW)/t(methanol) to 52 t(CW)/t(methanol).

The flow diagram of this advanced double effect distillation process is showed in Fig.1. Its advantages arise for limitations found in current units: (1) heat exchange network failed to fully recover energy causing large CW flow to remove extra heat; (2) for pressured column, bubble point reflux need larger reflux ratio than the sub-cooled; (3) the heat of distillate from pressured column was normally ignored; (4) the low concentration of methanol of feed inlet to recovery column leads to higher reflux radio, responsible for unsatisfied unit energy efficiency.

Fig.1 The double effect distillation process.

(Column1.Light ends column; Column 2.Pressured column;

Column 3.Atmospheric column; Column4.Recovery column)

As a solution, a new flow diagram is built up and showed in Fig.2. (1) HEN totally reorganized (not completely show); (2) sub-cooled reflux to pressured column caused 10% reduction to the reflux ratio, economizing steam consumption; (3) top product of pressured column give heat to crude methanol before feeding light ends column; (4) change product and by-product position of recovery column to give efficient unit energy consumption.

Fig.2 The reconstruction of distillation process.

(Column1.Light ends column; Column 2.Pressured column;

Column 3.Atmospheric column; Column4.Recovery column)

It is estimated that the production of methanol will be over 60 million tons in 2014. If this reconstruction could be applied in the whole methanol industry, 1.3 million tons of standard coal would have been saved. This is equivalent to reduce 5 million tons emission of carbon dioxide, so the environmental impact can be largely decreased. Field operation shows reconstruction perfect substitution worth to spread to the whole methanol industry, energetically efficient and low environmentally impacting.

443095 Experimental and Numerical Study of Transport Phenomenon of Falling Film Flow with Different Heating Methods

Abstract:

Falling film flow is a common phenomenon in chemical industry. It is widely used in chemical instruments, such as structured packed columns and falling film evaporators. An experiment was implemented to study the mass transfer phenomenon of binary-mixture (ethanol-water) falling film flows with different heating methods (uniform or concentrated heating). In order to study the detailed flow and mass transfer behaviors, a 2-D multiphase computational fluid dynamics (CFD) model based on volume of fluid (VOF) was developed. This model divides mass transfer into two parts: one is caused by the vapor-liquid equilibrium (VLE) and the other is caused by the heat flux. Results from simulation are inconsistent with the experimental data. Both experiments and simulations show that the ethanol concentration of the vapor with concentrated heating is 1.5% higher than the one with uniform heating.

443101 Thermodynamic Modeling of HNO_3-H_2so_4-H_2o Ternary System with Symmetric Electrolyte NRTL Model

Abstract:

Nitric acid concentration/sulfuric acid concentration (NAC/SAC) process has been widely used for concentrating dilute aqueous nitric acid and recovering spent sulfuric acid. Nitric acid is concentrated by employing sulfuric acid to bind water and break nitric acid-water azeotrope at approximately 68 wt. % nitric acid. To support heat and mass balance calculations and process simulation for NAC/SAC processes, we develop a comprehensive thermodynamic model for nitric acid-sulfuric acid-water ternary system based on previously published thermodynamic models of nitric acid-water [Wang et al., Fluid Phase Equilibria, (2015) (DOI: 10.1016/j.fluid.2015.04.015)] binary and sulfuric acid-water [Que et al., Journal of Chemical & Engineering Data, 56 (2011) 963-9772] binary systems developed with symmetric electrolyte NRTL (eNRTL) activity coefficient model. The present work takes into account one-step hydration of nitric acid molecules, two-step hydration of sulfuric acid molecules, and further hydration of hydronium ions. With up to three temperature coefficients for each eNRTL binary interaction parameter and up to two temperature coefficients for each chemical equilibrium constant, the ternary model is validated with boiling point data from 383.15 to 471.45 K and activities at 273 K. Furthermore, contour plots of boiling points, vapor phase composition, and specific heat capacity of the ternary system are generated for engineering use.

Keywords:

Nitric acid, Sulfuric acid, Electrolyte NRTL model, Vapor-liquid equilibrium, Boiling point, Heat capacity, Nitric acid concentration/sulfuric acid concentration (NAC/SAC)

443503 A Computational Analysis of Fire Scene for Safety in a Compartment

Abstract:

This article details numerical prediction of fire spread carried out to study the distribution of carbon monoxide concentration, smoke dispersion, soot concentration and temperature in laboratory compartment, using a commercial fire modeling software tool. The heat release rate curve predicted from the simulation was compared with literature values carried out in a road tunnel. The time variations of carbon monoxide transportation, smoke concentration and temperature spread in compartment room are reported to evaluate the effect on humans. The soot concentration was described and used throughout this study as an indicator of safety in the compartment room. The effects of vent dimension and ventilation capacity were analyzed by smoke detector response times. This work provides improvements derived from reinterpretation of actual fire scene considering the ability of mechanical exhaust system and ventilation in compartment room to mitigate the hazard of fire.

443640 Preparation and Characterization of (*FURCRAEA spp*) as Green Support of Iron Nanostructured Catalyst

Abstract:

PREPARATION AND CHARACTERIZATION OF (*FURCRAEA spp*) AS "GREEN" SUPPORT OF IRON NANOSTRUCTURED CATALYST

Karen G. Bastidas G[1], Anamaria Barrera B[1], Cesar A. Sierra A[2], Hugo R. Zea R[3]

[1] Department of Chemical and Environmental Engineering. National University of Colombia (Bogota).

[2] Associate professor. Director of the Research Group of Macromolecules. National University of Colombia (Bogota).

[3] Associate professor. Director of the Research Group of Materials, Catalysis and Environment. National University of Colombia (Bogotá).

Abstract.

The availability of natural water resources has declined over time, becoming one of the biggest problems today. Environmental risk in Colombia's rivers has increase due to the widespread use of harmful chemicals in a variety of processes, ranging from illegal mineral extraction to industry wastewater, e.g. Medellin's river, due to its strategic location along a textile industrial corridor might be subject to dye contamination; the gold and platinum mining zones in Choco and Antioquia regions are being heavily affected by the use of mercury in the extraction processes. In 2009, 520 tons of mercury were imported by miners working in this area, amount that exceeds several times the mercury used by regulated industry in previous years (Ministry of Mines and Energy of Colombia, 2007). Not only the environmental risk is a concern but also the economic conditions of many low-income families whose livelihood in many cases depends exclusively of the of the mining.

Due to the issues raised above, the main goal of this study is to research, develop and implement innovative technologies in the synthesis of highly efficient, low cost and environmentally friendly iron based catalysts capable of degradate pollutants, ensuring operational stability. A native fiber (Fique, FURCRAEA spp) has been selected as a solid support for these catalysts due to its abundance, low cost and special physical and chemical characteristics.

The combination of cellulosic fibers with iron oxide nanoparticles could provide exceptional biodegradable composite materials for the treatment of organic dyes and heavy metals present in wastewater. Likewise, the use of iron particles immobilized on fique fibers allows an easy separation catalyst-treated water, turning the biocatalyst in highly recyclable and reusable material.

Fique fiber was initially characterized by determining the percentage of ashes, total lignin, soluble and insoluble lignin, cellulose, hemicellulose, total solids and humidity, obtaining 0.94, 22.58, 0.5, 22.8, 36.3, 27.5 0.701 and 2.1%, respectively. Additionally, the isoelectric point (pzc), acidic and basic sites were measured.

Then, the fiber was subjected to a pretreatment in order to expose more active sites which on time serve as nucleation points for the deposition of nanoparticles. This was done by means of an ultrasound-assisted procedure, varying the sonication time (30, 60 and 90 minutes). The effect of the sonication was followed by measuring the contents of lignin, cellulose and hemicellulose within the fiber after each time. 60 minutes of sonication was found to be most beneficial time since the amount of lignin and hemicellulose present in the fiber was reduced, while conversely the amount of crystalline cellulose exposed increases.

In order to improve the chemical ability of the nucleation sites to interact with iron and then to form stable and size controllable nanoparticles, a cationization procedure was performed on the fiber; this was achieved by the exposure of the fiber to acidic and basic conditions promoting cationization. Three cationization times (1, 2 and 3 hours) were studied and their effect on the fiber chemical surface was evaluated using FTIR-ATR and XRD analysis. The results showed 3 hours as the optimal cationizing time.

Subsequent to the cationization of the fiber surface, a procedure of ionic/cationic exchange was implemented by exposing the fiber to a solution of iron salts in order to replace the cations in the fiber surface and anchoring the iron nanoparticles. Different conditions were studied in order to synthetize zero valent iron nanoparticles (nZVI) and others with different oxidation states. Finally, to track the impregnation process on the fiber, samples from the precursor solution were taken at different times and its iron concentration was determined by atomic absorption spectrophotometry (AAS). Complementary characterization performed by X-ray diffraction, BET surface area and electronic microscopy, among others.

Key words: Natural support (fique), iron nanoparticles, bionanocatalyst.

443764 Optimization of Heat Exchanger Network in Steam Assisted Gravity Drainage Oil Sands Surface Facilities

Abstract:

Steam assisted gravity drainage (SAGD) is the most viable and environmentally safe recovery technology for extracting heavy oil and bitumen. It is a thermal in-situ production process where parallel wells are drilled horizontally into an underground bitumen reservoir. Steam is produced at the central processing facility (CPF) and injected into the reservoir through the wells. The steam heats the bitumen to a point where gravity allows it to flow down to the lower well where the mixture of bitumen and water is then pumped to the surface. The water and bitumen are separated at the CPF. Typically, more than 90% of the energy consumed in producing a barrel of bitumen using SAGD is used to generate the steam that is injected into the bitumen reservoir. Fossil fuels is employed in the process, which causes GHG (greenhouse gas) emissions. The usage of hot and cold utility is reduced by increased heat exchange and optimization of the heat exchanger network, which results in reduction of GHG emissions. In this work, pinch analysis is performed to minimize energy consumption.

443774 Thermodynamic Anaysis of Dehydration of n-Propanol Using Heterogeneous Azeotropic Distillation

Abstract:

Heterogeneous azeotropic distillation is a widely popular process in chemical process industries for the separation of binary azeotropic mixtures into their constituent pure components. It involves the addition of an entrainer which is partially miscible and forms a heterogeneous azeotrope preferably with one of the original components. This is a powerful technique which combines several physical phenomena such as presence of azeotropes and liquid-liquid immiscibilities, and is used widely in industries to enhance difficult or intractable distillation separations. For the design of heterogeneous azeotropic distillation columns, a detailed study of the heterogeneous region is crucial to plan the distillation sequence, which requires experimental vapor-liquid-liquid equilibrium (VLLE) data.

The present study focuses on the investigation of the heterogeneous region of the well known industrially available azeotropic mixture n-Propanol-Water system using Propyl acetate as entrainer. The VLE data of the system was generated by means of modified Othmer still and the still is modified using a probe type ultrasonic homogenizer to account for the heterogeneity. Isobaric vapor- liquid- equilibria and vapor - liquid - liquid - equilibria data at atmospheric pressure have been obtained for the ternary system n-Propanol-water-propyl acetate and the constituent binary systems. The equilibrium data has been correlated using NRTL, UNIQUAC and UNIFAC models and compared with the experimental data.

443776 Study of Ionic Liquids As Entrainers for the Extractive Distillation of *N*-Propanol and *N*-Propyl Acetate Binary Azeotropic Mixture and the Isobaric VLE Studies of the System

Abstract:

Distillation is a widely accepted technique for the separation of binary liquid mixtures. Separation by simple distillation is not possible for binary mixtures that form azeotropes. One of the commonly used methods to separate azeotropes is extractive distillation. It is the process of adding an entrainer that can alter the relative volatility of the mixture and thereby breaking the azeotropic point. Ionic liquids on account of their physiochemical properties as well as the recyclability have been used as a preferred replacement for the conventional entrainers. Extractive distillation with ionic liquids has the advantages of both liquid solvent (easy operation) and solid salt(high separation ability). The corrosion and dissolution problems by using inorganic salts can be rectified by these non-volatile, chemically stable, high electrical conductive molten salts.

The azeotropic system under consideration in the present study is *n*-propanol - *n*-propyl acetate mixture. These compounds which constitute the products/reactants of esterification reaction of n-propanol form a minimum boiling azeotrope at 94.7°C. *n*-propyl acetate has been included in a list of volatile acetates as an ecologically benign CFC-free solvent for aerosol sprays.

Vapour liquid equilibrium(VLE) measurements of *n*-propanol and *n*-propyl acetate are performed at atmospheric pressure in a modified Othmer still which is a continuous distillation still in which vapour phase is recirculated. The most suitable ionic liquids for shifting of azeotropic composition of the system are selected using COSMO-RS model with suitable combinations of anions and cations. The isobaric VLE of the system with [EMIM][HSO_4] is obtained and compared with the existing data. The influence of alkyl chain length on the performance of ionic liquids is also analysed with regard to the previous studies. The effect of polarity is the most important factor on the behaviour of ionic liquid solvation in alcohols. Experimental VLE data were correlated using modified UNIFAC model.

443784 Measurement and Correlation of the Mutual Solubility of Diisopropylamine + Water and Triethylamine + Water Systems at High Pressure

Abstract:

The mutual solubility data of diisopropylamine + water and triethylamine + water systems were measured at the pressure up to 100 MPa and the temperature range was 280.2~316.2 K. Experiments were performed in a high-pressure autoclave with a moveable piston. The experimental data were correlated satisfactorily with NRTL activity coefficient model and the binary parameters were obtained.

443819 Improving High Shear Rate Friction Reduction Behavior of Polyacrylamide-Based Fracturing Fluids Using Xanthan Gum

Abstract:

The main component of slickwater in fracturing low permeability formations is a high molar mass polymer such as polyacrylamide (PAM), which acts as a drag reduction agent in turbulent flow. Polyacrylamide does not degrade in the environment and also suffers from mechanical degradation at high turbulent flow rates. Five different concentrations of PAM as well as mixtures of polyacrylamide/xanthan were prepared to investigate the possibility of improving shear stability of PAM. Drag reduction measurements were performed in a flow loop and friction factors at different flow rates were calculated using rheological parameters. Based on the observations, Xanthan (XG) is not as good a friction reducer as PAM, even at three-fold higher concentration. XG slightly improved the shear stability in mixed polymer solutions at low concentrations. But at high concentrations, XG significantly improves the friction reduction behavior of PAM at nonionic and ionic environments.

443836 Oxidative Dehydrogenation of n-Butane on Carbon Catalysts

Abstract:

Butenes and butadiene, which are one of the important industrial chemical products to produce synthetic rubber , are sythesized traditionally by oxidative dehydrogenation (ODH) of n-butane over various mixed metal oxides. Metal-free carbon nanomaterials has recently received much attention by researchers as a catalyst for the ODH reaction due to its environmental acceptability, corrosion resistance, and unique surface properties. For the reason, the reaction system was modelled in this study for the sythesis of butene and butadiene. For the modelling of the reaction system, the catalytic performance of several carbon catalysts, such as graphite and activated carbon, were initially tested. Then, the reaction mechanisms and kinetics were developed based on experimental results. Lastly, the mathematical modeling of the reaction system was optimized to maximize an overall yield of the main products. This approach allowed us to develop the optimized reaction system more efficiently.

This research was respectfully supported by Engineering Development Research Center (EDRC) funded by the Ministry of Trade, Industry & Energy (MOTIE). (No.N0000990)

443855 Modeling the Thermal Behaviors of an Ultracapacitor Module

Abstract:

Ultracapacitors, also known as supercapacitors, have the potential to meet the increasing power requirements of the sustainable energy systems for the grid-scale energy storage systems and electrified vehicles. As compared to batteries, ultracapacitors offer a higher power density, higher efficiency, and longer shelf and cycle life. Because the performance of an ultracapacitor depends on temperature, it is important to calculate accurately the thermal behavior of a single ultracapacitor cell and the module composed of multiple ultracapacitor cells for the efficient and reliable systems integration of an ultracapacitor in grid and transport applications.

In this work, modeling is performed to study the effect of the layout configugation of the ultracapacitor cells on the thermal behaviors of an ultracapacitor module. The ultracapacitor module is subject to the charge/discharge cycling with constant-current between the nominal voltage of the ultracapacitor module and the half of this voltage. The charge/discharge current values examined are 50, 100, 150, and 200 A. The validation of the modeling approach is provided through the comparison of the modeling results with the experimental measurements.

443856 Modeling the Electrolyte Flow Behaviors of a Zinc/Bromine Flow Battery Module

Abstract:

The zinc/bromine (Zn/Br_2) flow battery is a promising rechargeable power source for energy storage systems because of its inherent chemical simplicity, high degree of reversibility at the electrodes, good energy density, and abundant low-cost materials. The Zn/Br_2 flow battery system, intended for the grid-scale energy storage system, is to be made in modular form. A Zn/Br_2 flow battery module is composed of the multiple stacks consisting of bipolar carbon electrodes. Since the stacks require electrolyte circulation on both the negative and positive sides, two plumbing loops are provided. The flow loops contain an electrolyte storage tank and a circulation pump. A heat exchanger may be included for temperature control in the flow loop. The design of the electrolyte circulation plumbing determines the distributions of electrolyte flow and pressure within the stacks and the efficiency of the stack is dependent on the flow distribution throughout the stack. The modeling of the distributions of electrolyte flow and pressure within the module is needed to optimize the plumbing for electrolyte circulation and thereby to improve the performance of the module. In this work, modeling is performed to study the effect of the layout configuration of the plumbing for electrolyte circulation on the distributions of electrolyte flow and pressure within a 50 kWh Zn/Br_2 flow battery module composed of four stacks. It is seen that the plumbing layout for electrolyte circulation can have a significant effect on the flow rate distribution of electrolyte through the stacks.

443837 Blowdown Dynamic Modeling and Analysis of Pressure and Temperature

Abstract:

Vessel blowdown is one of the safety operation procedures that discharges flammable gas or liquid to flare system during emergency situation such as fire. When blowdown occurs, the liquefaction and freezing effect cause a pressurized vessel to be brittle, exposing it to possible hazard situation. Also the blowdown systems are recommended to be accordance with API(American Petroleum Institute) 521 code. API 521 requires pressure in vessel should be half of the initial pressure or be lower than 100 psig in 15 minutes after commencing blowdown in fire case. Therefore, accurate estimation over the change of temperature and pressure of the vessel during blowdown period is crucial for the design of blowdown system. For the reason, researchers have developed a few dynamic blowdown models so far. However, the model was found to be rather inaccurate under certain conditions. Therefore a new dynamic blowdown model was developed in this work. For the accurate analysis, the model algorithm considers both vapour and liquid phase in non-equilibrium condition that the temperatures of vapour and liquid phases are different. Futhermore, heat transfer from the vessel wall to discharge gas at either laminal or turbulent flow in the vessel was considered in the model to increase the accuracy of blowdown estimation. For thermodynamics, both Peng-Robinson and Soave-Redlich-Kwong equations of state were applied to the model, and the results were compared each other. Finally, the simulation results were compared with Haque et al's experimental and steady-state simulation results.

443974 Observer-Based Model Predictive Control for Time-Delay Systems

Abstract:

Model predictive control is one of the most widely used advanced process control technologies in industry.[1] It has many capabilities such as dealing with input/state constrains and uncertainties. Observer-based model predictive control was developed to estimate states information when the states are not measurable directly. Many chemical processes have time delays in either states or measurements, such as polymer production and some bioprocesses. [2] In this presentation, we are proposing an unknown input observer based model predictive control algorithm for time-delay systems. Unknown input observer is an observer design to eliminate certain types of unknown inputs in the system. Stability of unknown input observer-based model predictive control without time delay was provided by our previous study. Terminal cost and terminal region are designed for a class of non-linear time delay systems. The close-loop stability of the controller is provided. [3][4]

An example of bioreactor is provided which has uncertain time delays in measurements but with a known upper bound. Two measurements are provided which are online biomass and offline ethanol measurements. Model-plant mismatch is considered as unknown inputs in the system. State estimation and control are calculated using the observer and an optimization algorithm.

Reference:

[1] Marcus Reble et al., Model predictive control of constrained non-linear time-delay systems. IMA Journal of Mathematical Control and Information, 2010, 1-19

[2] Seung Cheol Jeong and PooGyeon Park, Constrained MPC algorithm for uncertainty time-varying systems with state delay, IEEE TRANSACTIONS ON AUTOMATIC CONTROL, 50, 2005, 257-262

[3] Mayuresh Kothare et al., Robust constrained model predictive control using linear matrix inequalities. Automatica, 32, 1996, 1361-1379

[4] Wook Hyun Kwon et al., A simple receding horizon control for state delayed systems and its stability criterion, Journal of Process Control, 13, 2003, 539-551

444020 Thermodynamic Modeling of Aqueous Lithium Sulfate and Its Solubility in the Presence of Na+, K+ and Mg2+ Sulfates

Abstract:

This work is an extension of the previous works done by our group towards development of a rigorous and comprehensive thermodynamic model for high salinity produced water that is essential for mass and heat balance calculations and simulation and design of novel treatment processes. Specifically this work presents a thermodynamic model that accurately represents all thermodynamic properties of aqueous lithium sulfate solution including the solubility of Li2SO4 in water and also in the presence of Na+–K+–Mg2+ sulfates. By explicit consideration of solution chemistry of each subsystem, symmetric electrolyte NRTL model is used to correlate available experimental data of all pertinent thermodynamic properties including activity coefficient, osmotic coefficient, excess enthalpy, and liquid heat capacity over the entire concentration range. This facilitates quantification of eNRTL binary molecule–molecule, molecule–electrolyte, and electrolyte–electrolyte interaction parameters. Also, solubility data are used to regress the thermodynamic constants required for modeling the liquid–solid equilibrium of lithium sulfate in water and in the presence of Na+–K+–Mg2+ sulfates. Displaying a significant prediction ability, the model accurately represents the thermodynamic behavior of the system from infinite dilution up to saturation and temperatures from 298.15 K to 573.15 K.

444044 Organic Solar Cells

Abstract:

This paper describes highlights the need of next-generation "organic solar cells". There is a continuing need for organic solar cells and arrays with very high specific power densities (1000-5000 kW/kg) for generating power in a new generation of ultra light space payloads example in NASA. An emerging technology with promise to meet these ambitious goals are organic solar cells based on very thin films of discrete layers. Theoretical efficiencies for multi junction solar cells are more ~70% as compared to ~30% for single junction devices. There is continuous development going on high mobility, band gap engineered organic donors and acceptors, as well as development of three junction devices and scaled up devices on lightweight flexible polymer substrates. The main interest in this paper is organic semi-conducting crystal: Pentacene, which has photo-conducting states. Organic solar cells manufactured today have low efficiency other than their stability issues. The efficiency of organic solar cells can be improved by either designing polymer materials that absorbs more of solar spectrum or by minimizing the charge collection efficiency by controlling the morphology of polymer films. Once fully developed, organic solar cells would revolutionary power the industry due to their flexibility and minimal weight

444055 A Refined Non-Random Two-Liquid Segment Activity Coefficient Model for Solubility Modeling

Abstract:

Accurate correlation and robust prediction for complex molecule solubility or solid-liquid equilibrium (SLE) in solvents and solvent mixtures have drawn much interest recently [Tanveer et al., *Chem. Eng. Prog.* **2014**, 110 (9), 37-47]. Solubility models such as the Non-Random Two Liquid Segment Activity Coefficient (NRTL-SAC) model [Chen and Song, *Ind. Eng. Chem. Res.* **2004**, 43, 8354-8362] are now routinely used in the industries. However, an important unfilled need is to model the solubility in oligomers/polymers/surfactants such as asphaltene nanoaggregates in petroleum fluids and drug molecules in lipid-based or solid dispersion formulation [Tung, *Org. Proc. Res. Dev.* **2013**, 17, 445-454]. This paper presents a refinement to the very successful NRTL-SAC model to better account for the contribution due to size differences of molecules. Specifically, the Flory-Huggins combinatorial entropy expression for conceptual segments in the model is substituted with the Staverman-Guggenheim expression for real molecules. We present the rationale for the refined model, the updated conceptual segment numbers for common solvents, and the newly identified conceptual segment numbers for common drug excipients. The model performance is illustrated by prediction of drug molecule solubility in various lipid solvents and solvent mixtures, compared to experimental data and previous work by other models. The refined NRTL-SAC model shows enhanced capability to correlate and predict drug molecule solubility in common solvents and excipients used in drug formulation.

444057 Explaining the Paradox of Asphaltene Precipitation with n-Alkanes

Abstract:

Explaining the paradox of asphaltene precipitation with n-alkanes
Md Rashedul Islam and Chau-Chyun Chen
Department of Chemical Engineering, Texas Tech University
Abstract
Paraffins, specifically alkanes, promotes asphaltene precipitation when they are blended with asphaltene-containing petroleum fluids. The alkane/oil volume ratio required for onset of asphaltene precipitation increases with the carbon number and yields a maxima. Experimental study indicates that the maxima lays at C7[1] or between C8 -- C10 [2]. However, precipitation yield decreases asymptotically with increase of the carbon number even if excessively high alkane/oil volume ratio is used [3]. For example, n-pentane be a better solvent for asphaltene than n-hexane in regard of both onset of precipitation and precipitation yield. However, high carbon number alkane beyond the maxima, n-hexadecane for instance, can be a better solvent at onset of precipitation point and a poorer solvent in regard of precipitation yield than n-pentadecane. A thermodynamic explanation for this paradox is necessarily important to find the compatibility of bitumen or crude oil with paraffinic hydrocarbons. In this work, this paradox is explained by the aggregation thermodynamics of asphaltene precipitation [4]. Both the enthalpy and entropy of mixing are investigated to explain the effect of paraffin carbon number on asphaltene precipitation. Athabasca and cold lake bitumen samples are studied.
References
[1] G. Hotier and M. Robin, "Action de divers diluants sur les produits pÃ©troliers lourds: mesure, interprÃ©tation et prÃ©vision de la floculation des asphalténes," Oil & Gas Science and Technology, vol. 38, pp. 101-120, 1983.
[2] I. A. Wiehe, H. W. Yarranton, K. Akbarzadeh, P. M. Rahimi, and A. Teclemariam, "The paradox of asphaltene precipitation with normal paraffins," Energy & fuels, vol. 19, pp. 1261-1267, 2005.
[3] D. L. Mitchell and J. G. Speight, "The solubility of asphaltenes in hydrocarbon solvents," Fuel, vol. 52, pp. 149-152, 1973.
[4] M. Wang, Y. Hao, M. R. Islam, and C. C. Chen, "Aggregation Thermodynamics for Asphaltene Precipitation," AIChE Journal, Submitted.

444086 Optimization of Layered Graphene Oxide Nanostructures through Controlling the Membrane Deposition Rate Towards Size-Sieving Separation

Abstract:

Single-layer graphene oxide (GO) flakes were coated on anodized aluminum oxide (AAO) substrate through a simple vacuum filtration process, the effect of the deposition rate on the nanostructures of obtained membrane was investigated though pure water flux and organic vapor diffusion, which indicated GO membrane assembled with slow deposition rate might have more ordered lamellar structure and, correspondingly, have smaller effective pore size. In contrast, under fast deposition rate, GO flakes assembled in a more random way. Therefore, through controlling the membrane deposition rate, the nanostructure of layered GO membrane could be optimized for size dependent separation. We found after thermal reduction, 100 nm GO membrane (prepared @ 0.02 nm/s) has effective pore size smaller than 6.1 Å, which exhibit great potential in n-hexane/2,2-dimethybutane (DMB) separation.

444142 Robust Design of a Radiant Wall Burner for a High Temperature Environment Utilizing Computational Fluid Dynamics Simulations

Abstract:

Radiant Wall Burners are often utilized in Ethylene Cracking Furnaces. The conditions and fuel gases for a specific unit can vary considerably within the unit and across similar units. With burner air preheated to high temperatures, premixing the fuel can result in dangerous and destructive flashback, while complete nozzle-mixing of the fuel can result in burner instability at startup or turn-down conditions. These challenges can be solved through the generation of a robust radiant wall burner design, which is facilitated by executing Computational Fluid Dynamics simulations of many different geometric configurations.

444161 Novel Microwave Synthesis Carbon Dots from PINE Bark and Its Application for Heavy Metal Sensing

Monday, April 11, 2016
Exhibit Hall E (George R. Brown)
Saheed Sanni, Chemistry, Vaal University of Technology, Vanderbijlpark, South Africa, Puthalapattu ReddyPrasad, Chemistry, Vaal University of Technology, Vanderbiljpark, South Africa and Augustine E. Ofomaja, Chemistry, Vaal university of Technology, Vanderbiljpark, WV, South Africa

NOVEL MICROWAVE SYNTHESIS CARBON DOTS FROM PINE BARK AND ITS APPLICATION FOR HEAVY METAL SENSING

Sanni S.O., ReddyPrasad P., Ofomaja A.E.

Biosorption and Waste water Treatment Research Laboratory, *Department of Chemistry, Faculty of Applied and Computer Sciences, Vaal University of Technology, P. Bag X021, Vanderbijlpark-1900, South Africa.*

INTRODUCTION

Microwave pyrolysis for the production of carbon has the advantages of being cheap, produces high yield of carbon and is absolutely a green synthetic method [1]. In recent times, the use of microwave pyrolysis have been extended to the preparation of carbon dots. Carbon nanodots (C-dots) falls under the class of carbon nanoparticles with size below 10 nm[2]. C-dots are carbon materials which processes high water solubility and strong fluorescence depending on their size, edge, shape, surface ligands, and defects. Since C-dots preparation can be achieved from difference carbonaceous sources and by varying a range of parameters during their synthesis, the properties of the final product may vary depending on these factors. C-dots have sp^2 characteristics and possess several functional groups on their surfaces that makes them highly hydrophilic and easily functionalized with various organic, polymeric, inorganic, or biological spices[3]. Several synthesis methods that have been explored includes electrochemical synthesis, combustion/thermal/hydrothermal/acidic oxidation, supported synthesis, microwave/ultrasonic, arc discharge, laser ablation/passivation, and plasma treatment[4]. C-dots processes excellent such as water dispersibility, chemical stability and photostability, ease of modification, excitation-dependent multicolor emission, low toxicity, and good cell permeability, great efforts have been paid to their potential applications in biosensing, bioimaging, drug delivery, and other biological related aspects.

METHODS

The carbon dots (C-dots) with green luminescence was synthesized from microwave assisted pyrolysis of pine bark waste our carbon source at 800, 1000 1200 W for 1 Hr. The dots produced were characterized using FTIR, TGA, XRD, UV-Vis and Raman spectroscopy. The dots were applied for heavy metal sensing.

RESULTS AND DISCUSSION

The functional groups present on the C-dots (**Fig.1**) includes -OH stretching band mode at 3413 cm^{-1}, C-H stretching at 2934 and 2845 cm^{-1}, stretching peaks of C=O, C=C at 1731, 1604 and 1437 cm^{-1}. The XRD pattern (**Fig 2a &b**) of produced C-dots depicts sharp peak at 20.9° and 26.8°, which is highly crystalline cores and possess amorphous nature of C-dots. This peculiar peak is ascribed to 002 of the graphitic carbon, which in line with inter-planar spacing of 3.77 Å with higher spacing than bulk graphite at 3.34 Å. The thermogravimetric and differential thermal analysis display (**Fig 3**) that the C-dots possess thermal stability more than 100 °C and mass losses occurs afterwards. The thermal analysis depicts 3 endothermic peaks at 331, 542 and 735 °C, which are ascribed to oxygenated groups such as carboxyl, hydroxyl, carbonyl and phenols. The optical properties of the synthesize C-dots were explored using UV-vis absorption and photoluminescence (PL) spectroscopy. The UV-vis spectra display (**Fig 4**)intense absorption peak in the range of 260 to 290 nm, which is ascribed to π-π^* transition of the aromatic sp2 domain and also presence of multiple polyaromatic chromophores. A shoulder peak at 330 nm which is ascribed to n-π^* transitions of C=O and structural defects of the C-dots. PL spectrum of the C-dots (**Fig 5a**)exhibits an intense maximum emission peak at 432 nm upon excitation at 330 nm, which also shows characteristic yellowish green color under sunlight and intense blue coloration UV lamp (365 nm). The PL properties of as prepared C-dots was varied at (**Fig5b**) different excitation wavelength from 300 to 405 nm. There was red-shifting in emission peak as the excitation peak was increased from 300 to 405 nm and also simultaneously decrease in fluorescent intensity, which is ascribed to quantum effect and surface defects. This behavior of the prepared C-dots is dependent on excitation-fluorescence, which is accordance with other report [13]. The quantum yield of synthesized C-dots was measured to be 17% at maximum excitation at 330 nm

Fig.6 shows the Raman spectrum of the PB-GQDs, which has been usually used to confirm the quality of the prepared PB-GQDs. There is two major Raman peaks at 1370 cm^{-1} (D band), 1580 cm-1 (G band). The peak of G band corresponds to the Sp^2 bonded carbon atoms in a two-dimensional (2D) hexagonal lattice and the peak of G band is associated with the vibrations of carbon atoms with dangling bonds in the termination plane of tangled graphite or glassy carbon. The D/G ratio band ratio is ID/IG: 0.68 is confirm the formation of graphene like structure from synthesized pine bark.

Fluorescence detection of Cu^{2+} ion

The detection of Cu^{2+} was performed at room temperature in phosphate buffer (PBS) solution pH 4.3. 1.0 mL of PB-CQDs solution was added into 0.5 mL of PBS buffer, followed by the addition of 1.0 mL different concentrations of Cu^{2+}. The fluorescence emission spectra were recorded after reaction for 5 min at room temperature (~28 °C).

Selectivity and sensitivity studies

Moreover sensitivity and selectivity is additional significant parameter to assess the performance of the sensing system. Consequently, we examined the fluorescence intensity changes in the presence of representative metal ions such as including Cu^{2+}, Hg^{2+}, Ca^{2+}, Al^{3+}, Ba^{2+}, Mn^{2+}, Ni^{2+}, Pb^{2+}, Co^{2+}, Zn^{2+}, Fe^{3+}, $Cs^{2+}$and Mg^{2+}, of each at a concentration of 15.0 µg mL^{-1} were reacted with a PB-CQDs solution. Fig. 7 A displays the relative change in PL intensity of PB-CQDs in the presence of various metal ions. Fe^{2+}, Mn^{2+} and Hg^{2+} ions cause the slight PL changes (defined as the relative change of PL intensity in 60–80% as compared to blank) which can be attributed to the nonspecific interactions between the carboxylic and/or amine groups and the metal ions. Among these metal ions, Cu^{2+} displays the strongest PL quenching effect on PB-CQDs, assigning to the special coordination between the Cu^{2+} ions and the phenolic hydroxyl and/or amine groups of PB-CQDs which has been widely used for the detection of Cu^{2+} ions or colored reactions in traditional organic chemistry. Such a specific fluorescence quenching effect may originate from the strong interactions between Cu^{2+} ions and the surface groups of PB-CQDs which transfer the photoelectrons from PB-CQDs to Cu^{2+} ions. These results clearly demonstrate that the PB-CQDs-based Cu^{2+} sensor is highly selective to Cu^{2+} over the other metal ions. Fig.8. Fluorescence responses of CCDs to the addition of different metal ions under acetate buffer (pH 4.3). The concentration of each metal ion is 15 µg $mL^{-1.}$ F_0 and F correspond to the fluorescence intensities of PB-CQDs at emission wavelength of 440 nm excited at 320 nm in the absence and presence of metal ions, respectively.

As shown in Fig. 3, the fluorescence intensity of PB-CQDs gradually decreased with the Cu^{2+} concentration. The fluorescence intensity displayed liner responses with the Cu^{2+} concentrations, with two linear ranges of 0.0□14.0 µg mL^{-1} and 15.0□25.0 µg mL^{-1} (Fig.3B). The linear equations are $F/F0$ =0.9827□0.8485C_{Cu2+} (R = 0.993) and $F/F0$ = 0.7565–0.0033 C_{u2+} (R = 0.996), respectively.

REFERENCES

[1] W. B. Lu, X. Y. Qin, S. Liu, G. H. Chang, Y. W. Zhang, Y. L. Luo, A. M. Asiri, A. O. Al-Youbi, X. P. Sun, Economical, green synthesis of fluorescent carbon nanoparticles and their use as probes for sensitive and selective detection of mercury(II) ions, Anal. Chem. **2012**, 84, 5351-5357.

[2] J. Jiang, Y. He, S. Li, H. Cui, Amino acids as the source for producing carbon nanodots: microwave assisted one-step synthesis, intrinsic photoluminescence property and intense chemiluminescence enhancement. Chem Commun **2012**, 48, 9634–6.

[3] Y. Wang, A. Hu, Carbon quantum dots: synthesis, properties and applications, J. Mater. Chem. C **2014,** 2, 6921–6939.

[4] C. Liu, P. Zhang, X. Zhai, F. Tian, W. Li, J. Yang, et al., Nano-carrier for gene delivery and bioimaging based on carbon dots with PEI-passivation enhanced fluorescence, Biomater. **2012,** 33, 3604-13.

Figures:

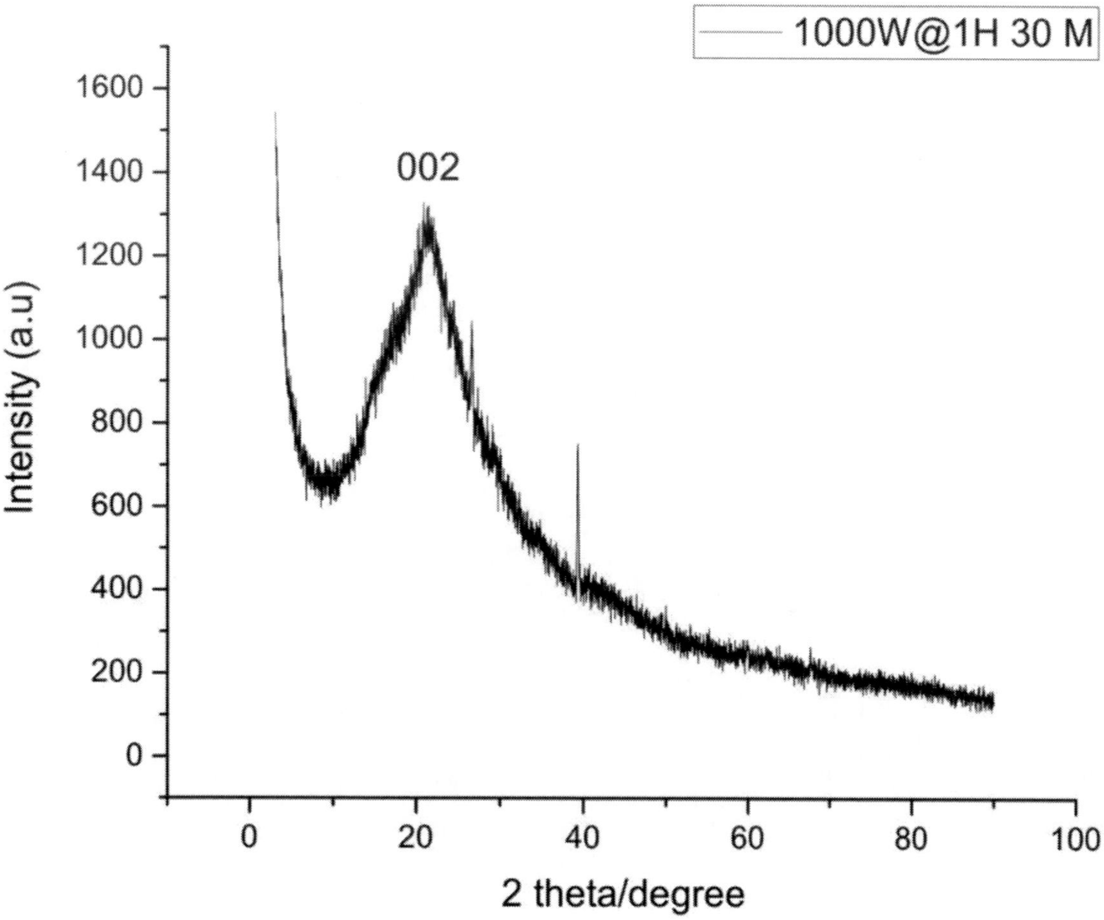
1000W@1H 30 M
002
Intensity (a.u)
2 theta/degree
1600
1400
1200
1000
800
600
400
200
0
0
20
40
60
80
100

Fig 1. FTIR spectra of as-prepared C-dots from Pine Bark

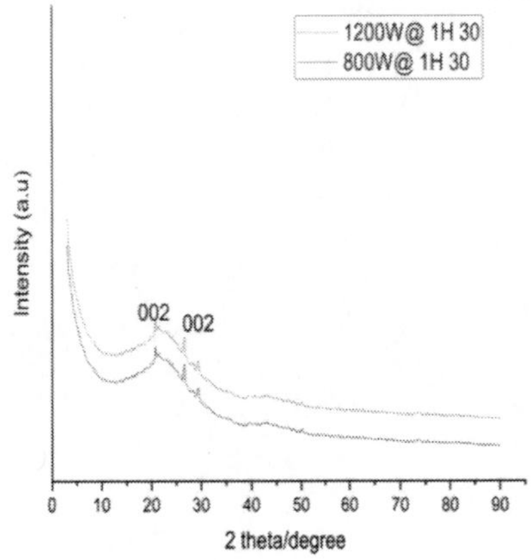

(a)

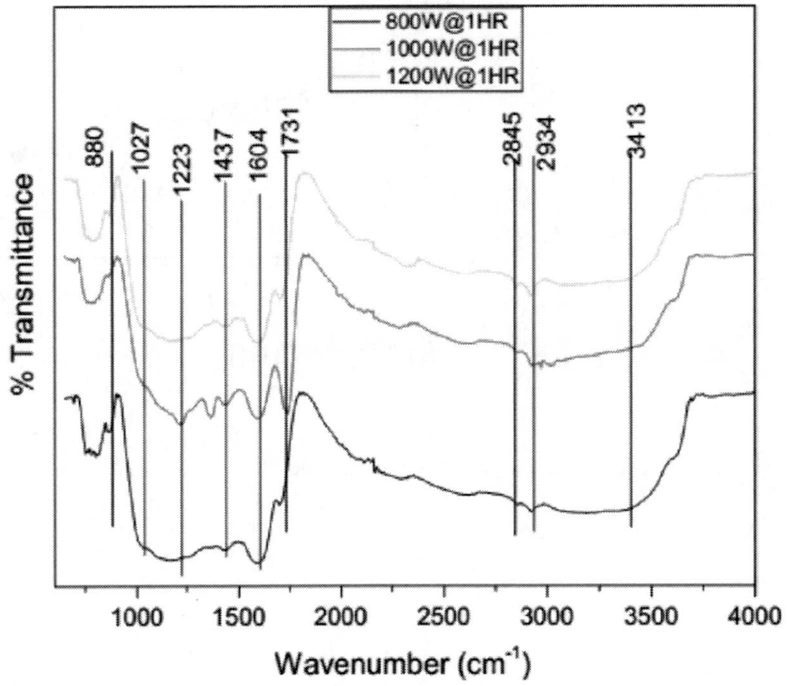

(b)

Fig. 2a & b. XRD pattern of C-dots.

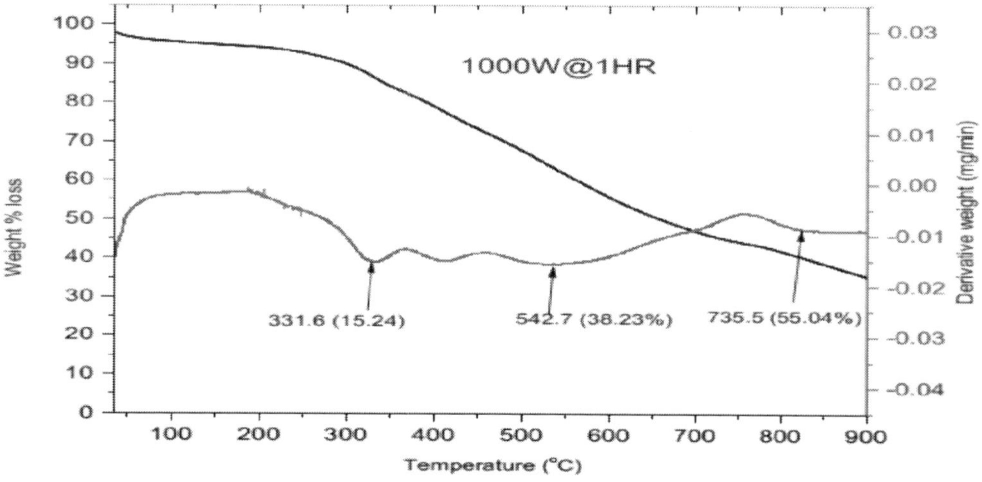

Fig 3. Thermal curves for C-dots under N_2 atmosphere.

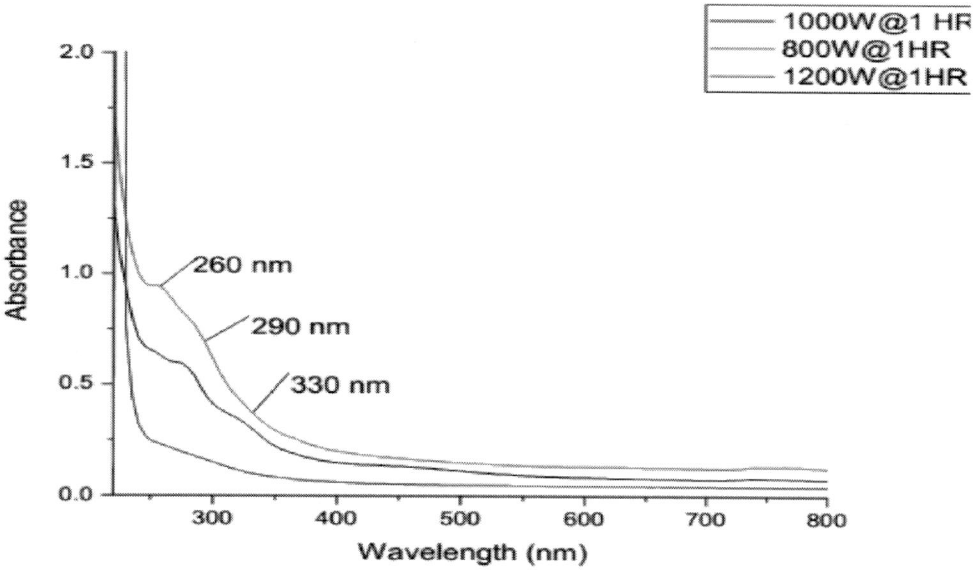

Fig 4. UV-vis absorption of C-dots.

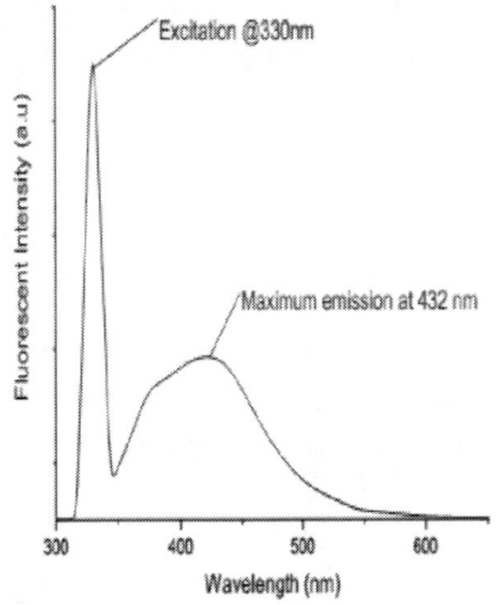

(a)

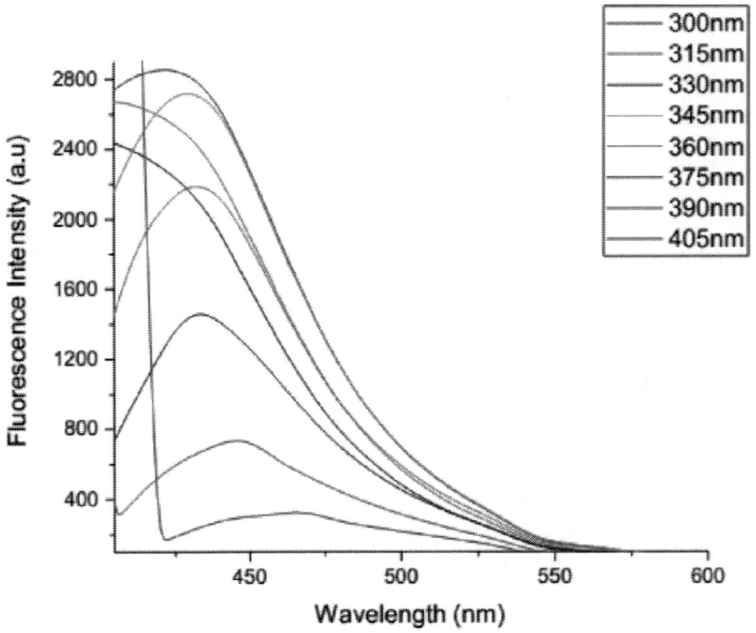

(b)

Fig 5a. PL excitation and emission wavelength (b) PL emission spectra at different excitation wavelength from 300 to 405 nm in 15 nm increments.

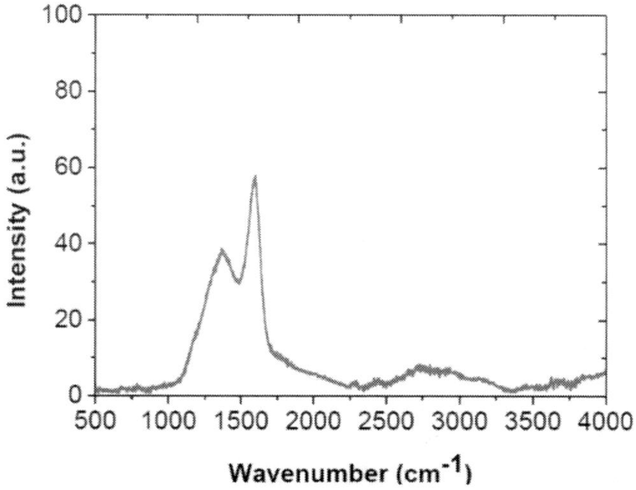

Fig. 6. Raman Spectra of PB-CQDs

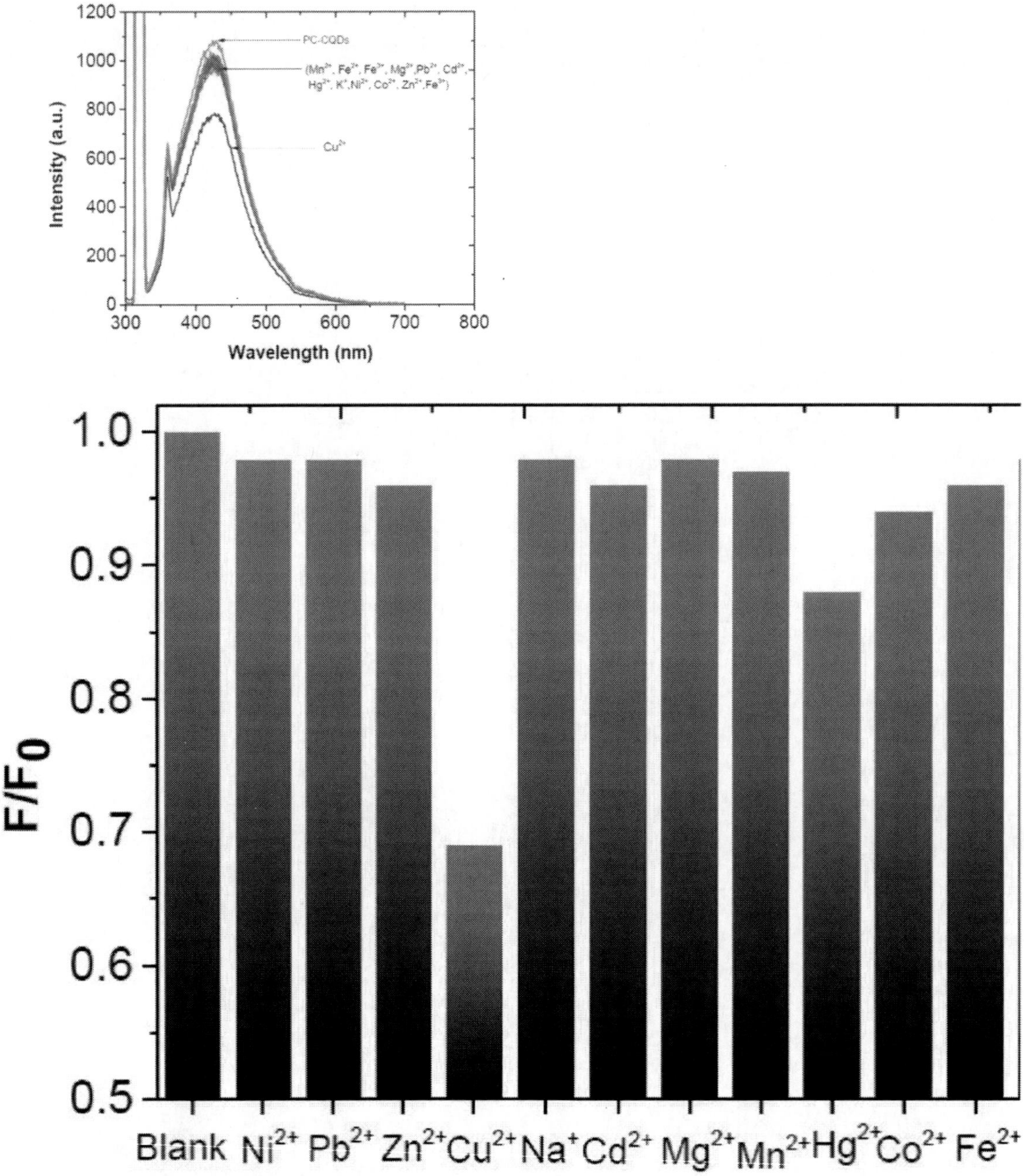

Fig.7. Fluorescence responses of CCDs to the addition of different metal ions under acetate buffer (pH 4.3).

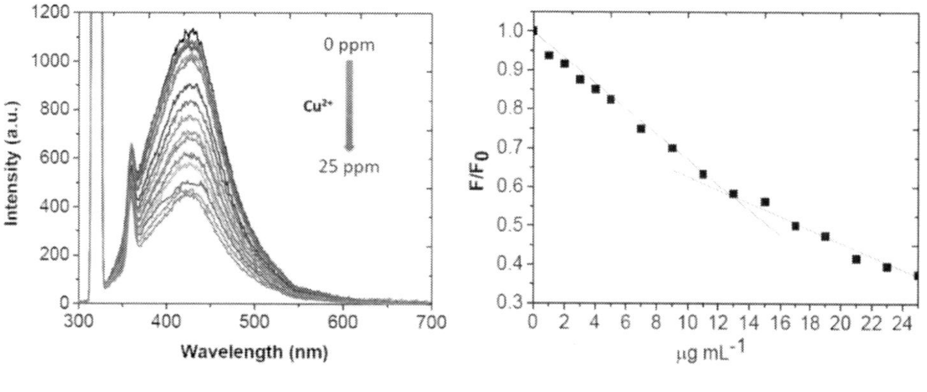

Figure 8. Fluorescence titrations of the y-CDs with Cu^{2+} from 0 to 25 μg mL^{-1}. (Influence of Cu^{2+} concentration on the fluorescence intensity of PB-CQDs (top to bottom: 0, 1, 2, 3, 4, 5, 7, 9, 11, 13, 15, 17, 19, 21, 23 and 25.0 μg mL^{-1})

444269 Functionalization of OH-Terminated Nanoparticles By Cyclic Azasilanes

Abstract:

Colloidal suspensions of functionalized nanoparticles have been suggested for enhanced oil recovery, formation damage remediation, flow assurance, and water treatment operations. Surface modified nanoparticles can also be used as tracers in reservoir studies, active media in thermal management fluids, and fillers in high performance elastomers. Conventional functionalization of nanoparticles with alkoxysilanes and chlorosilanes is limited by needs for catalytic amount of water for the reaction to proceed, requirement of multiple hours to fully saturate the surface, and release of by-products upon reaction completion. Monolayer deposition with a high density of functional groups represents a particularly difficult challenge.

Our ongoing studies indicate that application of cyclic azasilanes that have a strained ring structure with adjacent Si and N centers provides highly efficient functionalization of hydroxylated silica, alumina, magnetite, and diamond nanoparticles which is catalyst-free and by-product-free. The data on synthesis and characterization of functionalized nanoparticles as well as possible reaction mechanism of functionalization will be discussed.

444093 New Development on Hoist Scheduling with the Relaxation of Cyclic Precondition

Abstract:

Multi-Stage Material Handling (MSMH) processes are designed to manufacture materials requiring a series of treatments and in industry, they are broadly used due to their capability to handle large work orders and compatibility with materials with different recipes. A typical example is electroplating process. A hoist handles the transportation of each job among units (stages) according to its processing recipe. The hoist movement plan is predetermined through optimization and this plan is repeated to perform continuity. Obviously, the ultimate goal is to maximize the productivity within each movement plan, which is equivalent to minimize the cycle time to finish one work order. The optimization process is generally called Cyclic Hoist Scheduling (CHS), since every movement plan is a cycle due to its repeatability. This precondition in term of hoist operation and unit usage is general for most cases. However, it could block the really optimal results.

In this study, this precondition has been relaxed for multi-capacity units, thus the unit usage can be more flexible. Thus, the real optimality can be found through CHS. To deal with the cycle connection issue from the relaxation of the precondition, a new module Cycle Connection Identification (CCI) is proposed. Obtained hoist schedule will be checked. Once a schedule with cycle connection problems is identified, a new cycle will be scheduled forcing the smooth connection with the first cycle. The objective of scheduling a new cycle is to match the beginning of the first cycle and the ending of the second cycle. If the two-cycle group is still with cycle connection issue, more cycles will be scheduled till the cycle group is free from cycle connection problems. Each cycle is called a sub-cycle and the constructed cycle group without cycle connection problems is called a full cycle. Note that in each sub-cycle, the work order and all transportation operations are completed.

In this paper, hoist scheduling models with/without such a precondition have been developed and proposed as mixed integer linear programming. The performance of the two models is tested through case studies and the difference is compared and demonstrated in detail.

444359 Development of Novel Emulsified Acid Using Waste Oil

Abstract:

In carbonate reservoirs, hydrochloric acid (HCl) is commonly used for stimulation purposes. The main disadvantage for using HCl is its high and rapid reactivity allowing for less control on acid reaction with formation. In addition, HCl may react with the tubing causing severe corrosion and damages. Several slow-reacting acid systems, though HCl-based, have been developed to retard the HCl reaction and control its reactivity to allow deeper penetration in carbonate reservoirs before it breaks out and reacts with formation rock. However, the selection of the hydrocarbon phase of emulsified acid is critical to the creation and stability of the emulsified acid system. Diesel oil, crude oil and solvents such as xylene have been reported to be successful in the field.

In this work, alternative and less expensive source of oil is developed to replace the current practices. Emulsified acid is a water-in-oil emulsion where oil is the continuous phase and acid is the dispersed one. Industrial and filtered waste oil was used to prepare the emulsified acid of this work. Extensive work has been carried out to study the chemical composition of the waste oil, namely SARA and to select the proper emulsifier. Emulsions were prepared using 15% HCl with emulsifier concentration ranging from 0.5 to 2.0 vol% and 70:30 of acid-to-oil ratio. Droplet size distributions were measured and used to benchmark with other emulsions. Emphasis was given to the stability of the emulsion especially at high temperatures. In order to apprehend the effect of corrosion inhibitor on the emulsion stability, its concentration was varied (up to 0.3 vol%). Later, rheological analyses of the waste oil emulsion were compared to the commonly used diesel-based emulsified acid.

Results showed the potential use of waste oil to prepare high temperature emulsified acid (up to 100°C). The addition of corrosion inhibitor to the emulsified acid caused a negative effect on stability resulting in a less stable emulsion. On the other hand, addition of cationic emulsifier helped to formulate a more stable emulsion. Apart from that, the optimum concentration of the emulsifier was at 2.0 vol%. The rheological data confirms the stability of the new emulsified formulation at higher temperatures. This paper summarizes the findings of using waste oil emulsion and recommends it for field applications. The newly proposed emulsified acid introduces waste oil as a replacement of diesel.

The novelty of using waste oil will help to significantly reduce the expenses of formulating emulsified acid. Since waste oil is basically waste it has a good impact on the environment, that being said, instead of throwing it, the industry will be able to recycle it.

444364 Molecular Simulation of Controlling Ions and Their Role in Low Salinity IOR for Carbonate Reservoirs

Abstract:

Carbonate geology represents around 50% of hydrocarbon-bearing reservoirs in the world and more than 70% in some areas such as the Middle East. Water flooding has not been very efficient in carbonate reservoirs due mainly to the low water wetness. Modifying the ionic composition and salinity of the injected seawater demonstrated that the oil recovery from these carbonate reservoirs could be improved. Although the underlying mechanisms are still a subject of debate, the improved oil recovery has been mostly attributed to the increased water-wetness of the rock through the interactions of specific ions like Ca^{2+}, Mg^{2+}, SO_4^{2-}, and Cl^- , which are present in seawater with the rock surface under favorable conditions

The study of hydrated gas-phase metal ions provides a link between intrinsic chemistry of the isolated ion and its chemistry in solution. The structures and binding enthalpies of a variety of gas-phase and solution mono- and divalent ion hydrates containing up to 12 water molecules have been studied computationally. A DFT and molecular dynamics simulations were employed to study hydrated structures of the form $[\mathbf{M}(H_2O)_n]^{2+}.mH_2O$ for calcium and magnesium and $[\mathbf{M}(H_2O)_n]^{+}.mH_2O$ for sodium, in which there are n water molecules in the first coordination shell and m water molecules in the second coordination shell. The hydration, dehydration, and interactions of metal ions with Cl^{-1} and SO^{-2}_4 have been studied to better understand the mechanism of low salinity IOR. Density functional (DFT) theory using the B3LYP functional, and 6-311++g** basis set has been used. The charge of the metal ion decreases monotonically as the number of water molecules increases in the complex. Extensive calculations of the energetic contribution of individual structural characteristics as well as the energetic trends in binding energy are developed, which will lead to better understanding of how low salinity water IOR works.

Keywords: Hydration, carbonate reservoir, DFT, binding energy, Low Salinity

444278 Insight into the Mechanism of Toxic Chromium (VI) from Wastewater Using Magnetite Coated PINE Powder

Abstract:

INSIGHT INTO THE MECHANISM OF TOXIC CHROMIUM (VI) FROM WASTEWATER USING MAGNETITE COATED PINE POWDER

Pholosi A., Ofomaja A., Naidoo E.B.

Biosorption and Wastewater Treatment Research Laboratory, Department of Chemistry,

Faculty of Applied and Computer Sciences, Vaal University of Technology, P. Bag X021, Vanderbiljpark, 1900 South Africa.

INTRODUCTION

A knowledge of the interaction between pollutant molecules or ions and active sites on an adsorbent surface is an important phenomenon in many industrial processes such as catalysis, adsorption and other separation techniques. The use of lignocellulosic materials for Cr(VI) adsorption has gained interest over the years **(Miretzky and Cirelli, 2010)**. The adsorption properties of lignocellulosic materials has been attributed to acidic functional groups on their surfaces which can take part in ion-exchange, complexation, chelation and hydrogen bonding with pollutant species in solution **(Vithanage et al., 2015)**. Two mechanisms for lignocellulosic adsorption of Cr(VI) have been proposed; (i) an electrostatic mechanism in which at low pH, the protonated acidic functional groups on the biomaterial carrying a positive charge attracts the negatively charged Cr(VI) species **(Demirbas, 2005)** and (ii) the adsorption-coupled reduction mechanism in which Cr(VI) is reduced to Cr(III) either in solution (Direct reduction) or after being adsorbed (Indirect reduction) by electron-donor groups of the biomaterial that have lower reduction potential values than that of Cr(VI) **(Gao et al., 2008)**. The Cr(III) ions formed in solution may then form complexes with Cr(III)-binding groups on the biomaterial surface or released into bulk solution by repulsion between Cr(III) and other positively charged groups on the biomaterial **(Miretzky and Cirelli, 2010)**.

METHODS

Sodium hydroxide treated pine coated with magnetite (NTP-NC) was prepared by co-precipitation method. The effect of solution pH was performed in a batch system in which 0.5 g of the adsorbent was contacted in six 250 cm^3 beakers containing 75 cm^3 of 75 mg/dm^3 Cr(VI) solutions set at pH 1, 2, 4, 6, 8 and 10 at 26 °C and agitated at 200 rpm for 2 h and the clear solution analysed for Cr(V/I) left using UV-Vis spectrophotometry at 460 nm

RESULTS

The results of the effect of initial solution pH on the adsorption of Cr(VI) onto NTP-NC is shown in Fig. 1. To confirm the operating mechanism, the change in $[H^+]$ concentration at each initial pH, the oxidation/reduction potential before and after adsorption and the amounts of Total Cr, Cr(VI)and Cr(III) present in solution at each pH was monitored for NTP-NC and the results displayed in Fig1 . The change in $[H^+]$ concentration ($\Delta[H^+] = [H^+]_{final} - [H^+]_{initial}$) at the end of the adsorption process for each initial solution pH was recorded and plotted against the initial solution pH. The results showed that the change in $[H^+]$ concentrations at the end of the adsorption process where higher for lower initial solution pH's (where lower amounts of Cr was found in solution) than for higher initial solution pH's (where higher amounts of Cr was found in solution) and the change in $[H^+]$ concentration became small and almost constant after initial solution pH 4. This is consistent with the assumption that the removal of Cr from solution is associated with $[H]^+$ion consumption and that as the concentration of $[H^+]$ ions reduces in solution, the Cr removal reduces. Cr(VI) ion in acidic solution has a positive redox potential value (above 1.3 V at standard conditions) and in the presence of an electron donor becomes unstable and can be reduced to Cr(III) **(Park et al., 2008)**. The results show that the initial ORP values (before adsorption) for the Cr solutions at pH's of 1, 2, 4, 6, 8, and 10 were 339, 280, 165, 55, -53 and -172 mV for NTP-NC respectively. The positive values of ORP at lower solution pH's indicates higher possibility for reducing Cr(VI), while the lower and negative ORP values indicates lower possibility for reducing Cr(VI). On addition of the adsorbents into the solutions, a change in ORP were observed . The values of ORP at the end of the adsorption process became 330, 268, 30, 9 -29 and -61 for NTP-NC respectively. A reduction in ORP at the end of the adsorption stage signifies that reduction has occurred in the system. **Yu et al. (2014)** showed the amount of Cr(VI) removed from aqueous solution in contact with zero-valent iron (nZVI) could be related to the change in ORP and pH of the solution. The authors observed a decrease in ORP from 400 to 100 mV and a corresponding 10 unit pH increase as Cr(VI) was removed from solution.

Finally, analysis of the prepared Cr solutions at the various solution pH's showed that the concentration of Total Cr obtained from ICP and that of Cr(VI) using UV-vis measurements were very similar, suggesting that Total Cr was almost equal to Cr(VI) in solution. After contacting the various solutions with the adsorbents, the concentration of Total Cr in the pH 1 solution was observed to have reduced to 21.56 mg/dm^3 for NTP-NC, while the Cr(VI) concentrations was 10.66 mg/dm^3 for NTP-NC. With solution pH 2, the concentration Cr(III) left in solution was lower than that of solution pH 1 for both samples. The reason is that at pH 1 there is an abundance of H^+ ions in solution which accumulates on the functional groups on the adsorbent making the surfaces of the adsorbent positively charged and repelling away the positively charged Cr(III) ions from the adsorbent surfaces **(Suksabye et al., 2009)**. At solution pH 2, it is believed that there is less accumulation of H^+ ions at the adsorbent surfaces which led to reduced electrostatic repulsion and increased adsorption of Cr(III) from solution. This increase in Cr(III) adsorption is also confirmed by the reduction in Total Cr in solution between solution pH 1 and 2. The concentrations of Cr(VI) left in solution at pH 2 was higher than at pH 1 for NTP while for NTP-NC the Cr(VI) concentrations were lower at pH 2. For solution pH's higher than 2 and up to 10, the residual Cr(III) in solution becomes very low and almost constant. This is consistent with the small change in $[H^+]$ in solution which is almost constant at this pH range.

FTIR Evidence for the adsorption of Cr(III)

When Cr(III) is formed in solution via adsorption-coupled reduction mechanism it may undergo hydrolysis and/or complexation depending on the solution pH and the total Cr(III) concentration. Hydrolysis may lead to the formation of various Cr(III) species with an increase in solution pH which gives rise to a reduction in its solubility and increasing the chances of precipitation of the Cr(III) species. At solution pH 1 – 4, Cr^{3+} and $CrOH^{2+}$ are the major species which can be adsorbed by negatively charged ligands on the biomaterial surface. On the other hand, at solution pH range of 4 – 10, Cr(III) species (hydroxo complex), such as $CrOH^{2+}$, $Cr(OH)_2^+$ and $Cr(OH)_3$, are the dominant forms in the aqueous system and at high pH is readily transformed into soluble complex, $Cr(OH)_4^-$. When NTP-NC was contacted with Cr(VI) solution for 2 hr, the final solution pH was increased to pH 2.53. The FTIR spectra for NTP-NC (Fig. 2) shows the presence of characteristic Fe-O peaks of bulk Fe_3O_4 at 553.26 cm^{-1}, 564.36 cm^{-1}, 570.21 cm^{-1} and 576 21 cm^{-1}. Functional groups associated with the pine includes the broad band at 3250.99 cm^{-1} representing –OH of cellulose, 2905.17 cm^{-1} representing C-H stretching, while the peaks at 1627.69 and 1602.87 cm^{-1} shows the COO^- and C=O bonds respectively. After 2 hr contact with Cr(VI) solution, the peak at 564.34 cm^{-1} was observed to have split into several new peaks (Fig. 3b (ii)). These emerging peaks are characteristic of Fe-O bonds of maghemite (γ-Fe_3O_4), for example the characteristics peaks of maghemite was observed at 595.21, 685.65 cm^{-1} and 582.70, 634.91 cm^{-1} suggesting a change in the iron oxide phase (Fe_3O_4 to γ-Fe_2O_3) due to the conversion of Cr(VI) to Cr(III) by magnetite nanoparticles. According to **Peterson et al. (1997)**, Cr(III)(hydr)oxide produced is precipitated on the magnetite surface. In this study, α-Cr_2O_3 peak were observed at 604.63 cm^{-1} indicating that Cr(III) was formed during the adsorption process. Finally it was also observed that the carboxylate group at 1635.55 cm^{-1} was shifted to 1637.73 cm^{-1} while a new peak was observed at 1737.72 cm^{-1} suggesting that complexation between Cr(III) and Carboxylate ions also occurred.

REFERENCE

Demirbas, A. Energy Source 27 (2005) 1449–1455.

Miretzky, P., Cerelli, A.F. A review. J. Hard. Mater. 180 (2010) 1-9.

Vithanage, M., Rajapaksha, A.U., Ahmad, M., Uchimiya, M., Dou, X., Alessi, D.S., Ok, Y.S. J. Environ. Manage. 151 (2015) 443-449.

D. Park, C.K. Ahn, Y.M. Kim, Y.-S. Yun, J.M. Park, J. Hazard. Mater. 160 (2008) 422–427.

P. Suksabye, A. Nakajima, P. Thiravetyan, Y. Baba, W. Nakbanpote.. J. Hazar. Mater. 161 (2009) 1103–1108.

R.F., Yu, F.H., Chi, W.P., Cheng, J.C. Chem. Eng. J. 255 (2014) 568–576

M.L. Peterson, A.F. White, G.E. Brown, G.A. Parks Environ. Sci. Technol. 31 (1997) 1573–1576.

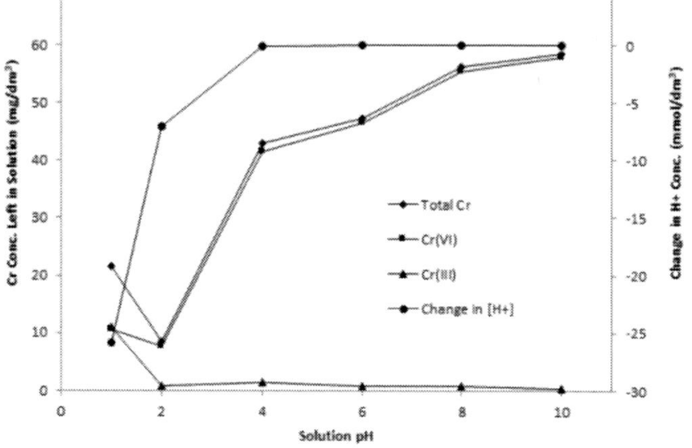

Fig. 1 Effect of solution pH on the adsorption of Cr(VI)

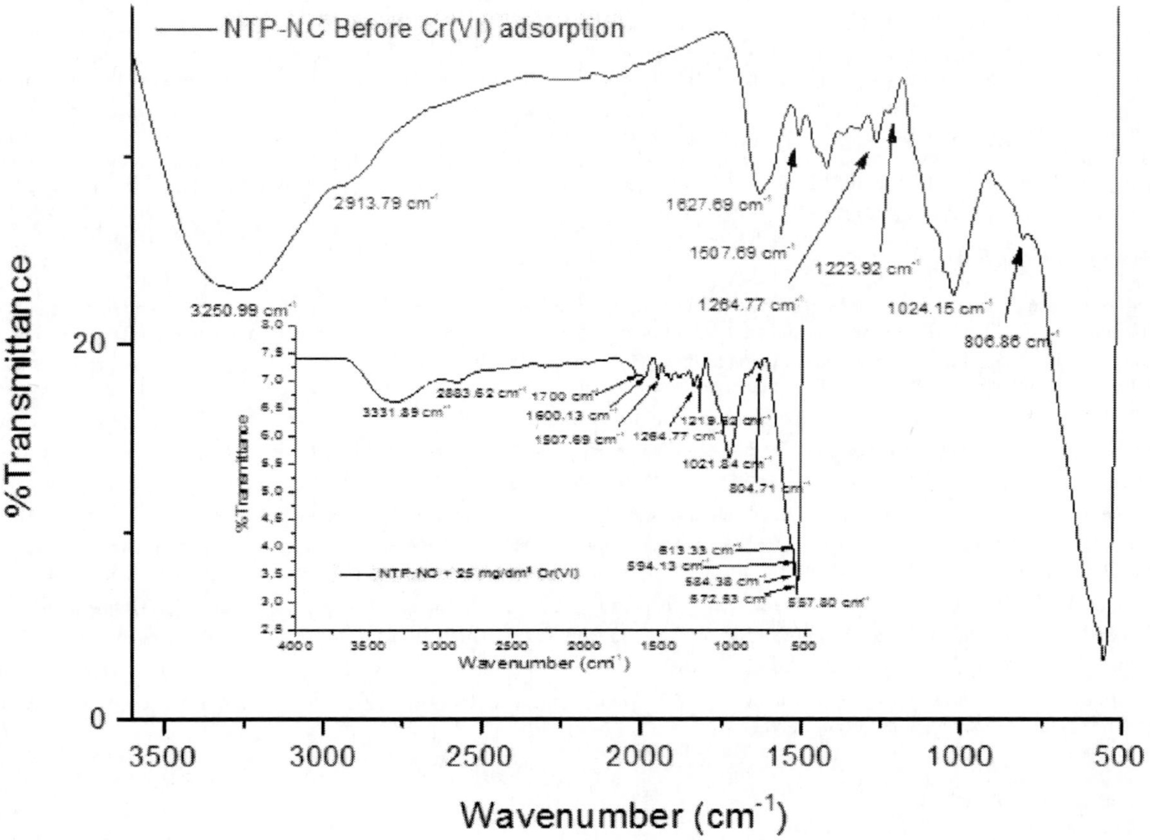

Fig, 2: FTIR of NTP-NC

444280 Copper-Phyllosilicate Core-Shell Nanocatalysts with Balanced Active Sites for Carbon-Oxygen Hydrogenolysis Reactions

Abstract:

Copper-Phyllosilicate Core-shell Nanocatalysts with Balanced Active Sites for Carbon-Oxygen Hydrogenolysis Reactions

Hairong Yue, Weimao Fang, Changmin Liu, Jiankang Ying

College of Chemical Engineering, Sichuan University, Chengdu, China

Email: hryue@scu.edu.cn

Abstract

Hydrogenolysis of carbon-oxygen (C¨CO) bonds (e.g., esters, ethers, furfural, and CO_2) has emerged as a versatile synthetic tool in organic methodology, as it could produce a variety of products (e.g., chemicals, fuels, and polymers). Copper-based catalysts have been intensively explored for hydrogenation reactions as the copper sites account for the selective hydrogenation of carbon-oxygen bonds and relatively inactive for the hydrogenolysis of carbon-carbon bonds. Work on understanding the active sites of copper catalysts for hydrogenation reactions indicted that both Cu^0 and Cu^+ species were crucial to the activity of Cu-based catalysts. However, researchers were not able to establish a relationship between activity and Cu^0/Cu^+ active species since the conventional Cu-based catalysts is the deactivation by metal particle growth and unstable surface Cu^0 and Cu^+ active species in the strong reductive H_2 and oxidizing carbon-oxygen atmosphere.

We renctly present on efficient approaches for fabrication of a series of copper-phyllosilicate core-shell nanocatalysts and nanoreactors with balanced and stable Cu^0 and Cu^+ active species. We chose the hydrogenation of dimethyl oxalate, ethylene cyclicarbonate and as probe reactions for a better understanding of properties of active sites and the structure-activity relationship. The results indicated a synergy between copper and the oxide components and the balanced surface Cu^0 and Cu^+ species can greatly improve the catalytic C-O hydrogenolysis performance of the catalysts. These nanocatalysts with balanced and stable Cu^0 and Cu^+ active species, confinement effect, intrinsic high surface of Cu^0 and Cu^+ and unique tunable tubular morphology, has potential applications in the high-temperature hydrogenation reactions.

Keywords: core-shell nanocatalysts, Hydrogenation reaction; copper catalyst

References

1. Yue, H.; Zhao, Y.; Ma, X.; Gong, J.: Ethylene glycol: properties, synthesis, and applications. *Chem Soc Rev* 2012, 41, 4218-4244.

2. Gong, J.; Yue, H.; Zhao, Y.; Zhao, S.; Zhao, L.; Lv, J.; Wang, S.; Ma, X.: Synthesis of Ethanol via Syngas on Cu/SiO2 Catalysts with Balanced Cu^0-Cu^+ Sites. *J Am Chem Soc* 2012, 134, 13922-13925.

3. Zhao, S.; Yue, H.; Zhao, Y.; Wang, B.; Geng, Y.; Lv, J.; Wang, S.; Gong, J.; Ma, X.: Chemoselective synthesis of ethanol via hydrogenation of dimethyl oxalate on Cu/SiO_2: Enhanced stability with boron dopant. *J Catal* 2013, 297, 142-150.

4. Yue, H.; Ma, X.; Gong, J.: An Alternative Synthetic Approach for Efficient Catalytic Conversion of Syngas to Ethanol. *Acc Chem Res* 2014, 47(5):1483-1492

5. Yue, H.; Zhao, Y.; Zhao, L.; Lv, J.; Wang, S.; Gong, J.; Ma, X.: Hydrogenation of dimethyl oxalate to ethylene glycol on a Cu/SiO2/cordierite monolithic catalyst: Enhanced internal mass transfer and stability. *AIChE J* 2012, 58, 2798-2809.

6. Yue, H.; Zhao, Y.; Zhao, S.; Wang, B.; Ma, X.; Gong, J.: A copper-phyllosilicate core-sheath nanoreactor for carbon¨Coxygen hydrogenolysis reactions. *Nat Commun* 2013, 4, 2339.

Stimuli-responsive Rubber in Oil and Gas Applications

Rostyslav Dolog, Sankaran Murugesan, and Valery Khabashesku

Baker Hughes, 14990 Yorktown Plaza Dr, Houston, Texas, 77040, USA

ABSTRACT

A problem that exists today in the oil & gas industry is the loss of mechanical properties by reactive rubber seals due to swelling, and the limited space in downhole well applications to accommodate larger swellable packing elements.

This paper presents cost-effective, reinforcing, stimuli-responsive fillers, such as certain classes of fly ash, for use in rubber composite materials in various oil and gas applications. Apart from cost effectiveness, these fillers can also enable stimuli-responsive effects, as well as improved mechanical properties (e.g. modulus) for rubber composites when swelled in water. Fly ash (Class C) is a common component of Portland cement which was used in the literature to achieve a stimuli-responsive effect when added to Hydrogenated Nitrile Butadiene Rubber (HNBR) as filler. Exposure to water brines can cause a filler to promote an increase in pH which enables the hydrolysis of HNBR nitrile groups to carboxyl groups, thus creating an ionomer. This process is followed by neutralization of the new ionomer with metal ions, and formation of ionic crosslinks, further improving mechanical properties of the rubber.

This paper demonstrates that incorporation of fly ash as a reinforcing filler allows development of swellable seals that can retain, or even increase, their modulus and stiffness upon swelling. These characteristics allow avoiding the loss of mechanical properties by seals due to swelling and also eliminating the need for large swellable packing elements that occupy additional space downhole.

KEYWORDS: Elastomers, stimuli-responsive, fly ash, composites

INTRODUCTION

There is a growing demand for high performance elastomers in the oil & gas industry. Rubber compounds are used for various sealing applications, packer elements, stators for pumps, etc. In general, rubber formulations are comprised of elastomer matrix, curing system, antioxidants, plasticizers, along with reinforcing filler, which provides the compounds with their strength. The reinforcing filler is usually represented by carbon black and, to a lesser extent, silica. Along with other factors, filler mechanical properties and concentration generally define hardness, modulus and tensile strength of the elastomer compound. Throughout the years, many different materials have been tried as alternative fillers for rubber composites, including nanomaterials (CNTs, graphene, nanofibers), recycled rubber particles, and ash/fly ash. In the case of ordinary carbon black, the filler serves not only for reinforcement, but also as a cost reducer, because raw elastomer is often a more expensive component. Mechanical properties of the composites are affected by the aspect ratio of the filler, quality of the dispersion, and interface between blend components. The latter determines the effectiveness of load transfer from polymer matrix to the reinforcing filler. High performance compounds can be designed by incorporating well dispersed fine particles with a high aspect ratio and good compatibility with the polymer matrix.

In this context, the fly ash is a cheaper, low reinforcing filler alternative to carbon black, which will help reduce cost of the compounds for "non-demanding" sealing applications requiring large amounts of raw materials. Fly ash is produced in large quantities as a byproduct of coal burned at power plants, and can be considered as a recycled material. Fly ash costs several cents per pound, while carbon black, depending on the grade, is usually one order of magnitude more expensive. Hemmings et al[1] demonstrated that fly ash can be used as filler for polymeric composites with enhanced mechanical properties. Adoption of this technology in the oil and gas industry could change the design approach for reactive rubber compounds. Further functionalization of fly ash through a cost-efficient process could ensure compatibility with the rubber matrix, while promoting filler dispersion and ensuring an efficient load transfer. Such functionalization's include, but are not limited to, surface modification with fatty acids, dicarboxylic acids or phosphoric acid. The reasoning for surface modification is to attach functional groups with aliphatic chains to fly ash particles to make them compatible with aliphatic rubber chains, or to attach acidic groups on their surface that can interact with nitrile groups of HNBR. Initial attempts towards chemical functionalization of fly ash have been made by Seal et al for treatment of offshore oil spills.[2]

Apart from cost effectiveness, composites with fly ash can show stimuli-responsive properties. It has been shown that adding Portland cement to the HNBR matrix improves its mechanical properties in the swollen state by water uptake.[3] Exposure to water causes Portland cement to grow in volume and mass, and promotes an increase in pH which enables the hydrolysis of HNBR nitrile groups to carboxyl groups, thus creating an ionomer. This process is followed by neutralization of the new ionomer with metal ions and formation of ionic crosslinks, further improving mechanical properties of the rubber. Fly ash (Class C) is a common component of Portland cement, which contains a large variety of metal oxides, and is capable of increasing the

[1] Filler Comprising Fly Ash For Use In Polymer Composites. Hemmings et.al., US 6,916,863 B2, July 2005.
[2] Hydrophobic High Surface Area Zeolites Derived from Fly Ash for Oil Spill Remediation, Seal et.al., Environ. Sci. Technol., 2013, 47 (11), pp 5843–5850
[3] Stimuli-Responsive Cement-Reinforced Rubber, Musso et. al., ACS Appl Mater Interfaces. 2014 May 14;6(9):6962-8.

pH of the water solution by hydrolysis, which has a potential to achieve a stimuli-responsive effect when added to HNBR as filler.

Degradation of tensile properties is a common problem for swellable packers and other sealing elements. By selectively using certain classes of fly ash as effective reinforcing filler, one can develop swellable seals that can retain or even increase their modulus and stiffness upon swelling. Therefore, there is a potential to minimize or even avoid the loss of mechanical properties by water swellable reactive packing elements due to swelling. This provides an opportunity to decrease the size of these elements, which are often represented by rather long sections, and hence occupy a significant amount of space downhole. In this paper we have studied the reinforcing effects of fly ash fillers with HNBR compounds as well as stimuli responsive properties of these compounds.

EXPERIMENTAL SECTION

Rubber compounds T1, T2, T3, and T4 are elastomers based on a custom-built peroxide curing system. Formulation F2 was prepared by adding 15 wt. % of fly ash class C to compound F1, which is a commercial swellable NBR sulfur-cured compound.

T series compounds were prepared using a C.W. Brabender® internal mixer operated at 60 RPM for 20 minutes at ~160°F and then molded into sheets and cured at 340°F for 30 minutes in a hot press. Compound F2 was prepared by mixing fly ash with compound F1 in the Brabender mixer for 6 minutes at 60 RPM at ~100°F. Compound F1 is a commercial water swellable formulation that was used as received. F series compounds were cured at 300°F for 60 minutes in a hot press.

Tensile properties of rubber samples were tested using dog bone specimens, according to the ISO37 standard, at room temperature and at 350°F using an MTS® testing system.

Swelling tests in 3% KCl brine and LVT 200 oil were performed to assess the effect of fillers on the chemical resistance of elastomers. Standard one-inch rubber buttons were tested at 200°F at ambient pressure for 6 days. Thermal properties were studied using TGA Q50 and DSC Q2000 instruments.

To analyze the stimuli-responsive effect introduced by fly ash fillers, rubber buttons were aged for 10 days in 3% KCl in the oven at 200°F and then removed from the solution, dried for 30 minutes in the oven and then for 3 hours at room temperature. Samples were later tested in the MTS testing system in compression mode.

Viscoelastic properties of stimuli-responsive samples, based on water swellable compounds, were accessed using the RSA G2 dynamic mechanical analysis (DMA) instrument. This experiment was conducted in a temperature ramp mode of 2°C/minute at 1Hz frequency and 0.2% amplitude strain. A Nicolet 6700 Fourier Transform Infrared (FTIR) spectrometer was used to confirm neutralization of carboxylic groups by metal cations.

RESULTS AND DISCUSSION

The reinforcing properties of fly ash were evaluated at room temperature (Table 1) and at 350°F (Table 2) and compared with N550 carbon black. Sample T3, reinforced with fly ash, demonstrated inferior tensile properties and hardness compared to a compound filled with carbon black. However, at room temperature, fly ash was capable of delivering a reinforcing effect to the rubber compound without sacrificing its elasticity, e.g. the ultimate elongation of elastomer compound remained virtually unchanged after incorporation of 50 parts per hundred rubber (PHR) of fly ash, while its modulus and tensile strength have increased. Therefore, potentially, one can substitute carbon black with high loadings of fly ash in certain applications, thus further decreasing the cost of the rubber product. The summary of mechanical properties is provided in Table 1, E_x denotes modulus of a compound at certain strain (x), i.e. stress at strain.

Sample	N550 CB (PHR)	Fly ash Class F (PHR)	Tensile strength (psi)	E_{25} (psi)	E_{50} (psi)	E_{100} (psi)	Elongation (%)	Hardness Shore A
T1	0	0	829	71	97	121	472	48
T2	50	0	3402	165	273	744	290	68
T3	0	50	1176	94	127	166	467	54

Table 1. Room temperature tensile properties and hardness of elastomer samples

The relatively low increase in modulus and hardness of T3 compared to T1 at room temperature can be explained by a low aspect ratio of fly ash and weak interactions between blend components. The latter problem can be solved by functionalization of fly ash particles.

The general data trends from the high temperature tensile tests closely resemble the results of the tensile test conducted at room temperature. However, at high temperature, rubber specimens break at much lower elongation, which leads to a great decrease in the measured tensile strength. This behavior is typical for elastomers and limits their application to a certain maximum service temperature and is also due to the fact that high temperature tests are conducted at conditions that push the limits of the material. The modulus values at 15% (E_{15}) and 20% (E_{20}) strain are not provided herein, because half of the tested specimens were broken before they reached such strains.

The superiority of carbon black as an efficient filler became even more pronounced at high temperature. However, incorporation of fly ash doubled the tensile strength and increased the elongation at break by one-third when compared to the unfilled elastomer.

Sample	E_5 (psi)	E_{10} (psi)	E_{15} (psi)	E_{20} (psi)	Tensile strength (psi)	Elongation (%)
T1	31	46			91	23
T2	61	130	194	261	832	39
T3	42	73	98	119	173	30

Table 2. Summary of high temperature (350˚F) tensile data for elastomer samples

Thermal properties of samples T1-T3 were assessed with Differential Scanning Calorimetry (DSC) and Thermogravimetric analysis (TGA). TGA allows evaluating thermal stability of samples and to verify the actual composition of a sample from the ash content. The results of the DSC and TGA analysis are summarized in Table 3. During TGA, samples were heated to 600°C at 10°C/minute rate in a nitrogen (N_2) environment, and subsequently heated at 10°C/min to 600 °C in air to burn off carbon-based fillers. This allows decomposing of organic matter first by applying heat, and then burn the remaining carbon-based material in air atmosphere.

The initial degradation of the polymer started at 200°C and progressed to complete decomposition of aliphatic chains in the temperature range of 420-500°C with a highest rate of decomposition at ~460°C. Sample T2 incurred the second rapid mass loss after 600°C due to burning of the carbon black, which comprised most of the ash content at 600°C in a N_2 environment. The DSC experiment demonstrated the glass transition temperature of the elastomer remained largely independent of the type of filler - the respective DSC heating scan was conducted with a 10°C/minute temperature ramp after precooling the sample to -50°C.

Sample	W_{600} (%)	W_{800} (%)	$T_{deg}1$, °C	$T_{deg}2$, °C	T_g, °C
T1	2.4	2.4	452.8	-	-29
T2	29.8	2.7	458.7	616.5	-29
T3	32.8	32.7	447.3	-	-28

Table 3. TGA/ DSC results for elastomer samples

Table 3 contains ash contents (W) at 600°C and 800°C, degradation temperatures (T_{deg}) which represent the temperature of the highest rate of decomposition during TGA experiment, and glass transition temperature (T_g) of the samples, as determined by DSC scan.

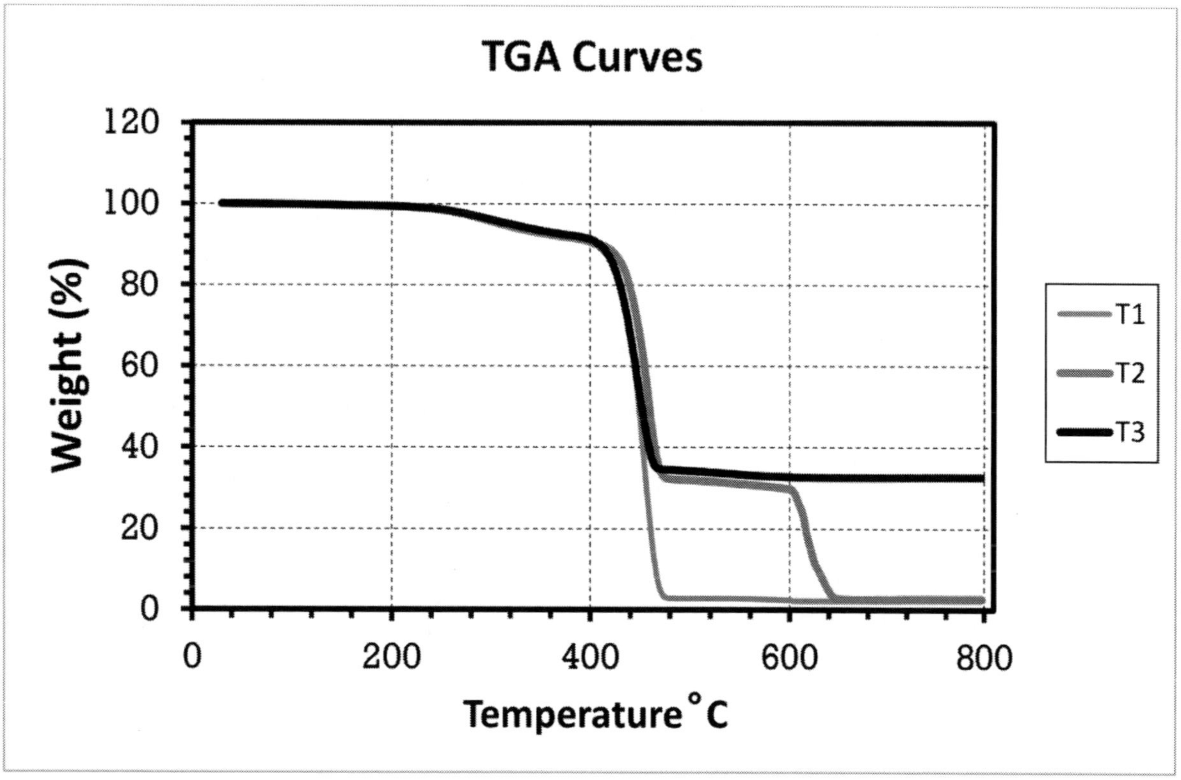

Figure 1. TGA Curves for samples T1, T2, and T3

Sample \ Testing fluid	3% KCl brine	LVT 200	Specific gravity
T1	0.6	13.0	0.96
T2	0.3	11.6	1.13
T3	0.8	9.1	1.18

Table 4. Volume increase (%) of elastomer samples due to swelling for 6 days at 200°F

Musso et. al. developed a stimuli-responsive HNBR composite reinforced with Portland cement that was increasing its stiffness and modulus when swelled in water[3]. Fly ash (Class C) is a common component of Portland cement, which contains the components necessary to achieve a similar effect. Therefore, one can avoid the loss of mechanical properties by water swellable packing element seals upon activation. To test this hypothesis, we studied the effect of aging in water brines on the properties of elastomers with fly ash fillers.

Rubber samples were molded and cured into standard size rubber buttons (cylinders of ~0.25" height and ~1" diameter). These buttons were aged for 10 days in 3% KCl brine in the oven at 200°F, removed from the solution, dried for 30 minutes in the oven and for 3 hours at room temperature. The swelling of both compounds was close to ~1% by volume (Table 4). Samples were later tested in an MTS machine in compression mode. Each button was compressed to 70% strain, buttons remained intact during the experiment, and visually all appeared to have a certain level of compression set at the end of the experiment. The results of the experiments are summarized in Table 5 and Figure 2. Only one button was tested for each sample. The provided compression modulus values (C_n) represent values of stress at a percentage strain (n), e.g. C_5 is stress at 5% compression strain.

Compound T4 was reinforced by fly ash Class C, which contains larger weight fraction of CaO than fly ash Class F (~25% vs ~10-15% respectively) and is expected to produce a higher pH and closely resemble cement that was used to prepare stimuli-responsive rubber in the original paper by Musso et.al.[3] The results demonstrate that exposure to water (3% KCl) at high temperatures can improve the modulus for the compounds filled with fly ash Class C (T4), while the control sample (T1) showed the opposite behavior. After aging, the compressive modulus at 50% strain (e.g. stress at 50% strain) of control sample T1 decreased by 19%, as compared to the unaged sample, while compressive strength of sample T4, reinforced by fly ash, increased by 15%. The difference in modulus values is even more drastic at 70% compression strain (-35% and 31%, respectively). This is in agreement with the hypothesis stated earlier.

Sample	Filler	Compressive modulus (psi)				
		C_5	C_{10}	C_{25}	C_{50}	C_{70}
T1	None	19	42	127	468	1810
T1 swollen	None	16	36	109	380	1185
T4	Fly Ash class C (50 PHR)	21	46	134	434	1628
T4 swollen	Fly Ash class C (50 PHR)	29	57	153	501	2125

Table 5. Summary of compressive modulus properties after 3% KCl soaking and 200°F drying

To further investigate the stimuli-responsive property of fly ash filler, the latter was incorporated into a water swellable compound, which provided for significantly more interaction of filler with water to occur. Sample F2 comprises unmodified commercial water swellable NBR formulation F1 mixed with 15 wt.% of fly ash Cass C. Sample F2 was processed into a 0.04" slab, then a rectangular sample was cut from the slab and swollen in 3% KCl solution until 100% volume increase. The F2 sample was then dried overnight in the oven, which lead to de-swelling of the sample due to solvent evaporation. Mechanical properties of the F2 sample were assessed with DMA prior to the swelling test, in a swollen state, and after drying - the data summary is shown on Figure 3.

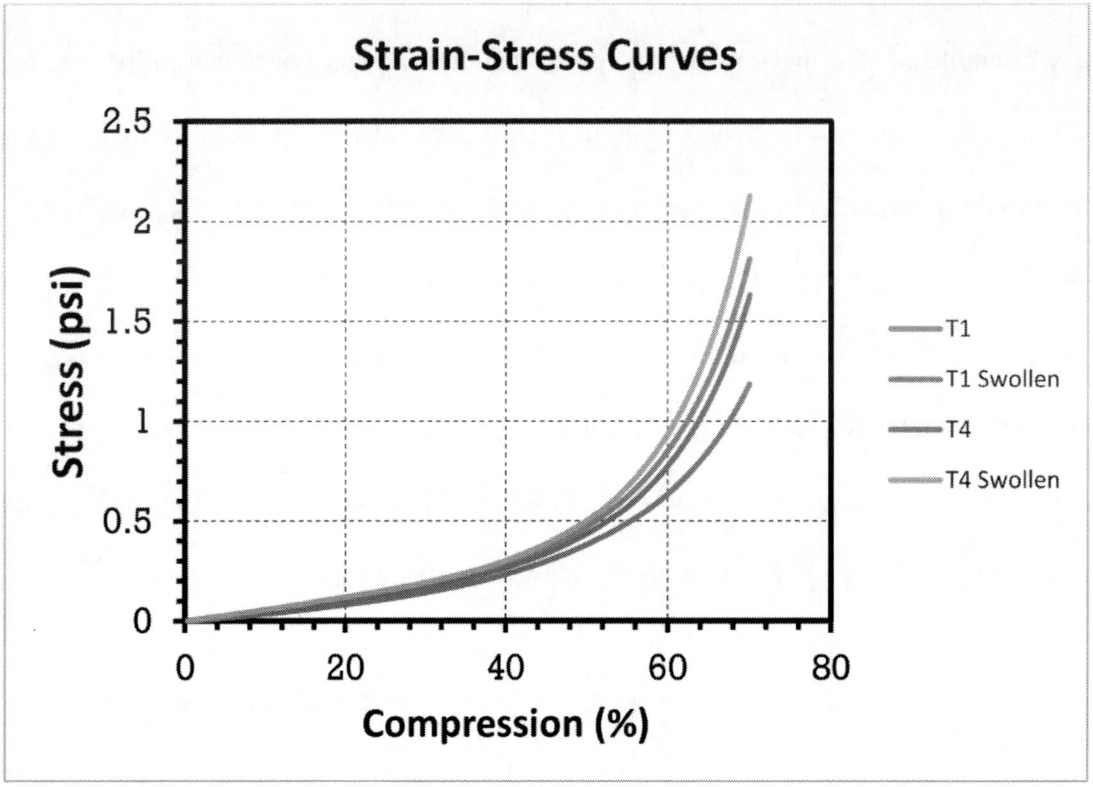

Figure 2. Stress-strain curves

The storage modulus of sample F2 after swelling and subsequent drying was about an order of magnitude higher than that of original sample F2 measured prior to swelling experiment. The storage modulus of the dried F2 sample was also less sensitive to the increasing temperature, contrary to the weakening original sample F2. The increase in the storage modulus of swollen F2 with temperature is simply due to the drying of the sample. The obtained result demonstrated that fly ash can essentially act as a stimuli-responsive material.

The significant improvement in the modulus is believed to be due to the neutralization of the swellable component of the commercial F1, and potential hydrolysis with subsequent neutralization of NBR elastomer. This is confirmed by the shift in T_g to a lower temperature range after drying as illustrated in Figure 3.

Carboxylic groups were detected in the original sample F2 prior to swelling, hence, they cannot be attributed solely to hydrolyzed HNBR, but rather to the super-absorbent polymer present in water swellable rubber compound (see Figure 4). FTIR confirms neutralization of carboxylic groups because a significant increase of the peak near ~1540 cm^{-1} and a lowering of peak intensity at ~1720 cm^{-1} have been observed. This neutralization is believed to promote the formation of ionic clusters in the compound with fly ash considered to be the source of cations. The neutralization of carboxylic groups could as well be attributed to K^+ cations present in the KCl brine solution, however, in that case, the respective peak in FTIR spectra would appear at higher wave numbers, approaching ~1600 cm^{-1}.

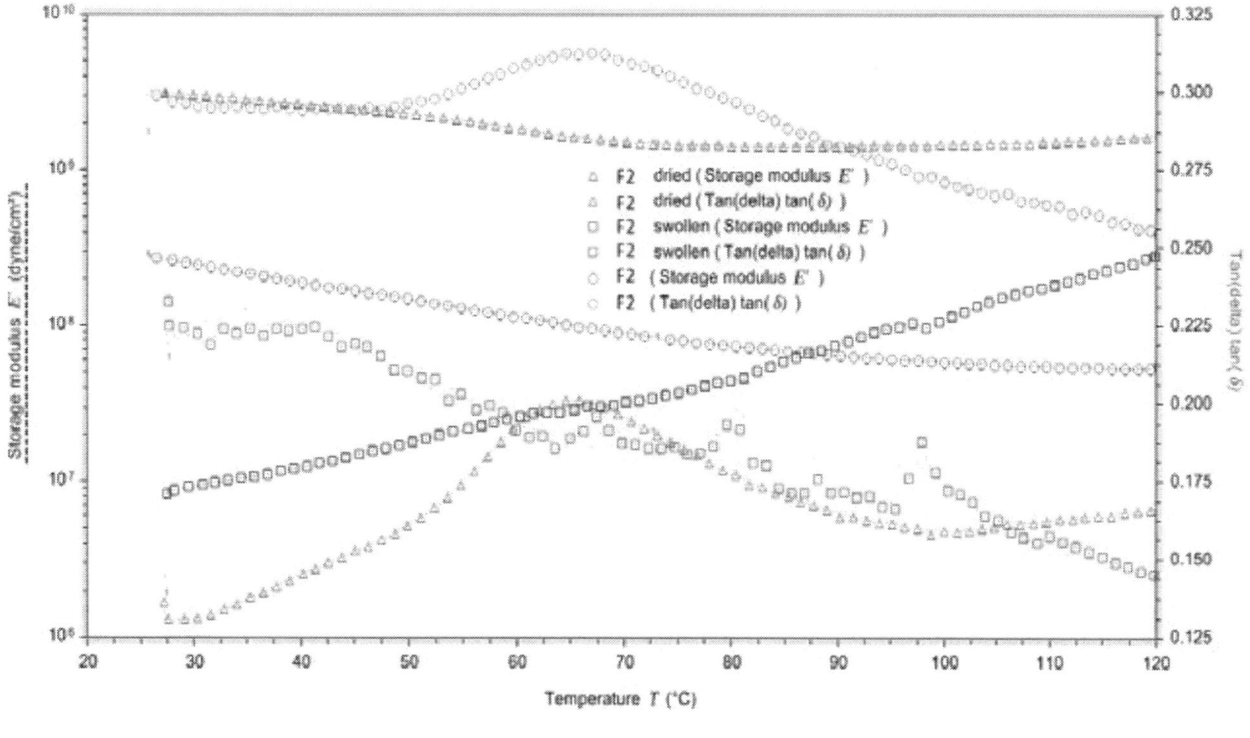

Figure 3. DMA data on F2 sample

It is important to mention that sample F2 exhibited a certain degree of brittleness after drying, which is undesired in rubbers, but this can be managed by incorporation of common ionomer plasticizers.

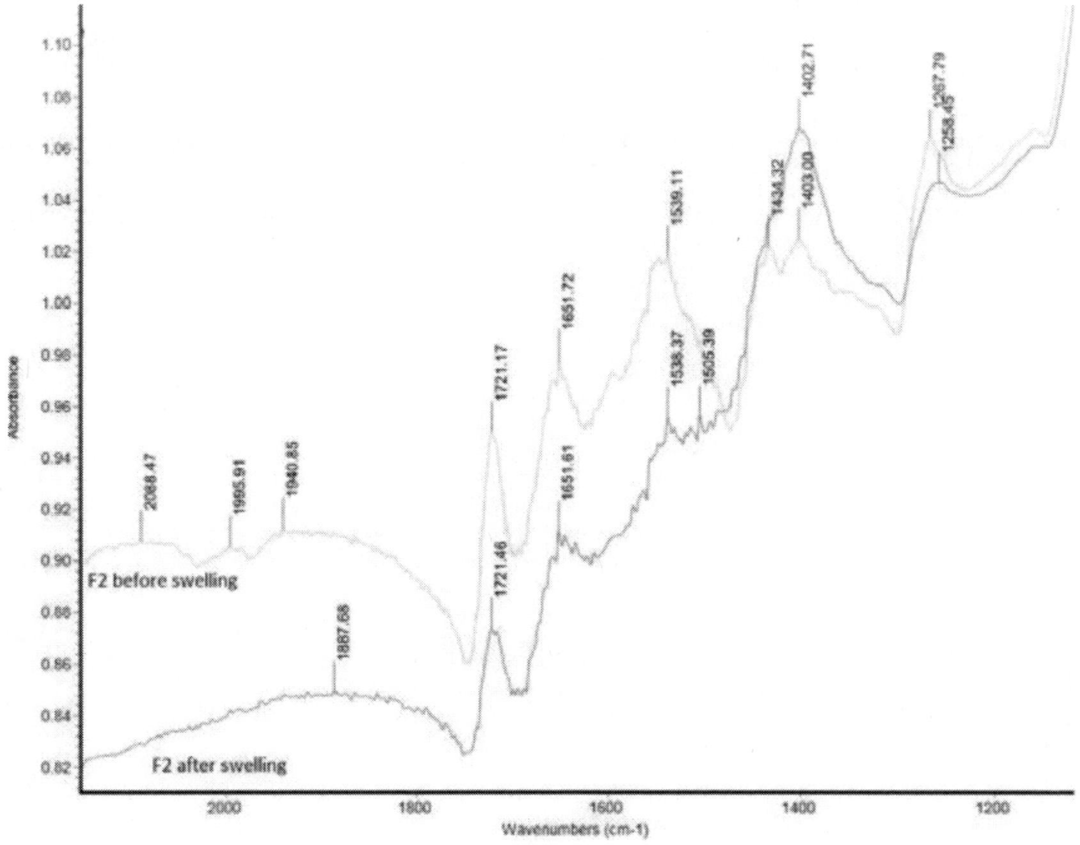

Figure 4. FTIR spectra of F2 sample taken before and after swelling

Conclusions

Fly ash demonstrated a great potential for consideration as cost effective, low reinforcing filler for downhole oil and gas applications. Although the reinforcing properties of fly ash are inferior to carbon black, incorporation of fly ash can improve the mechanical properties of HNBR elastomer compared to the unfilled raw crosslinked elastomer at both room temperature and at 325°F.

Furthermore, composites of HNBR with fly ash Class C demonstrated a stimuli-responsive effect by significantly increasing their modulus upon swelling in water brine. This effect was even more pronounced in composites of fly ash with swellable water compounds, which provided an easy access of filler to water upon swelling. A ten-fold increase in modulus was achieved in such fly ash reinforced sample, after swelling in 3% KCl water brine and drying in the oven overnight. The FTIR confirmed neutralization of carboxylic groups present in the investigated system, which contributed to overall improvement of mechanical properties of the compound upon swelling.

444310 Formal Modeling of Accidents and Automatic Generation of Tutoring Scenarios for Intelligent Safety Training of Process Operators

Abstract:

With the emphasis of accident-free operations and prevention of catastrophic events, safety-training demands for the chemical plants have been on the increase globally. However, the research results on modeling of accidents and involving human operators and the advanced theory and technology for (safety) operator training have been shared limitedly among domain experts and commercial developers. Even with the recent trend of simultaneously delivering the operator training system with the construction of a new process plant, in commercial operator training systems (OTS), safety training used to implement quite limited and obvious scenarios (and the burden was almost up to the trainer), partially due to the difficulty in modeling propagations of probable accidents. By the regulatory requirements or proactive investment by companies, effective scenarios for training of safety and abnormal situation management are necessary to be improved in the aspects of the size of volume as well as the training depth. Thus, considering the emergence of enhanced virtual reality technologies and inexpensive gadgets like HMD, supporting the immersive training environment, the future efficiency of the training heavily relies on the quantity and quality of safety scenarios, which would be essential in guiding the entire training process.

In this research we first overview a government-sponsored research project developing an enhanced safety training system for chemical process operators, using haptic devices in addition to enhanced audio and visual equipment. The initial version of the system runs with a fixed set of training scenarios developed for predetermined accident scenarios. However, with a limited set of training scenarios, the trainees simply memorize and rote learn the steps, required to be mastered to proactively solve the problems with an advent of possible accidents. In addition to that, without formal modeling, the system cannot give timely feedbacks and the performance of the trainee cannot be measured in real-time. Then to enhance the contents and usefulness of the training, the knowledge of process operations is collected and systematically stored as a form of knowledge base, in addition to the accident scenarios obtained by executing Fault Tree Analysis and HAZOP study.

By using a formal modeling language like UML and Petri nets, the accident scenarios and operators' problem-solving tasks are modeled hierarchically and classified into distinguishable subtasks. Then a new training scenario is generated on-line through automatic synthesis based on planning using the knowledge base. Safety training scenarios are generated intelligently for the tutoring of a single operator's job function as well as effective collaboration among a group of operators for real-scale malfunction diagnosis and intervention at the initial propagation stage of an accident. As case study, safety trainings for city gas governing stations are executed based on systematic scenario generation and analyzed for improved effectiveness and field applicability of the system. Reduction of time and cost in the training scenario production and real-time customized training based on new operator's level of knowledge and skill are also discussed in the framework of intelligent tutoring systems.

444376 Ozone Source Apportionments of Local Anthropogenic Emissions

Abstract:

Ozone is one of six principal air pollutants designed by US EPA. NOX and VOC emissions from anthropogenic sources are the most important factors to control ozone pollutant.

In this report, photochemical dispersion models are used to estimate the effect of ozone from local anthropogenic emission sources. EKMA-type diagrams are generated for NOX- and VOC- effects. Furthermore, APCA (Anthropogenic Precursor Culpability Assessment) technology is used to estimate ozone source apportionments.

EKMA-type diagrams show that 10-30% emission reductions could reduce ozone concentrations to meet standard. Further APCA results show the average ozone contributions from regions and emission categories. APCA results explain different effects for different monitoring stations.

444397 Flare Minimization and Emission Characterization during Start-Ups of Chilling Train in Ethylene Plants

Abstract:

Flaring is a crucial safety operation to protect chemical plant personnel and equipment during plant start-up. However, it generates huge amounts of off-spec products, and causes negative environmental and social impacts, as well as results in tremendous raw material and energy losses which could be saved to generate much more needed products. Thus, cost-effective start-up flare minimization strategies which is a double-win practice, simultaneously benefiting both the industrial and environmental sustainability, are becoming more and more important and attractive to the industries.

The chilling train is a very important section for the ethylene plants and accounts for more than 50% of the total start-up time and flare emissions of the whole ethylene plants. Within this section, the charge gas from the upper stream process is refrigerated to a very low temperature (lower than -160 °C) to separate hydrogen and methane. Meanwhile, the left charge gas is liquidized for cryogenic separation. The chilling train system consumes over 40% of the refrigerating capacity of an ethylene plant. During the chilling train start-up, different operational conditions with different compositions and flowrate of precooling process will affect both product separation efficiencies, start-up stability time and flare emission.

Traditional start-up scenarios are developed based on experience and improved by trial and error, which is inefficient and dangerous. In this study, a general methodology on flare minimization for chilling train start-up operations via rigorous dynamic simulation is presented. The methodology starts with setup and validation of chilling train steady-state and dynamic simulation models. The initial state of start-up in the real plant is achieved through systematically transforming from the validated dynamic model, and several operation conditions with different compositions and flowrate of precooling stream are examined to identify an optimal initial state of start-up based on the obtained initial state. Thereafter different flare minimization strategies for the chilling train start-up are conducted on the optimal initial state model. Also, emission source characterization under different flare minimization strategies for chilling train start-up is investigated. Deep insights of the start-up strategies, emission source distribution and dynamic emission profiles are provided. This study provides detailed technical support for both the industry and environmental agencies on evaluating and developing cost-effective flare minimization strategies in the future.

Experimental Study on Najas Minor of Producing Bio-Oil

Abstract: Due to the diminishing of petroleum reserves and the environmental degradation caused by consumption of fossil energy , many attentions have been paid to the application of biomass energy which is acceptable as an abundant renewable energy resource. Najas minor grows in dense clusters with highly branched stems of which a small fragment could propagate into a thrive one.In this study, we researched the properties of Najas minor to make biofuels .Firstly,the najas minor after heavy metal ions degradation were fed into pyrolyser to prepare biofuels. Significant differences of the product was observed compared with only Najas minor pyrolysed.

KEY WORDS: Najas Minor; catalytic pyrolysis; bio-oil

1.Introduction

Energy demands and environmental problems caused by fossil energy have been a matter of growing concern.Bio-oil has been considered to be a promising way for producing renewable bio-fuels or chemicals[1-3].In recent years,many countries pay attention to the importance of biomass that is an abundant renewable energy resource.Biomass can be defined as any organic material mainly consisting of carbon, hydrogen,oxygen, nitrogen, and some other components in small proportions[4].Although the biomass has been studied for nearly a century,most studies mainly concentrated in Lemna minor, Lemna gibba, Spirodela polyrhiza and Landoltia punctata.We still know little about the biological characteristics and potential application value of other varieties of biomass[5].Najas minor,a submersed herb,grows in dense clusters with highly branched stems of which a small fragment could propagate into a thrive one.Najas Minor is native to Europe, Asia and North Africa from the Netherlands to Morocco east to Japan and the Philippines,including China,Siberia, Central Asia, Iran, Turkey, Ukraine, Germany, France Italy and a host of other Countries.

Catalysts used for catalytic cracking are basically divided into the following categories: molecular sieve, alkali metal salt, metal oxide, etc.Different catalysts have different catalytic effects.The use of metal nanoparticles has shown to be very important in recent industrial applications[6].The application of precious metal nanometer materials in catalysis,which has caused the extensive concern of researchers in recent 30 years,has become a hot research topic in the field of catalysis.Gold nanoparticles have received a great deal of attentions due to their special optical and electrical performances, catalytic properties and remarkable quantum size effect, surface effect etc[7]. The synthesis of metal nanoparticles can be classified broadly into wet and dry methods[8]. The cost of production is relatively high both materially and environmentally. There is an unequivocal need to develop more cost-effective and environmental benign methods to alternate these existing methods[9].Researchers in this field turn to use organisms to produce inorganic nanostructures extracellularly[10].

Although there has been a large volume of work on the synthesis of gold nanoparticles[11-13], and obtainning bi-oil by abundant biomass, there are relatively fewer attempts to combine these two methods.In this work, gold nanoparticles were obtained by aqueous chloroauric acid solution with the of Najas Minor at room temperature. The direct pyrolysis and catalytic pyrolysis experiments of Najas Minor were carried out to observe the effects of catalysts on pyrolysis process of Najas Minor.

2. Materials and Methods

2.1. Experimental samples

The Najas Minor collected from Tianjin city and the solutions of Au(III) were used in this study.The stock solutions of Au(III) were prepared by dissolving analytical grade chloroauric acid

($AuCl_3 \cdot HCl \cdot 3H_2O$, Kewei Reagent, China) in deionized water.

2.2.Experimental apparatus and procedure

Pyrolyzer with GCMS,shown in Fig. 1,is mainly used for analyzing the composition of polymer.Pyrolyzer with GCMS includes three parts :quartz pyrolysis tube,pyrolyzer and GCMS.

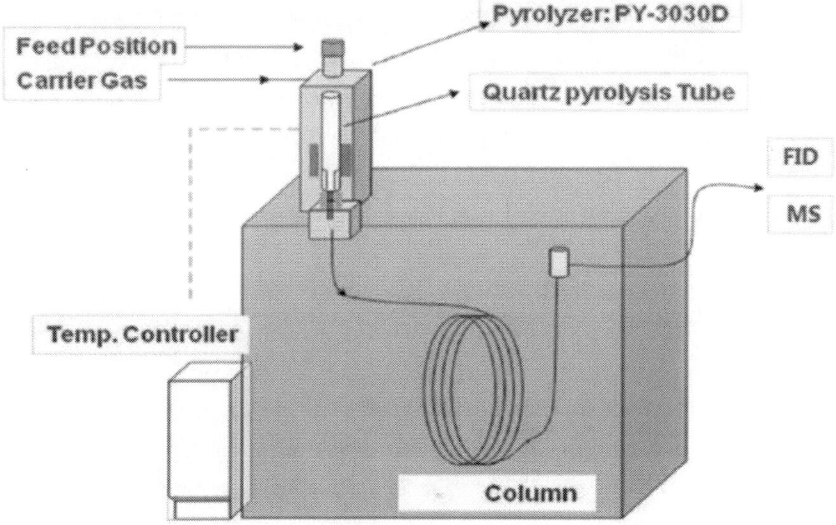

Fig. 1 Diagram of Pyrolyzer with GCMS

The quartz pyrolysis tube (shown in Fig.2) length is 12cm.The sample end diameter is 5 mm,and the pyrolysis products end diameter is 3mm.This special structure is beneficial to the fixation of the reactant in the tube.The top and bottom of the reactants are both supported by quartz wool to prevent reactants blowed by the carrier gas ,or fine particles into the chromatographic column.All pyrolysis experiments were carried out on the EGA/PY-3030D pyrolyzer(Frontier Lab,Japan) .The samples were filled in the special quartz pyrolysis tube to reduce the amount of valuable catalyst and reduce the cost of research work.

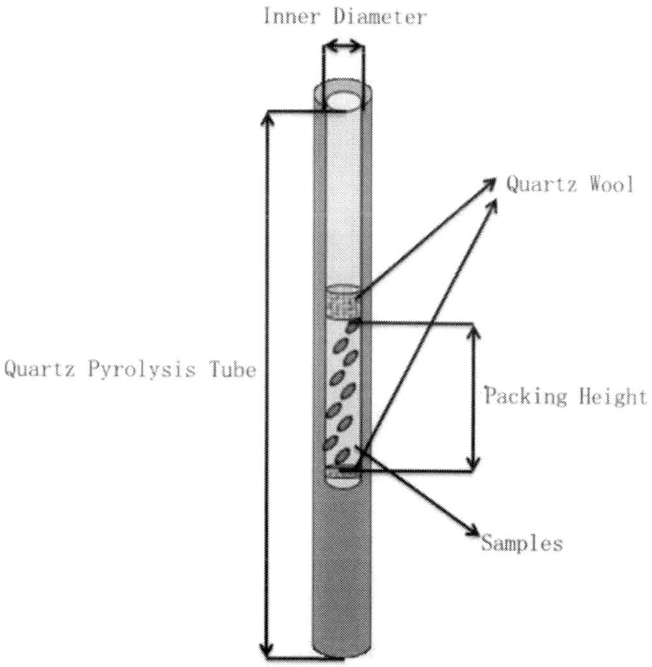

Fig. 2 Diagram of Quartz Pyrolysis Tube

Approximately 0.5g Najas Minors were used in each experiment.The samples were impregnated in 50ml deionized water and 50ml solution of chloroauric acid(0.0793mmol/L) for a night,followed by drying at 80℃ until the weight didn't change.Each experiment used 2.5mg Najas Minors as reactants.Each sample was heated, under the control of heating controller, from ambient temperature to 600 ℃ repeatly and kept at 600 ℃ for 2 h.

3. Results and discussion

3.1.MS Analysis of the Products during the Pyrolysis

Under the same pyrolysis conditions, bio-oil obtained by direct pyrolysis and catalytic pyrolysis were analyzed by GCMS,and the main chemical compounds were shown in Table1 and Table2.

Table 1 Indentification and quantification of main chemical compounds by direct pyrolysis

NO.	peak area%	Formula	Compound
1	12.67	$C_3H_6O_2$	2-Propanone, 1-hydroxy-
2	3.67	C_7H_8	Toluene
3	3.28	C_7H_8O	p-Cresol
4	3.81	$C_5H_{10}O$	Pentanal
5	3.53	C_8H_8O	Benzofuran, 2,3-dihydro-
6	2.17	C_8H_8N	Indole
7	5.93	$C_{20}H_{40}O$	3,7,11,15-Tetramethyl-2-hexadecen-1-ol
8	2.52	$C_{22}H_{42}O_2$	Phytol, acetate
9	5.25	$C_{16}H_{32}O_2$	n-Hexadecanoic acid
10	8.99	$C_{18}H_{30}O_2$	9,12,15-Octadecatrienoic acid, (Z,Z,Z)-

Table 2Indentification and quantification of main chemical compounds by catalytic pyrolysis

NO.	peak area%	Formula	Compound
1	6.01	C_7H_8O	p-Cresol
2	1.75	$C_7H_7NO_2$	3-Hydroxypyridine monoacetate
3	4.58	C_8H_7N	Indole
4	3.05	$C_{10}H9N$	1-Phenyl-1-cyclopropanecarbonitrile
5	3.90	$C_{12}H_{22}O_{11}$	.alpha.-D-Glucopyranose, 4-O-.beta.-D-galactopyranosyl-
6	8.04	$C_{20}H_{40}0$	3,7,11,15-Tetramethyl-2-hexadecen-1-ol
7	2.82	$C_{20}H_{40}$	2-Hexadecene, 3,7,11,15-tetramethyl-, [R-[R*,R*-(E)]]-
8	3.99	$C_{15}H_{28}$	1-Pentadecyne
9	2.12	$C_{15}H_{28}$	3,7,11-Trimethyl-2,4-dodecadiene
10	8.75	$C_{16}H_{32}\ 0_2$	n-Hexadecanoic acid

3.2 discussing

According to the chemical category and common characteristics, the products were divided into hydrocarbons, alcohols, acids, ketones, nitrogen compounds and other compounds.Nitrogen compounds detected by GC/MS analysis ,mainly come from the pyrolysis of amino acids in Najas Minor.Table 3 showed the distribution of pyrolysis products.As shown in the table,the hydrocarbons ,alcohols and nitrogen compounds in biological oil of catalytic pyrolysis were significantly increased compared with direct pyrolysis.The content of hydrocarbons increased from 12.4% to 34.6%.The content of alcohols increased from 9.1% to 16.6%.The gold nanoparticles loaded in surface of Najas Minor, was conducive to the generation of hydrocarbon and alcohol in pyrolysis.By analyzing the GC/MS results, the main hydrocarbons in bio-oil by catalytic pyrolysis were long chain alkanes,and olefins.The main components in gasoline and diesel is hydrocarbons. As a result, whether the biomass can be effective as future power fuel oil, depends largely on the hydrocarbon content of bio-oil after further treatment.

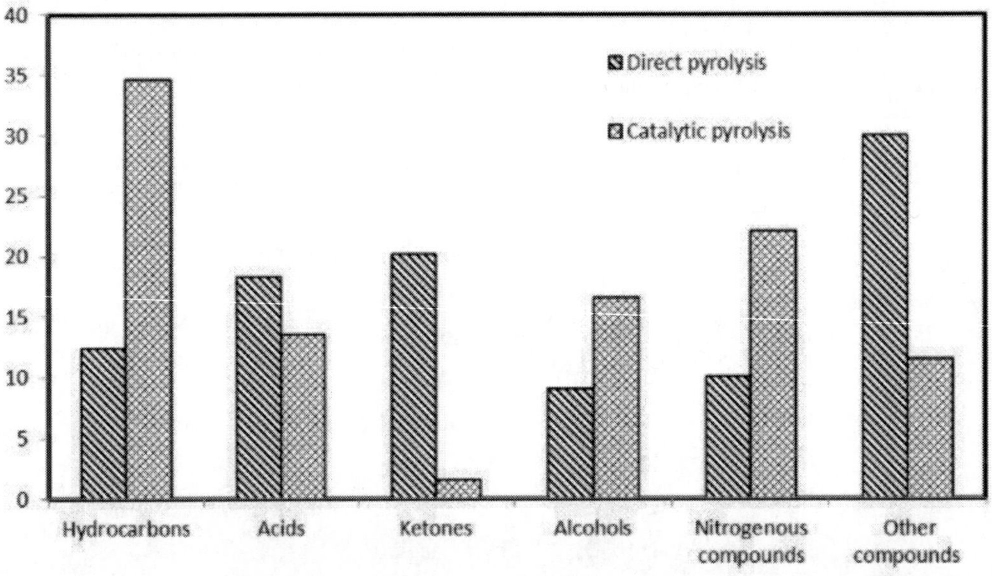

Fig.3 Distribution of pyrolysis products

4.Conclusion

Biomass was explored as a potential biomass energy resource. The experimental study on the direct and catalytic pyrolysis of Najas Minor had been presented, using Pyrolyzer with GCMS,to observe the catalytic effects of gold nanoparticles.In this study,remarkable catalytic effects for pyrolysis were observed.The pyrolysis of Najas Minor with gold nanoparticles,provided more hydrocarbons, alcohols and nitrogen compounds.

参考文献:

[1]Gong F, Yang Z, Hong C, et al. Selective conversion of bio-oil to light olefins: Controlling catalytic cracking for maximum olefins[J]. Bioresource Technology, 2011, 102(19):9247-54.

[2]Zhang Z, Bi P, Jiang P, et al. Production of gasoline fraction from bio-oil under atmospheric conditions by an integrated catalytic transformation process[J]. Energy, 2015, 90:1922-1930.

[3]Hui S, Yuqi D, Jinzhao D, et al. Transesterification of sunflower oil to biodiesel on ZrO2 supported La2O3 catalyst.[J]. Bioresource Technology, 2010, 101(3):953–958.

[4]Yaman S. Pyrolysis of biomass to produce fuels and chemical feedstocks[J]. Cheminform, 2004, 35(31):651-671.

[5]Hejný S. Ellias Landolt The family of Lemnaceae—A monographic study—Vol.1 biosystematic investigations in the family of Duckweeds (Lemnaceae) (vol. 2)[J]. Folia Geobotanica Et Phytotaxonomica, 1993, 28(1):50-50.

[6]Gardea-Torresdey J L, Tiemann K J, Gamez G, et al. Gold Nanoparticles Obtained by Bio-precipitation from Gold(III) Solutions[J]. Journal of Nanoparticle Research, 1999, 1(3):397-404.

[7]Marie-Christine D, Didier A. Gold nanoparticles: assembly, supramolecular chemistry, quantum-size-related properties, and applications toward biology, catalysis, and nanotechnology.[J]. Chemical Reviews, 2004, 35(16):293-346.

[8]Liu B ,, Xie J ,, Lee J Y, et al. Optimization of high-yield biological synthesis of single-crystalline gold nanoplates.[J]. Journal of Physical Chemistry B, 2005, 109(45):15256-63.

[9]Raveendran P, Fu J, Wallen S L. Completely "Green" Synthesis and Stabilization of Metal Nanoparticles[J]. Journal of the American Chemical Society Jacs, 2003, 125(46):págs. 13940-13941.

[10]Absar Ahmad , Satyajyoti Senapati , MIK, et al. Extracellular Biosynthesis of Monodisperse Gold Nanoparticles by a Novel Extremophilic Actinomycete, Thermomonospora sp.[J]. Langmuir, 2003, 19(8):3550-3553.

[11]Selvakannan P, Mandal S, Phadtare S, et al. Water-dispersible tryptophan-protected gold nanoparticles prepared by the spontaneous reduction of aqueous chloroaurate ions by the amino acid[J]. Journal of Colloid & Interface Science, 2004, 269(1):97-102.

[12]Gittins D I, Caruso F. Biological and Physical Applications of Water-Based Metal Nanoparticles Synthesised in Organic Solution[J]. Chemphyschem, 2002, 3(1):110 - 113.

[13]And A G, Murphy C J. Seed-Mediated Synthesis of Gold Nanorods: Role of the Size and Nature of the Seed[J]. Chemistry of Materials, 2004, 16(19):3633-3640.

444353 A Combined Rheology and Stability Study on Emulsified Acid for High Temperature Applications

Abstract:

HCl emulsion has been a popular formulation to stimulate carbonate reservoirs due to the various advantages that emulsified HCl has over ordinary HCl; including but not limited to deeper penetration and less corrosion damages. Emulsified HCl has even more potential advantages if utilized properly. The reaction kinetics, stability and rheology of emulsified HCl plays a significant role on achieving successful acidizing treatment.

The objective of this paper is to experimentally study the factors affecting the stability of emulsified HCl. The emulsion was prepared using 15wt% HCl and diesel with cationic emulsifier. The acid solution-to-diesel ratio was 70:30. The effect of stirring intensity, concentration of the emulsifier, addition of corrosion inhibitor, temperature, addition rate of acid phase to oil phase and mixing time of the emulsion have been studied. Stirring intensity was varied from 600 to 1500 rpm. Emulsifier dosage ranged between 5 to 20 gpt. Temperature used for heating, to mimic reservoir condition, was as high as 120°C.

The results showed that there is an optimum condition for emulsifier dosage and for mixing time of the emulsion. The optimum emulsifier dosage was found to be 10 gpt with 5 minutes of mixing time. As for the addition rate of acid phase to oil phase, 1.4 ml/min of drop-wise addition was found to give a very stable emulsion. The effect of stirring intensity on the emulsion stability was proportional, the higher the rpm speed the better the stability. Corrosion inhibitor, on the other hand, played a role as a destabilizer. With regards to heating temperature, it was found out that the higher the temperature, the faster the emulsion breaks.

The novelty of this study is the comprehensive approach in combining stablity and rheology of emulsified acid for high temperature applications. The particle size analysis and solution morphology help in better understanding HCl emulsion and how it works.

COMPARATIVE ANALYSIS OF MOLECULAR STRUCTURE IDENTIFIABILTY BASED ON SIGNATURES AND DESCRIPTORS

Zelimir F. Kurtanjek
University of Zagreb, Croatia

Abstract

This work is focused on molecule computer representation aimed for molecule predictive model (QSAR) and inverse modelling. Two different approaches are compared, i.e. the molecule structure coding based on graph theory (Faulon et al.[1] extended valence) and the chemical based descriptors[2]. Applied are 1-3rd level molecule graph fragments and 630 2D molecular descriptors. The molecule test sets are alkanes, alkenes, acetones, aromatics, organic acids and halogenated hydrocarbons in the range of C_1-C_{12}, selection of biomolecules (including aminoacids) and a set of binary ionic liquids (cations: imidazole, pyridinium, quinolinium, ammonium, phosphonium). The comparative comparison of molecule description based prediction potential are statistically evaluated by octanol-water (PWA) partition coefficients (for hydrocarbons), viscosity of ionic liquids and biological impact on acetycholinesterase inhibition. The evaluation is based on two distinct levels. On the first level the comparative analysis based on internal and external correlation between "scores" in multidimensional spaces of the separate first, second and third order valance spaces, and the molecular signature space. Firstly, the potential of information content applicable for inverse modeling is tested on ability for molecule classification by use of linear discrimination (LDA) based on the singular value decomposition (SVD) of corresponding molecular fragments and descriptors. For chemical-biological property inverse modelling proposed is application of constrained genetic algorithm (GA). Tested are linear structure multivariate (chemometric) models such as principal component regression and partial least squares and nonlinear decision trees (random forest). Applied are singular value decompositions for reduction of the molecule signature and the descriptor spaces up to the significance level of p=0.05. The selected level of significance is presumed to be in coherence with experimental data of molecule physical and chemical properties. The main focus is on extraction of the main molecular fragments (partial graphs extracted from molecular signatures) and specific molecule descriptors. Importance of application of open source software modules provided by Bioclipse[3] and R software packeges[4] for computational-statistical numerical algorithms is emphasized.

Key words: Molecule signatures, descriptors, inverse modelling, decision tree

Introduction

Advancement of high-throughput experiments have lead to „big data" databases applied in pharmaceuticals, medicine, new material science, catalysis, and new industrial processes. The role of chemoinformatics as a science for computer encoding of molecule properties and algorithms for data mining and inference are the essential steps in the development. Computer encoding of molecular properties has been a part of chemical engineering from the beginning of process design. It is mostly based on molecule group contributions, but has been extended by application of molecular descriptors (1-2-3 D), molecule signatures, and application of molecular quantum mechanics. Modeling is faced with the problem of „high dimension curse". For example, about 5 000 various molecular descriptors can be determined for each molecule, although most of them are highly correlated. It results in an ill-conditioned problem of modelling with large uncertainties in parameter and poor model prediction accuracy outside modelling training sets. Great improvement has been achieved by application by use of model ensembles, like random forest of many individual decision trees. These models are highly flexible, can be trained on discontinuous and nonlinear functional relations, can encompass various data formats like numerical and categorical variables, and due to large number of randomly selected individual predictors in principle do not have tendency to overfitting. However, when the object of modelling is extraction of important molecular features, due to autocorrelation between descriptors random forest, produced lists of importance are large sets of poorly distinguishable variables. The focus of this work is to apply feature regularization for optimization and reduction of list of variable importance. Compared are two methods for structure encoding, molecular descriptors and molecular signatures of first and second order.

Modeling and Factor Regularization

In this work are selected two distinct data sets of organic molecules from open source data bases (J. Sangster, 1989) and ionic liquids ILs (The UFT/ Merck and NIST, 2015), and lists of the selected molecules are given in Table 1. As target properties for organic molecules are selected the corresponding molecular classes and partition octanol-water (POW) coefficients, and dynamic viscosity in the range of 30-80 ^{0}C of ILs is selected The molecules are coded into SMILES and used as input lists into software for evaluation of molecular signatures. Most popular open source software tools are DRAGON (R. Todeschini, et al., 2009) and PADEL (Chun Wei Yap, 2010). Here are used 1-2 D descriptors generated by PADEL. For the same list of molecules are determined first and second order molecular signatures or extended valances (J.P. Faulon, 2003). The signatures are two dimensional graph based structural molecular fragments. The vertices are atoms and the edges are corresponding bonds (lengths or heights), and are here only used of the first and second level. A molecule is characterized by a complete set of signatures as input data for modelling. Since the models are based on large number of variables (molecular features) possibly lager than number of samples (molecules), usually denoted as p>>n models. They are intrinsically „ill conditioned" leading to good predictions on train data but very poor predictions on test data. Also, produced list of variable importance are very dense "clusters are not distinguished" with large uncertainties (broad confidence intervals). This makes the problem of inverse modelling aimed to predict a set of candidate molecules for given set of target properties, such as in a case of high dimension combinatorial task for selection of ILs cations and anions, very unreliable. In order to reduce model dimensionality and improve robustness and parsimonious of important molecular features here is applied modelling with feature regularization. The model is based on ensemble of random trees (CART models, L. Breiman et al. 1998) with included feature regularization.

Data sets of organic molecules	Range of number of atoms				
Molecular classes	C	H	O	Cl	F – I - Br
Alkanes	5-12	12-30			
Alkenes	4-9	6-18			
Acetones	6-24	6-26			
Halogenated hydrocarbons	3-10	6-22	1		
Organic acids	1-16	2-32	2		
Esters	4-11	6-22	2		
Aromatics	1-12	1-21		3	2 – 2 - 1
Data sets of ionic liquids (ILs)					
Cations	**Anions**				
1,3-Dimethylimidazolium (+)	PF_6 (-) Cl (-) CH_3COO (-)				
1-Methylpyridinium (+)	BF_4 (-) $MeSO_4$ (-) $EtSO_4$ (-)				
1,1-Dimethylpyrrolidinium (+)	Tf_2N (-) CF_3SO_3 (-)				

Table 1. The molecular groups of the organic and ionic liquid molecules selected for random forest models and extraction of the most important molecular features. The properties for the organics are obtained from J. Sangster (1989) and for the ionic liquids from Merck and NIST data bases (2015). Molecular descriptors and signatures of ILs ions are recorded separately and adjoined as a single record.

The modelling is formulated as an optimization (functional and multivariable minimization) problem with two parts of the objective function (T. Hastie et al., 2009).

$$\min F_{object.}(X,T,\omega) = SSE(y,\hat{y}) + \phi(T,\omega) \tag{1}$$

The first part is the sum of square errors (SSE) based on difference between observed y and prediction $\hat{y}$ based on input data by an ensemble T of interrelated CART models (L. Breiman et al. 1998), The second part is a selected metrics ϕ for the model complexity depending the sum of leaf probabilities for population of individual CART models and their total number N_T .

$$\phi(T,\omega) = \gamma \times N_T + \frac{1}{2} \times \lambda \times \sum_i^{N_T} \omega_i^2 \tag{2}$$

The ensemble output prediction is a sum of predictions of individual T_i tree CART models:

$$\hat{y} = \sum_i^{N_T} T_i(X) \tag{3}$$

A very efficient open source code Extreme Gradient Boosting XGB (T. Chen, 2016) is available in R language (R Developm-ent Core Team, v. 3.2.3, 2015). The optimization algorithm is consecutive, starting with first CART model which is appended with a next tree boosted by errors in predictions of previous tree. At each step optimization is solved by linearization of the objective function to quadratic terms. Number and structure of trees as well as the parameters for regularization need to be found by trial and error and minimal number of trees by folding and cross validation error on the training set. The critical step is validation on untrained set to have acceptable model error and avoid model overfitting with minimal model complexity.

Results and Discussion

Tested are models for prediction of the seven molecular classes (Table 1.) based on molecular descriptors and molecular signatures (first and second level). Results for molecular descriptor feature importance are shown on Fig. 1. The effect of regularization is very pronounced. Out of hundreds of descriptors for the molecular classification needed are only few, the most important being BCUTw.1h. On contrary, when an ensemble of independent CART trees is applied, i.e. random forest model, level of accuracy of prediction is the same, but the list of variable importance is densely populated with numerous descriptors.

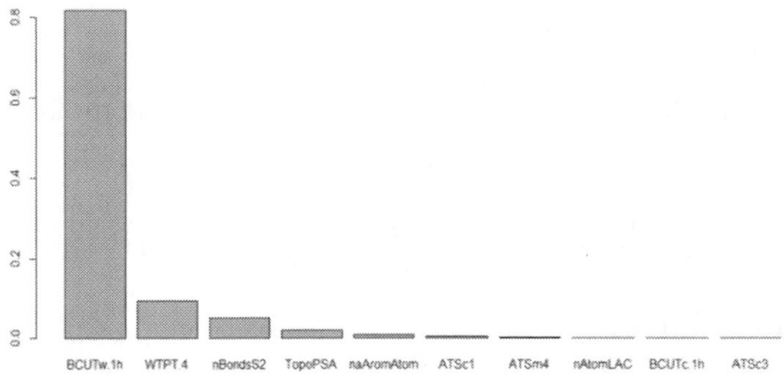

Fig 1. Variable importance molecular descriptor factors for the inverse modeling of the selected molecular classes. The first five descriptors are: BCUTw.1h (based on a weighted version of the Burden matrix which takes into account both the connectivity as well as atomic properties of a molecule), WTPT.4 (Randić's weighted path), nBondS2 (bond count), and naAromAtom (number of aromatic atoms).

The same classification problem was tested by the first and second order molecular signatures. The important signatures are depicted on Fig. 2-3. Accuracy of prediction on trained sets of molecules compared to descriptors is about the same, though the importance factors are not as distinctive as in case of the descriptors.

significantly better when molecular descriptors are applied. The

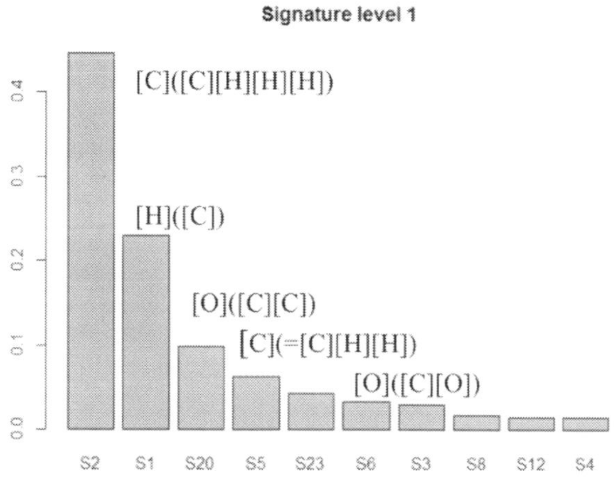

Fig 2. Relative importance of the individual first level signatures for inverse prediction of the selected molecular classes. The first five most important signatures are coded according to SMILES (Simplified Molecular Input Line Entry System).

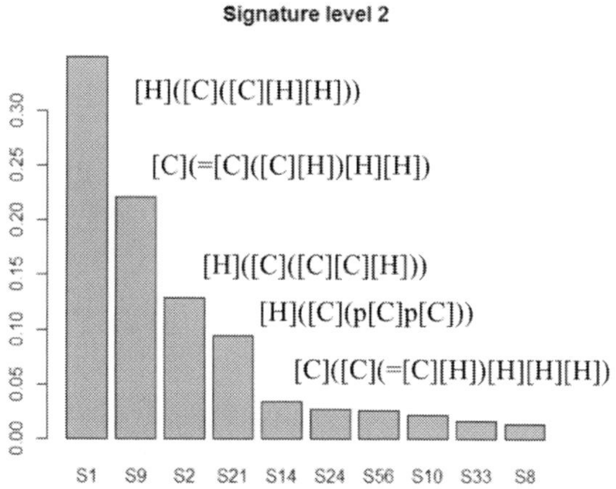

Fig 3. Relative importance of the individual second level signatures for inverse prediction for the selected molecular classes. The first five most important signatures are coded according to SMILES (Simplified Molecular Input Line Entry System).

However, model accuracy on untrained set of molecules, depicted in Fig. 4, is significantly lower for the model with molecular descriptors compared to signatures. The average error with descriptors is about 5% while for signatures about 11%. The "box and whiskers" plots are obtained by error evaluation of ten randomly generated untrained sets.

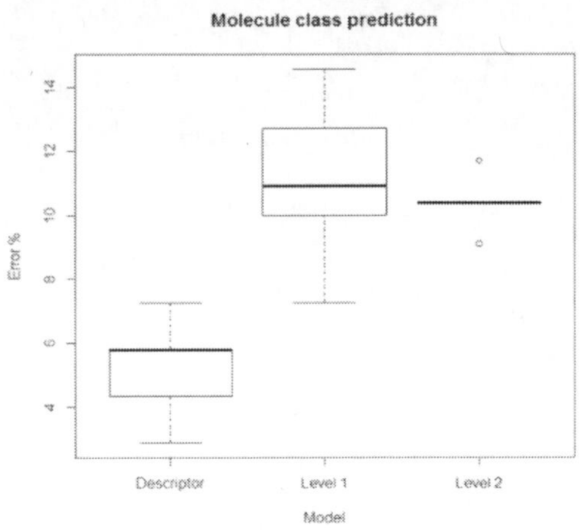

Fig 4. "Box and whiskers" plot for the model validation (relative errors) for inverse prediction of the molecular classes based on the set of untrained molecular descriptors and the first and second level of molecular signatures.

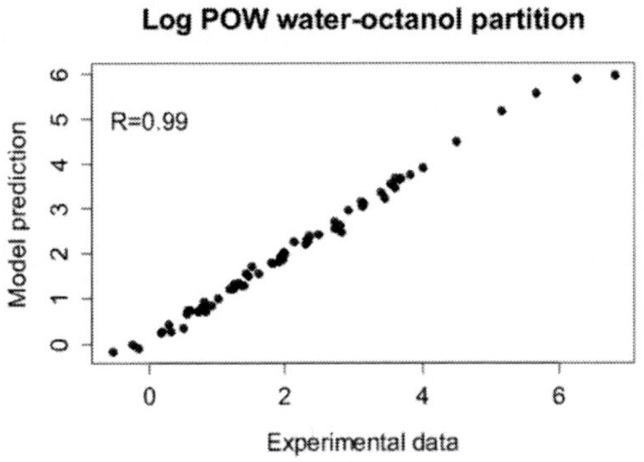

Fig 5. Predictions of the partition octanol-water coefficients POW, for the specified molecular classes of organic molecules, are based on the random forest model with the molecular descriptors.

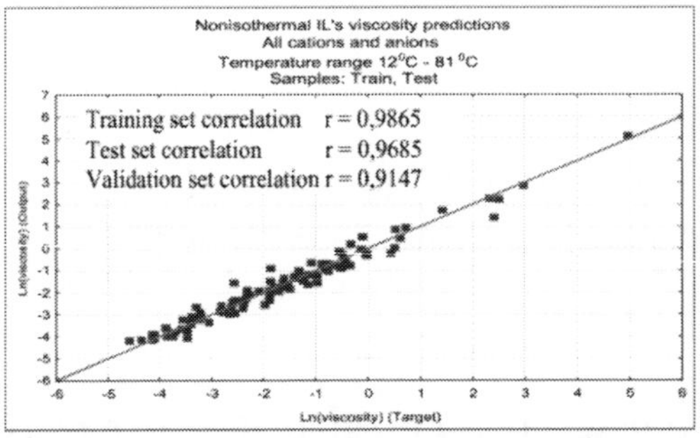

Fig 6. Predictions of dynamic viscosities for the specified groups of ionic liquids (ILs) based on the random forest model with the molecular descriptors. The viscosities cover the temperature range from 12 – 81 ^{0}C and five orders of magnitude.

In Fig. 4 can be observed that dispersion of errors (quantiles) for the second level molecular signatures is negligible, though two outliers are present.

For prediction of molecular properties applied are random forest models based on molecular descriptors. For all of the organic molecules, included in the seven classes, evaluated are octanol-water partition coefficients POW, shown in Fig. 5. The range of POW values is very high, on log scale fro 0 to 6, but the obtained accuracy of the model predictions is high (Pearson correlation R=0.99). The same approach is applied for prediction of ionic liquid viscosity data in temperature range from 12 to 81 ^{0}C. Since temperature effects on viscosity are intrinsically nonlinear, they represent a modelling challenge. However, the decision trees are very efficient for accounting nonlinearities, though nonlinearity is reflected in non-orthogonal decision planes, rendering trees with significantly increased number of branches. For the range of 6 order of magnitude the accuracy of prediction on trained set is R = 0.98 and on untrained set R = 0.91.

Conclusions

Compared are ensemble decision trees models for molecular feature class and property prediction based on extensive list of molecular descriptors and first and second level of extended valances or signatures.

The aim was to test effects of regularization on ensemble of CART decision tree models. Proven is effectiveness of the extreme gradient boosting algorithm in cases with extensive list of (molecular) features.

Obtained is increased accuracy on untrained sets of data, and more importantly, obtained are parsimonious lists for ranking of variable importance lists. This enables further significant reduction in development of robust and simple decision tree models.

The reduction in number of molecular features applied for decision tree models may greatly simplify the challenging problem of inverse modeling based on molecular descriptors and/or signatures.

References

1. J. Sangster, J. Phys. Chem. Ref. Data, **18**(3)(1989) 1111-1227
2. The UFT/ Merck Ionic Liquids Biological Effects Database, www.il-eco.uft.uni-bremen.de
3. NIST, Ionic Liquids Database Standard Reference Database #147, <ilthermo.bouder.-nist.gov /ILThermo/mainmenu.uix>
4. R. Todeschini and V.Consonni: "Molecular Descriptors for Chemoinformatics", (2 volumes), WILEY-VCH,Weinheim (Germany) 2009.
5. Chun Wei Yap, PaDEL-Descriptor: An Open Source Software to Calculate Molecular Descriptors and Fingerprints, J. Comp. Chem., (2010):**32**: 1466-1474.

6. J.-L. Faulon, D.P. Visco Jr., R.S. Pophale, The signature molecular descriptor. 1. Extended valence sequences vs. topological indices in QSAR and QSPR studies, J. Chem. Inf. Comput. Sci. **43** (2003) 707–720.
7. J.-L. Faulon, C.J. Churchwell, J.D.P. Visco, The signature molecular descriptor. 2. Enumerating molecules from their extended valence sequence, J. Chem. Inf. Comput. Sci. **43** (2003) 721–734.
8. L. Breiman, J. H. Friedman, R. A. Olshen, C.J. Stone, „Classification and Regression Trees", CRC, Boca Raton, 1998.
9. T. Hastie, R. Tibshirani, J. Friedman, The Elements of Statistical Learning, Data Mining, Inference and Prediction, Springer Verlag, Second Edition, New York, 2009.
10. T. Chen, T. He and M. Benesty (2016). xgboost: Extreme Gradient Boosting. R package version 0.4-3. https://CRAN.R-project.org/package=xgboost
11. R Development Core Team, , R: A language and environment for statistical computing. R, Vienna, Austria,www.R-project.org 2015.

444791 Dissolution and Precipitation of Minerals Under Different Alkali Concentrations in a Reactive CO_2 Batch System

Abstract:

The primary focus of this research is to study the dissolution and precipitation of minerals and salts from Berea sandstone and brine samples taken from the Great Salt Lake in a reactive CO_2 batch system. Various experiments were constructed to measure dissolution and precipitation reaction rates under different alkali concentrations. Changing the alkali concentrations of the Great Salt Lake brine samples produced changes in dissolution and precipitation of minerals from the Berea sandstone and the Great Salt Lake brine samples. Data from ICP-MS was used to quantify mineralogical changes in the samples and Micro-CT was used to measure the porosity change in the Berea sandstone samples. The results from these analytical instruments showed patterns of mineralogical dissolution and precipitation in the post-experimental fluid that correlate to the change in alkali concentrations.

444825 Significance of Energy Intensity and GHG Emission Reduction Towards Sustainability of a Petrochemical Plant

Abstract:

Sustainability is an important pillar in today's petrochemical industry for improving process efficacy by reducing specific energy consumption, greenhouse gas (GHG) emissions, and wastewater. SABIC (Saudi Basic Industries Corporation) like other giant petrochemical companies has set stringent sustainability targets and investing huge resources to transform its plants from the present position to the first quartile in each category.

SABIC is the world largest producer of ethylene oxide and ethylene glycol. For each ton of glycol produced, 0.2 to 0.5 tons of CO_2 is emitted as GHG. The significance of sustainability "in general" while the methodology adopted to reduce energy intensity, carbon footprint, and wastewater "in particular", will be discussed thoroughly in our talk.

444827 Syngas to Ethylene Glycol: A Technology Overview

Abstract:

SABIC is the world largest MEG producer petrochemical company and most of its MEG is being produced through the hydrolysis of EO. Ethylene oxide (EO) is a simplest form of epoxides and due to its molecular structure, acts as a versatile intermediate for a variety of chemicals such as; monoethylene glycol (MEG), di- and triethylene glycols, ethoxylates, ethanol amines, and glycol ethers. The growing demand of MEG owing to its downstream utilization in polyester industry has prompted the research of an efficient, cost effective, and environmentally benign EO production process.

The major feedstock for MEG are ethylene and syngas. The heavy dependence of MEG producers on ethylene and its scarcity in some parts of the world promoted the need of alternate feedstock for highly demanded MEG. The technological developments utilizing syngas as a feedstock will be discussed along with the commercial analysis and future growth in EO/EG business.

444834 Stability and Rheology of Emulsified Acid in Waste Oil: Effect of Emulsifier Concentration, Mixing Speed, Temperature and Droplet Size

Abstract:

Acid in oil emulsion is used in the industry for stimulation purposes. In particular, HCl as the acid phase and diesel as the oil. Emulsifying HCl has numerous benefits over regular HCl. Perhaps the main reason is due to deeper penetration. Apart from that, less corrosion damages are caused since the external phase is hydrocarbon. Several studies showed the success of replacing diesel as an external phase with other hydrocarbon oil, such as crude oil and xylene.

This work utilizes the extra hydrocarbon left unused –or sometimes dumped- from refineries. This unused hydrocarbon is referred to as waste oil. First the chemical composition of waste oil was studied using HPLC, SARA fractionation as well as FTIR. This helped in selecting the proper emulsifier agent for the waste oil. The HCl-waste-oil emulsion was prepared using 15wt% HCl with a ratio of 70:30 of acid-to-oil. Extensive work was done to ensure reproducibility of data. This study was done to experimentally investigate the effects of distinct important variables on thermal stability and apparent viscosity of the HCl-waste-oil emulsion. These variables are; emulsifier concentration, mixing speed and the addition of corrosion inhibitor. All experiments were conducted at a high temperature 120˚C. To assess the quality of the HCl-waste-oil emulsion, droplet size was measured as a function of temperature (up to 100 ˚C).

The results showed that the HCl-waste-oil emulsion is a shear-thinning fluid. Power-law model was applicable seamlessly to all of the apparent viscosity data for all of measured temperatures. The optimum conditions were found to be; 0.5vol% and 600 rpm for emulsifier concentration and mixing speed, respectively. Apart from that, the emulsions prepared with corrosion inhibitor were less stable compared to the ones without it. However, the addition of corrosion inhibitor tended to generate monodispersed emulsion but bigger size droplets for temperatures 25˚C and 50˚C and polydispersed droplets for higher temperatures. Droplet size measurements showed an average size of 1.47 to 3.09 µm. Overall, the lab results showed a promising potential for the HCl-waste-oil emulsion to be used in the field.

This work takes into account; reducing the budget of acidizing job and at the same time environmental concerns. It makes use of the unwanted hydrocarbon from refiners. And, it saves the environment by using the waste oil instead of dumping it.

444994 Multiscale Simulation of Proteins Motion in the Microchannel

Abstract:

Protein adsorption is an important phenomenon, which plays a prominent role in chemical, industrial and medical areas. In this work, the dynamic behavior of the egg-white lysozyme has been investigated in a micro-channel with a multiscale modeling approach. The Molecular Dynamic simulation, free energy computation and Langevion Dynamics (LD) simulations have been combined to study lysozyme adsorption process on Au (111) surface. The gold surface has strong attraction to proteins. The protein-surface adsorption or desorption free energies is calculated. From the case studies of LD simulations, the surface has effects on the protein dynamic adsorption process. By the comparison of LD, MD and the experiments, the time step is concluded as an important factor to study protein adsorption behavior at the water-surface interface. Fine time step at the scale of 1 picosecond is necessary to provide enough accuracy to resolve proteins' dynamics trajectories.

445004 Grand Challenges in Future Energy and Chemicals: Climate Change and Sustainability As Drivers for New Technology and Innovation

Abstract:

Global demand for energy will double by 2050, driven by increasing population and affluence especially in emerging economies. While fossil fuels will continue to underpin the energy supply for the near future, concern over impact on climate change continues to grow, driving interest in expanded use of renewables and development of carbon capture and sequestration or re-use solutions. In addition to CO_2, water use and air emissions are critical issues to be addressed in future and current energy supply, as is energy storage for effective use of renewables. This presentation will provide an overview of what industry is doing today to address these issues, including the factors and metrics used in decision making.

445029 Electrospun Nanocomposite Fibers for the Removal of Heavy Metals in Water

Abstract:

Uniform electrospun nanofibers with small diameters were successfully prepared by systematically optimizing the electrospinning parameters including polymer concentration, salt loading, applied voltage, tip-to-collector distance, and flow rate. These parameters are shown to affect the fiber morphology and the addition of iron (III) nitrate salt was found to be able to reduce the fiber size. Due to the conductive nature of iron salt in polymer solution, even a relative lower voltage of 6 kV could generate a stable fluid jet and yield the nanofibers. The thermal properties of the composite fibers were investigated by thermal gravimetric analysis (TGA) and differential scanning calorimetry (DSC). Rheological studies were also conducted. The composite nanofibers were further carbonized to carbon nanofibers. Applications of the nanofibers in environment remediation have shown promise.

445030 Heavy Metal Detection in Treated Water Via a Fabric Nanocomposite Sensor

Abstract:

We present the conception of an electrochemical sensor for on-site/portable water analysis. This flexible sensor was developed to quantify the amount of heavy metals (in real-time), alleviating the wait time, large sample size, possible contamination, and expensive analytical equipment associated with current procedures for water analysis. This sensor is composed of a multiwalled carbon nanotube (MWNT) functionalized polymer mat, for a flexible yet conductive sensor. The MWNTs are then functionalized with a cyclo-oligomeric calixarene, which is shown to selectively form a supramolecular complex with metal ions. Upon complex formation, the charge carriers are drawn away from the carbon layer and the detection is prevalent at appropriate levels. The optimization of MWNT and calixarene functionalization, as well as the selectivity and sensitivity of the sensor will be discussed.

447119 Development R&D of Manoir Industries and Emphasis on Coating

Abstract:

The Senior VP Technical Support & Projects will present the development R&D of Manoir Industries and especially the coating (it is our lattest development) as well the services we can provide such as metallurgy expertise, advice, convector cleaning, site welding, welding supervision in order to increase the furnace productivity. So it is 60% R&D and 40% services. Manoir Industries manufactures cast components for the petrochemical market and develops high temperature alloys such as the Manaurites.

444478 Biogas Reforming with Addition of Carbon Dioxide to Syngas over Ni-Based Bimodal Pore Catalyst

Abstract:

Methane dry reforming of (MDR) has been widely studied due to its important applications in producing of syngas ($CO + H_2$), removal of two greenhouse gases, and upgrading of biogas (mainly composed of CH_2 and CO_2) into value added chemicals. Compared to the noble metal catalysts, nickel-based catalysts are suitable for industrial scaled-up MDR due to their high activity and low cost. However, nickel-based catalysts have been encountering a significant challenge that is the active metal sintering and coke deposition, leading to the catalyst deactivation and then the poor stability. Nevertheless, it has been reported that the deactivation of nickel catalysts can be suppressed by adding promoters like strong Lewis bases with the enhancement of chemisorb CO_2, such as MgO and CaO, and like lanthanide elements with the capacity of oxygen storage and release, such as CeO_2 and La_2O_3. Al_2O_3 is a commonly applied catalyst carrier own to its good pore size dispersion, high specific surface area and high mechanic strength.

Herein, we report that the bimodal NiCeMgAl catalysts with different NiO-loading were synthesized by the refluxed co-precipitation method and were evaluated for MDR performance by using real biogas from anaerobic digestion of food wastes. The sample containing 15wt% NiO-loading (Ni_{15}CeMgAl) was found to be active enough at 750 °C with a high CH_4 conversion of around 90%. Ni_{15}CeMgAl catalysts kept their bimodal porosity after reduction and MDR process. The evolution of the Ni_{15}CeMgAl catalyst before and after MDR process was investigated by SEM, TEM, XRD, and TGA techniques. The correlation between the structure evolution and catalytic performance change was also discussed.

444551 Continuously Adjustable, Molecular-Sieving "Gate" on 5A Zeolite for Distinguishing Small Organic Molecules By Size

Abstract:

Zeolites/molecular sieves with uniform, molecular-sized pores are important for many adsorption-based separation processes. Pore size gaps, however, exist in the current zeolite family. This leads to a great challenge of separating molecules with size differences at ~0.01 nm level. Here, we report a novel concept, pore misalignment, to form a continuously adjustable, molecular-sieving "gate" at the 5A zeolite pore entrance without sacrificing the internal capacity. Misalignment of the micropores of the alumina coating with the 5A zeolite pores was related with and facilely adjusted by the coating thickness. For the first time, organic molecules with sub-0.01 nm size differences were effectively distinguished via appropriate misalignment. This novel concept may have great potential to fill the pore size gaps of the zeolite family and realize size-selective adsorption separation.

The Principle of Intrinsic Reality
Solving the World's Woes through Chemical Engineering and Economics

Jed Moyano Bellen, SCPM, PMP
Hyundai Engineering & Construction Co. Ltd.
9 Temasek Blvd., #27-02 Suntec Tower 2, Singapore 038989
jed.bellen@hdec.co.kr

Prepared for Presentation at
American Institute of Chemical Engineers
2016 Spring Meeting
11th Global Congress on Process Safety
Houston, Texas
April 10-14, 2016

UNPUBLISHED

The Principle of Intrinsic Reality
Solving the World's Woes through Chemical Engineering and Economics

Jed Moyano Bellen, SCPM, PMP
Hyundai Engineering & Construction Co. Ltd.
9 Temasek Blvd., #27-02 Suntec Tower 2, Singapore 038989
jed.bellen@hdec.co.kr

Keywords: Principle of Intrinsic Reality, Chemical Engineering, Economics, Sustainability

Abstract

Human beings in profound rationality always seek for explanations of the phenomena around them. To understand these phenomena, human beings have developed systematized body of knowledge as a practical way of dealing with life's challenges. Among the systematized body of knowledge developed are chemical engineering and economics.

An avant-garde way of looking at each of these disciplines leads to the Principle of Intrinsic Reality. This principle relates the fundamental laws of the two disciplines to glean insights on the constraints imposed by conservation laws on economic systems at both the macro and micro levels.

It can be argued that if mass and energy in the universe are constant, then logically diminishing returns and marginal utility are bound to be observed. This logic is quite evident in a microorganism culture which follows the Monod growth model.

It can also be thought that conservation laws imply scarcity of resources. As such, it enjoins everyone to find optimal and efficient use of resources in the context of trade-offs, production decisions and sustainability. It even calls for the development of new technologies to produce more outputs from less inputs (i.e. shifting of production possibility frontier to the right). In this way, the Principle of Intrinsic Reality leads to the imperative for sustainable optimization.

Furthermore, it is deemed that the conservation laws can explain the phenomenon called "polarity of wealth." As mass and energy when harnessed create wealth, then the constancy of mass and energy implies boundary for wealth creation. In effect, wealth can neither be created nor destroyed but can only shift from one entity to another thereby creating polarity in wealth distribution. This assertion can be the key towards more equitable restructuring of economic systems to close income gap and thereby result to stable and resilient populace.

In conclusion, this paper envisions for a paradigm shift from traditional money-centric view of chemical engineering and economics to a synergistic convergence of disciplines which can powerfully tackle the world's woes and thereby effect sustainability.

1. The World in which We Live Today

The past few years saw unprecedented global challenges which have put into question existing economic models and scientific paradigms. Among these challenges are the dystopian global economy, climate change, environmental degradation and terrorism. The magnitude of these challenges has grown to a profound level that newer ways of thinking and perspectives become necessary. With that end in mind, this paper attempts to look at the fundamental laws of chemical engineering and economics in new ways to recognize our intrinsic reality from which these pressing challenges can be tackled at large.

2. The Principle of Intrinsic Reality

It is an observable fact that advancement in science increasingly blurs boundaries of various disciplines in tackling proactively the pressing global issues or problems. Within this purview, two seemingly diverging disciplines namely chemical engineering and economics can be taken together to glean insights on our intrinsic reality and what it means for our economic and ecological systems on a larger scale.

In the author's contention, intrinsic reality is grounded on the fundamental laws of chemical engineering and economics. These laws are the Conservation Laws (Chemical Engineering) and Law of Diminishing Returns (Economics). It can be argued that the Law of Diminishing Returns is observable because of the constraints imposed by the Conservation Laws. As such, this leads to the Principle of Intrinsic Reality which can be stated thus;

> *In any system, both at the macro- and micro-levels, the limiting factor for growth is the constancy of potential inputs.*

Potential inputs mean not the actual inputs to processes but the constant material or energy available which can be harnessed to create wealth. The reasoning or proof of this principle is presented in subsequent discussions.

2.1 Proof Based on the Monod Growth Model

The best illustration to understand the Principle of Intrinsic Reality is the Monod Growth Model. Monod in his 1949 seminal paper presented an empirical relation between exponential growth rate and concentration of essential nutrient. The data was fitted into hyperbolic equation:

$$R = R_K\left(\frac{C}{C_1 + C}\right)$$

In this equation, R is the growth rate during the exponential phase; R_K is the rate limit for increasing concentrations of C; C is the concentration of the nutrient; and C_1 is the concentration of nutrient at which the rate is half the maximum.

Figure 1: Graphical Representation of Monod Growth Model

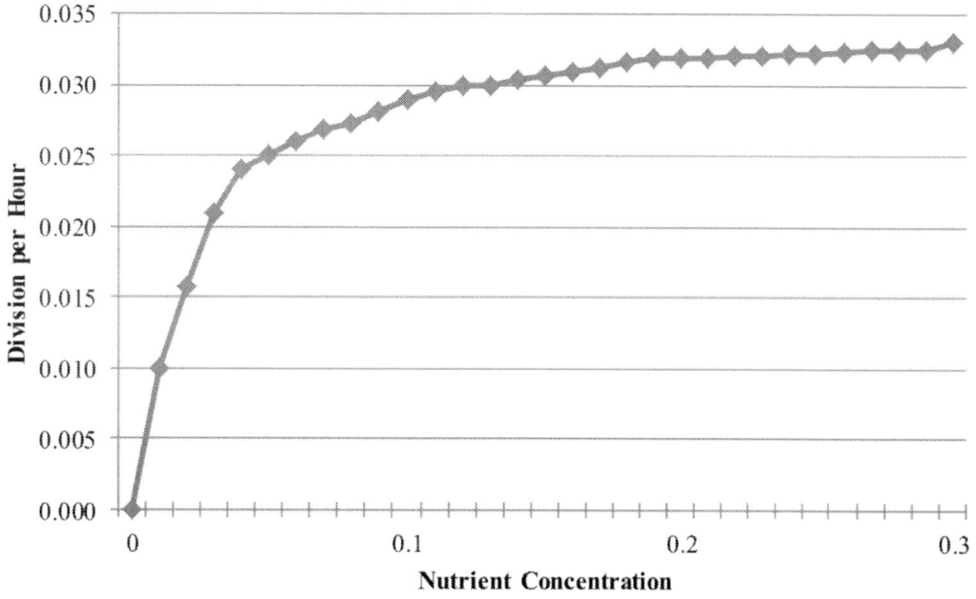

Monod (1949) further pointed out that the limiting factors commonly found are exhaustion of nutrients, accumulation of toxic metabolic products and changes in ion equilibrium.

The foregoing discussion has been observed at the micro-level with the bacterial culture. At the macro-level (i.e. the world at large), analogous phenomena are also observed to occur because of the same constraint imposed by the constant quantity of resources (i.e. material and energy). One example of these phenomena is that human beings cannot grow indefinitely. If they do, then how can the world sustain that growth with limited available resources (e.g. foods) to support it?

Another analogous phenomenon is in population growth. Thomas Malthus (1789) predicted that when population growth exceeds agricultural production, the population level is forced to return to subsistence-level conditions. This phenomenon as Malthus (1789) believed is due to the restriction of available resources supporting population growth, which others have claimed to be exploding as the earth becomes over populated.

In a surprising manner, a similar phenomenon can be observed in production processes in which the increase in a single factor of production assuming all other factors constant, results not to more marginal output but to a reduced one. This was attributed by classical economists, Thomas Malthus and David Ricardo, to the decreasing quality of inputs while neoclassical economists attributed this to the disruption in the entire production process as additional units of input are added to a fixed amount of capital. But in view of the Principle of Intrinsic Reality, the phenomenon is viewed intrinsic to the production system in which case infinite amounts of mass or energy cannot be processed with perfect efficiency. Hence, the law of conservation of mass and energy comes into the fore.

It can be deduced from the discussion that there is an underpinning intrinsic reality to the more obvious phenomena in the world from the micro-scale (bacterial growth) to the macro-scale (human growth and production processes).

2.2 Proof Based on Diminishing Marginal Utility

Another way to prove the Principle of Intrinsic Reality is through the diminishing marginal utility observable among humans and perhaps other organisms as well. Marginal utility, as defined in economics, is the gain from an increase or the loss from a decrease in the consumption of goods or services. The diminishing marginal utility is an observable phenomenon in which the utility or satisfaction an individual gains diminishes as he or she consumes more of the same unit of goods or services. Graphically, this can be represented as in Figure 2.

Figure 2: Graphical Representation of Diminishing Marginal Utility

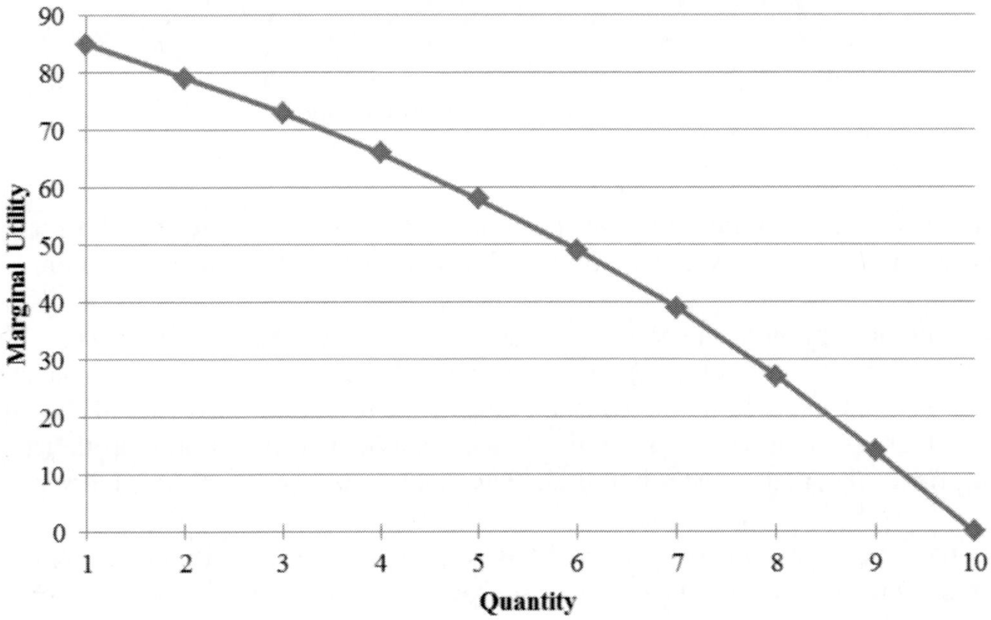

It can be deduced from this phenomenon and in support of the discussion in section 2.1 that human beings cannot grow indefinitely due to the fact that in our intrinsic reality, there is always a limit so as the constancy of available resources (i.e. mass and energy) can support our existence.

This means that in view of our intrinsic reality, we cannot have infinite greed otherwise an upsurge of unintended consequences are bound to happen (i.e. environmental degradation, terrorism and eventually demise of humanity).

Therefore, recognizing our intrinsic reality is the first step towards solving the world's woes and thereby resulting to sustainability and security.

2.3 Expanding the Production Possibility Curve through R&D

The caveat here is that the constant mass or energy (i.e. the limits) leads to trade-offs. Trade-offs is encountered when personal or business decisions are made. In making these decisions, one has to ensure that he or she becomes better off with his or her chosen option. We also realize in our life that we cannot get all that we want and we have to make trade-offs so long as it places us in a better or preferred position. This is more of a philosophical discourse, but it goes to show that the concepts of chemical engineering can be as useful in life and in business as it is in chemical processing.

To tackle these trade-offs, the Production Possibility Frontier in economics can be used. In view of the Principle of Intrinsic Reality where growth is limited by the fixed amount of material and energy, the Production Possibility Frontier needs to shift to the right as shown in Figure 3.

Figure 3. Graphical Representation of Expanding the Production Possibility Curve

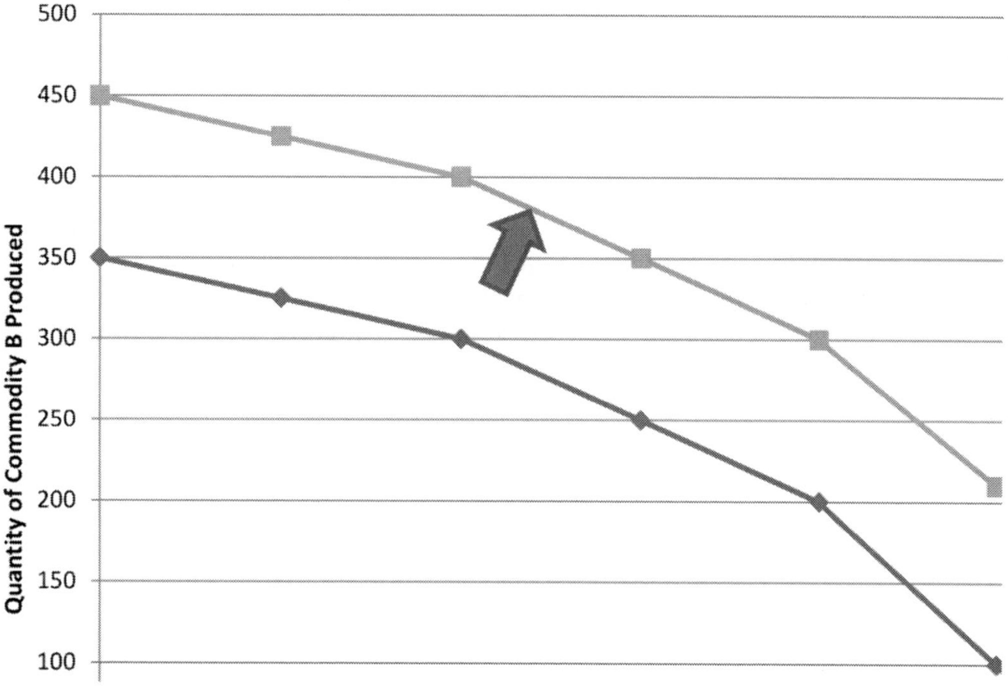

This shift has an implication on the practice of chemical engineering as Research and Development (R&D) is needed to achieve the said shift. R&D enables the economy to produce more of the two commodities (i.e. Commodity A and Commodity B) resulting in economic growth. The key challenge here to Chemical Engineering, particularly on Chemical Engineering R&D, is to look at the bigger picture where economic growth emanating from R&D output is pivoted on sustainability; considering its impact to the environment and the society at large.

3. Paradigm Shift in the Practice of Chemical Engineering

"Money" or "Profit Maximization" has been at the center of the practice of chemical engineering (Smith 2005). Various mathematical optimization methods are used to design systems, processes or components of processes with the end in mind of optimizing profits. This mental model however, will no longer work in our situation today as we face economic, ecological, social and spiritual crisis (Scharmer 2010).

To address this, Scharmer (2010) pointed out seven acupuncture points to shifting capitalism into a regenerative ecosystem economy. Within this purview, the practice of chemical engineering has a greater role to assume. For example, Scharmer (2010) suggested that the design of all production and consumption cycle be completely earth-to-earth, i.e. without necessitating for landfill and that it has to co-evolve with the natural ecosystem. In this regards, the way we do optimization should be revamped into sustainable optimization. By sustainable optimization, it means optimization of the entire system and adding in environmental as well as societal objectives to the equation.

In this definition, two things need to be considered. The first one is systems thinking which emphasizes the understanding of the entire system so the interaction among components will be known and unintended consequences mitigated (Senge 2006). This is a holistic approach to problem solving and finds enormous use in process design.

It should be noted however that the construction of processing plant, whether it is a greenfield or a brownfield project, is rooted on projected market demand. In some instances, economic volatility necessitates companies to experiment with multiple pathways to survive. In this circumstance and within the purview of systems thinking, design thinking comes into the fore. Design thinking is a collective principle emphasizing empathy for users, doing prototyping and tolerance for failure (Kolko 2015). This principle finds enormous use in chemical engineering designs such as in product development, process development, systems improvement and the like. This is also useful for simplifying or humanizing chemical engineering designs with the end in mind of contributing to sustainability. It also has to be noted that of paramount importance in using this tool is imagining the future with the view in mind of status quo and the likely outcome for sustainable development.

The second consideration is the inclusion of environmental and societal objective functions to the objective functions (economic criteria) of profit, cost, energy and yield (Edgar, et al. 2001) in optimization. The environmental objectives can be the impact or risk to environment and the societal objectives can be the impact or risk to society or common good.

These are just few examples of the needed paradigm shift in chemical engineering practice in consideration of the principle of intrinsic reality. But there can be more. A picture of the envisioned new paradigm for the practice of chemical engineering is shown in Figure 4.

Figure 4. New Paradigm in the Practice of Chemical Engineering

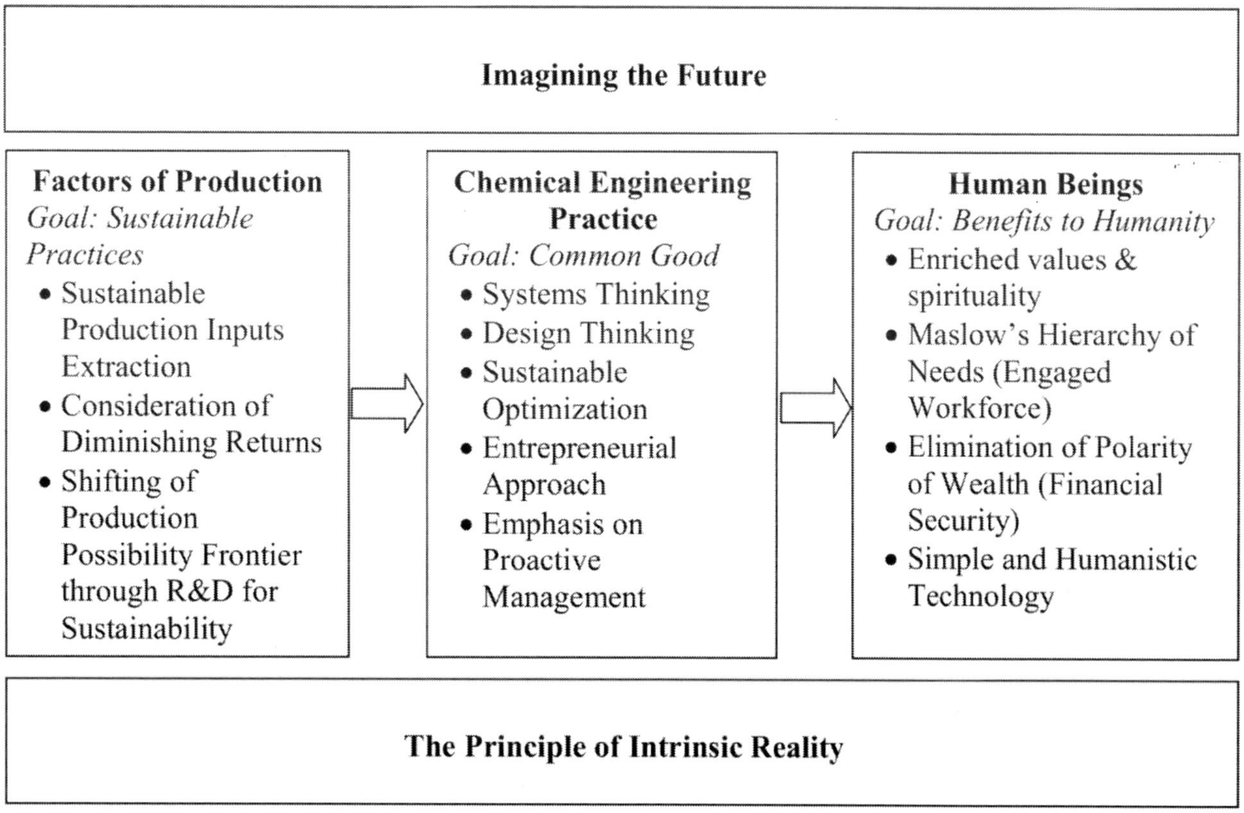

It should be realized that at the core of our economic systems is chemical engineering which can drive the economy into positive direction with the least unintended consequences. Related to that is capitalism in which upgrading of its operating systems can affect the practice of chemical engineering to a substantial degree. A discussion of this is provided in the following section.

4. Wealth Creation and Its Imperative for Sustainability

The development of this paper was inspired by the book "Freakonomics" in which the authors Steven D. Levitt and Steven J. Dubner (2005) asserted that economics can explain almost everything. It came to mind that if economics as a basic science can explain almost everything, then why not chemical engineering which was founded on chemistry, the central science?

Reflecting on it deeper, while explanations clear our doubts, we need more than that so we can tackle the world's woes. As such, this author thought that perhaps combining the two disciplines of chemical engineering and economics can be a potent force towards this end. How this can possibly be done is discussed in this section.

The starting point here is a review of the economic history of the world. In his book "A Farewell to Alms", Gregory Clark (2007) reviewed the economic history of the world in which he divided it into two periods; the period before 1800 and the period after. This is graphically represented in Figure 5.

Figure 5. Shares of Major Economies in World GDP through Time

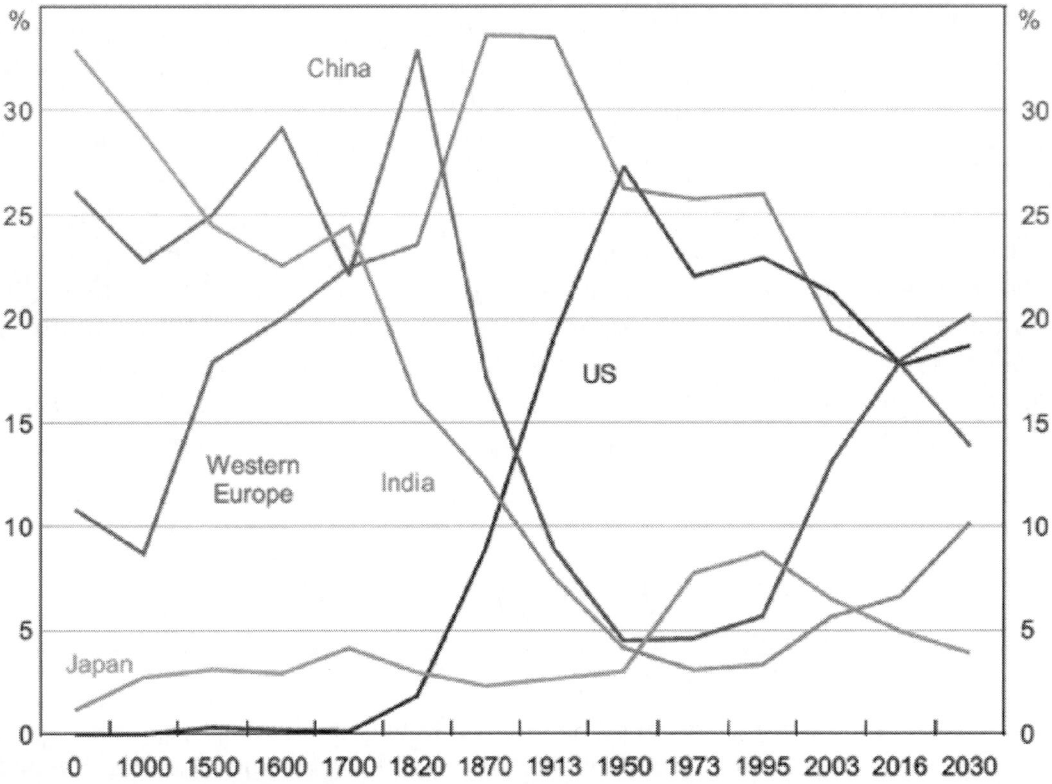

Note: Based on purchasing power parity (PPP) valuation of country GDP

Sources: International Monetary Fund (IMF), *World Economic Outlook* database, April 2011; Maddison (2006)

Thompson (2012) interpreted this graph by stating that "everything to the left of 1800 is an approximation of population distribution around the world and everything to the right of 1800 is a demonstration of productivity divergences around the world." In other words, everything to the left before Industrial Revolution depicts economic power by the sheer size of population and everything to the right after the Industrial Revolution shows the effect of mass production on country's economic output.

Clark (2007) on the other hand called the era before the industrial revolution, i.e. ca. 1800, as Malthusian era in which income per person varied across societies and epochs but growth is not observed. As Clark (2007) explained, this was due to a mechanism called Malthusian trap in which short-term increase in income through advancement in technology is dragged down by population growth. He further explicitly stated:

> "Economics as a discipline arose in the dying decades of the Malthusian era. Classical economics was a brilliantly successful description of this world. But the torrent of goods unleashed by the Industrial Revolution not only created extremes of wealth and poverty across nations, it also undermined the ability of economic theory to explain these differences."

In view of this statement, Figure 5 is reinterpreted in this paper. The contention of this author is that *polarity of wealth* can be gleaned from the graph. *Polarity of wealth* means the uneven concentration of wealth in the global economic system as some parts of the world are richer than others. The polarity can also be observed to shift as some richer parts of the world decline in economic power and some poorer parts of the world rise in economic wealth.

A historical example can be gleaned from Figure 5. India and China up until 1500 CE had 50 and 60 percent share of the world's economy, respectively. In 1 CE, China had a population of almost 60 million, while the area of the present day United States had a population of 680,000. In the late 18th century, the Industrial Revolution rendered countries with large population, just countries with the largest populations.

Looking at North America, it took them 1800 years to overtake China's economic output. However, by 1950, the US economic output was three times as great as China, even though its population is three times less. Comparing China with UK in 1913, the UK had a population of 45 million and China had 437 million but the economic output was almost the same. Analyzing what happened in Europe and later in the United States, Europe was on the lead even before 1800 (Nye 2012).

In 2010 according to data, China overtook Japan as the second biggest economy, though the numbers may not be strictly comparable. Japan's data are seasonally adjusted while those for China were not. Angus Maddison suggested that China and India were the biggest economies in the world for most of the past 2000 years. That is why economists find it a mystery why they fell so far behind and now they are flourishing ('China's economy' 2010). This may not be mysterious considering the *polarity of wealth* as was mentioned earlier and that this polarity can be argued to emanate from the Principle of Intrinsic Reality.

The reasoning here is that the world has constant mass and energy. Assuming that all these mass and energy can be harnessed to create wealth, it could mean that there is a limit for wealth creation. Theoretically, the limit can be expanded indefinitely through R&D but experience demonstrates that this was not achieved yet or getting anywhere near being achieved. As such, the constant mass and energy vis-à-vis the limit for wealth creation can be thought of as the reason for the polarity of wealth, i.e. wealth can only shift from one entity to another.

The implication of this assertion can be demonstrated by the fact that the 1% richest have accumulated more wealth than the rest of the world combined (Oxfam International 2016). The world has never seen 100% of its population to be wealthy. As such, time and again, equitable distribution of wealth is called for and mechanisms to achieve such_should be made a priority in government policies.

In a more concrete way, Scharmer (2010) suggested measures to address this growing inequality which is consistent with the mindset espoused by the Principle of Intrinsic Reality. These measures are as follows:

- Creation of a New Class of Common Property Rights

The creation of common property rights is intended to address the cost to society of private economic undertakings. For example, a private company is extracting natural resources from a certain place. While the private company can profit from that endeavor, the community in that place might suffer the undesirable consequences (e.g. flooding due to forest denudation). To ensure sustainability and institutionalize the voice of future generations in this case, a trust can be established which will own the rights and distribute funds (e.g. from cap-and-trade system for carbon emissions) to citizens (one share per person), communities and the government for investment in public infrastructure.

The trust is an effective means because it does not function like a company (driven by profits) or government (driven by special interest groups) but that it is accountable for long-term sustainability of the specific commons it manage for future generations.

- Creation of a Human Right to Basic Income

One way to ensure equitable distribution of wealth is to create a human right to basic income for every human being. This right when combined with free or affordable access to healthcare and education would create a playing field where everyone has a chance to pursue aspirations and dreams. In other words, it enables individuals to put their real creativity for the benefit of larger community. Another intention of this measure is to free up paid labor (a form of dependence) so that human beings can enhance their creativity and entrepreneurial skills.

- Inclusive Capitalism

To eradicate the *polarity of wealth*, capitalism should be made inclusive so that money and capital flows are directed to all sectors of the economic system. Currently, our money system leads to accumulation of oversupply of money and capital in high financial but low environmental and social gain areas. At the same time, there is an undersupply of money and capital in areas that address societal and community needs. It implies that there is too much money in areas where it is not needed but not enough on areas where it is needed. To address this imbalance, it is recommended to redesign the system so that the current extraction-based financial system is replaced by institutions which can support the real economy and thereby serve the ecology and community.

It should be noted that shifting of wealth is also observed in various geographical regions. When this occurs, geo-political instability can result. This goes to show that if we want lasting stability and security within the limits of our intrinsic reality, then we should not tolerate unabated greed. Otherwise, *polarity of wealth* will continue to persist and we cannot realize our full potential as individuals and collaborators for common good.

5. Way Forward...

Recognizing our intrinsic reality leads us to greater understanding of our very nature bounded by the constancy of mass and energy. It shed light on why growth stabilizes after addition of more inputs. Within this purview, trade-offs are also encountered which necessitates for R&D. R&D initiatives however should be underpinned on sustainability.

Looking at the practice of chemical engineering from the lens of intrinsic reality, the approach should be futuristic while considering relevant aspects related to factors of production and human beings.

The bigger picture shows that Engineers particularly Chemical Engineers are major players in driving the real economy. As Scharmer (2010) aptly puts it, "The essence of technology and knowledge creation, and the ultimate driving force of the economy, is the same: it is the human capacity to create." As such, if sustainability is to be achieved, Chemical Engineers have a greater responsibility to innovate for regenerative and social technologies. In this way, we can stem the polarity of wealth by providing affordable technology to enable disadvantaged individuals enhance their creativity, innovation and entrepreneurship. This can help them partake in the benefits of inclusive capitalism.

As demonstrated in this paper, the synergy emanating from the convergence of Chemical Engineering and Economics can powerfully tackle the world's woes. As such, there is a good reason for chemical engineers and economists alike to collaborate particularly on sustainability initiatives.

6. References

[1] 'China's economy overtakes Japan's in real terms' 2010, *The Economist*, 16 August, viewed 5 March 2016, http://www.economist.com/node/16834943

[2] Clark, G 2007, *A Farewell to Alms: A Brief Economic History of the World,* Princeton University Press, Princeton, NJ.

[3] Edgar, T, Himmelblau D & Lasdon, L 2001, *Optimization of Chemical Processes*, McGraw-Hill, New York.

[4] Huang, Y & Wang, B 2011, 'From the Asian Miracle to an Asian Century? Economic Transformation in the 2000s and Prospects for the 2010s', in H Gerard & J Kearns (eds), *Proceedings of the Reserve Bank of Australia annual conference*, Reserve Bank of Australia, Sydney, viewed 30 January 2016,

http://www.rba.gov.au/publications/confs/2011/huang-wang.html

[5] Kolko, J 2015, 'Design Thinking Comes of Age', *Harvard Business Review*, pp. 66-71, viewed 24 February 2016, https://hbr.org/2015/09/design-thinking-comes-of-age

[6] K.N.C. 2012, 'Mis-charting economic history: more 2,000 years in a single graphic', *The Economist blog*, web log post, 20 June, viewed 29 February 2016,

http://www.economist.com/blogs/graphicdetail/2012/06/mis-charting-economic-history

[7] Levitt, S & Dubner, S 2005, *Freakonomics: A Rogue Economist Explores the Hidden Side of Everything*, William Morrow, New York, NY.

[8] Malthus, T 1798, *An Essay on the Principle of Population*, J. Johnson in St. Paul's Church-yard, London.

[9] Maslow, A 1943, 'A Theory of Human Motivation', *Psychological Review*, vol. 50, pp. 370-396, viewed 24 February 2016, http://psychclassics.yorku.ca/Maslow/motivation.htm

[10] Monod, J 1949, 'The Growth of Bacterial Cultures', *Annual Review of Microbiology*, vol. 3, pp. 371-394, viewed 2 February 2016,

http://www.annualreviews.org/doi/abs/10.1146/annurev.mi.03.100149.002103

[11] Nye, J 2012, 'Fascinating new graph shows the 'economic history of the world since Jesus'', *MailOnline*, 23 June, viewed 15 November 2015,

http://www.dailymail.co.uk/news/article-2163610/Fascinating-new-graph-shows-economic-history-world-Jesus.html

[12] Oxfam International 2016, *An economy for the 1%, how privilege and power in the economy drive extreme inequality and how this can be stopped*, Oxford, viewed 5 March 2016,

https://www.oxfam.org/sites/www.oxfam.org/files/file_attachments/bp210-economy-one-percent-tax-havens-180116-en_0.pdf

[13] Samuelson, P & Nordhaus W 2009, *Economics*, McGraw-Hill Higher Education, New York, NY.

[14] Scharmer, O 2010, 'Seven Acupuncture Points for Shifting Capitalism to Create a Regenerative Ecosystem Economy', *Oxford Leadership Journal*, vol. 1, pp. 1-21, viewed 24 February 2016, http://www.oxfordleadership.com/journal/vol1_issue3/scharmer.pdf

[15] Senge, P 2006, *The Fifth Discipline: The Art & Practice of the Learning Organization*, Doubleday/Currency, New York.

[16] Smith, R 2005, *Chemical Process Design and Integration*, John Wiley & Sons Ltd., West Sussex, England.

[17] Thompson, D 2012, 'The economic history of the last 2,000 years in 1 little graph', *The Atlantic blog*, web log post, 19 June, viewed 29 February 2016, http://www.theatlantic.com/business/archive/2012/06/the-economic-history-of-the-last-2-000-years-in-1-little-graph/258676/

Acknowledgment

The author would like to thank Hyundai Engineering and Construction Co. Ltd. for the support in this endeavor. The author is also thankful to Ms. Sofia Ayu for meticulously editing this manuscript and Mr. Marcial N. Barroga for the poster design.

445978 Commercial Aspects of Liquefied Natural Gas (LNG) Projects

Abstract:

Historically, Liquefied Natural Gas (LNG) projects have been characterized as large base-load facilities that are highly capital intensive, with development time frames ranging from several years to several decades. Although construction, commissioning and start-up schedules are comparable to other processing systems, it is the activities leading up to a Final Investment Decision that drive overall development timing and ultimate success.

Capital costs for LNG projects are driven by many factors. Refrigerant compression horsepower, dedicated electric power generation and the need for materials suitable for cryogenic service, contribute significantly to the overall cost of LNG projects. Site development, environmental and local issues can also add significant costs and lengthen the development schedule.

The LNG industry has a track record of successful implementation of cryogenic liquefaction technology, in plants of ever-increasing capacity, resulting in safe, reliable and commercially viable ventures. Given the typical scope of investment these projects require, and the need for consistent natural gas pricing across multiple markets, LNG projects must have a robust commercial structure that guarantees reliable off-take with predetermined mechanisms for product pricing.

LNG commercial structures have been predicated on a few key attributes: a) coordinated supply/demand timing, b) long-term off-take agreements with credit worthy off-takers, c) a reliable and predictable feedstock supply, d) a strong internal balance sheet or third party financing, e) a willing host, and f) availability of a site and transportation capacity. The challenge is to implement all these activities, or "links" simultaneously.

Combining the links allows development of a commercial "chain", which forms the basis of the LNG development. Hence the term "LNG Value Chain".

In more recent years, some LNG industry paradigms have shifted in such a way that projects previously thought to be unlikely are now seen as the new generation for the industry. Changes in gas supply mechanics and sources, pricing basis, contract term, new base-load liquefaction capacity geography, capacity tolling arrangements and non-traditional developers characterize these new LNG projects.

NGL Recovery Plant Technology: Ethane Recovery or Rejection for Changing Market Conditions

Scott Schroeder
Advisian
Houston, TX
scott.schroeder@advisian.com

Prepared for Presentation at
American Institute of Chemical Engineers
2016 Spring Meeting
11th Global Congress on Process Safety
Houston, Texas
April 10-14, 2016

UNPUBLISHED

NGL Recovery Plant Technology: Ethane Recovery or Rejection for Changing Market Conditions

Scott Schroeder
Advisian
Houston, TX
scott.schroeder@advisian.com

Keywords: NGL, Natural Gas, Turbo-expander, cryogenic, LPG

Abstract

In the United States ethane has typically been recovered from natural gas due to its increased value as a petrochemical feedstock. The rapid increase in natural gas production from developing shale plays has resulted in a change from the norm. The increase in ethane supply is exceeding current demand, resulting in depressed ethane pricing; furthermore many of these new sources of natural gas are not well connected to the traditional ethane handling network resulting in high transport costs. In this new environment rejecting the ethane into the gas stream instead of recovering it will be a preferred option for some plants. The rapid development of shale gas production has also led to extensive use of pre-engineered "standard cryo plants" which offer a shortened delivery time compared to a custom engineered plant. These "standard cryo plants" are typically designed to be as flexible as possible, but may not be optimum economic solution for each case. This paper will examine the typical requirements for ethane rejection, the typical performance of "standard cryo plants" and some of the alternate designs that maybe economically attractive for specific cases.

1. Background – Natural Gas Processing

Natural gas from the wellhead is primarily methane but also contains heavier hydrocarbons – ethane, propane, butanes, pentane and c6+ and may also contain H_2S, CO_2 or N_2. Although other compounds can be found, this is typical most of the gas that is commercially produced, either from conventional formations or from the new shale gas formations that have led to a huge increase in natural gas production in the United States. The gas from the wellhead is also quite commonly saturated with water.

The natural gas from the wellhead must be processed to meet sales gas specifications. The exact specifications vary regionally, but can be generalized as follows:

- Water content – the water content is specified to avoid liquid water condensing from the gas in the transmissions network. Typically specified as a water dewpoint or in Lbs H_2O/MMSCF (mg/m^3) of gas.
- H_2S content – the hydrogen sulfide is highly toxic as well as corrosive. Typically specified as grains/100 scf (mg/m^3).
- CO_2 content – Carbon dioxide can be corrosive and reduces the heating value of the gas. Typically specified as a maximum volume percentage.
- N2 content – Nitrogen is inert and reduces the heating value of the gas. Typically specified at a maximum volume percentage, maybe combined value with other combustion inert compounds.
- Hydrocarbon Dewpoint – The hydrocarbon dewpoint is specified so the liquids do not condense in the pipeline transmission system. Typically specified as a maximum temperature at a specific pressure or pressure range.
- Heating Value – The heating value range is specified to allow interchangeability of the gas. Systems that combust natural gas are normally adjusted for a specific heating value and a large fluctuation in the heating value will cause problems with end users. Typically specified in BTU/SCF (MJ/m^3). An additional interchangeability specification such as a Wobbe Number may be used in some cases.

The values for these specifications are regionally dependant and generally have evolved over time due to technological advances, market development and regulations.

The processing of natural gas to separate the hydrocarbons heavier than methane is the area where evolution has occurred due to cycles of technology and market developments. During the operation of the first natural gas pipelines it was discovered that the removal of heavier components was required to prevent liquid slugs from accumulating in low spots in the pipelines. This spurred the development of early dew-point control plants. The development of the automobile increased the demand for gasoline and processes were developed to recover increasing amounts of "natural gasoline" which consists mainly of pentanes and heavier. The processing technology continued to advance resulting in higher recoveries of butanes and then propane. The recovery of propane and butane allowed the development of the LPG (Liquefied Petroleum Gas) market. LPGs are used as a portable fuel since they can be maintained as a liquid under moderate pressures.

Starting in the 1960's, the introduction of the turbo-expander and the development of plate-fin or brazed aluminum heat exchangers allowed for the next step in the technology/market development cycle. This new technology made high levels of ethane recovery economically possible. However, as a pure product ethane is notably different than propane and heavier compounds. Ethane is not easily portable as a liquid and has not developed as a fuel product. Ethane is used almost exclusively as a feedstock to the petrochemical industry, primarily for the production of ethylene by steam cracking.

The growth of the petrochemical industry's use of ethane as a feedstock, especially in North America, has led to deep ethane recovery being standard practice. The typical recovery process is a turbo-expander cryogenic process which recovers an ethane plus mixture. The ethane plus mixture is often referred to as "Y-grade" or raw NGL mix. This ethane plus mixture is transported by pipelines to fractionation plants that separate the ethane, propane, butanes, and natural gasoline into purity products. These fractionation plants typically have petrochemical facilities located nearby to allow the ethane to be delivered by dedicated ethane pipelines. The fractionation facilities also often are located near salt caverns that allow cost effective storage of large volumes of hydrocarbon products.

2. The Shale Gas Revolution

The Shale Gas Revolution, as it has been called, is the rapid increase in production from what are called "unconventional" wells. These "unconventional" wells are drilled in relatively impermeable shale formations. Unlike conventional wells that are drilled vertically and typically flow freely, commercial production from the shale formations requires extended horizontal drilling followed by hydraulic fracturing (fracking). Although horizontal drilling and hydraulic fracturing are not new, it was not until about 2006 that the right conditions of improving technology combined with high natural gas prices and favorable regulation ignited the revolution. The drop in natural gas prices at the end 2008 didn't stop the growth in shale production; with the large improvements made in production efficiency and continued high oil prices, production moved to formations with high liquids content and high productivity to counter the lower gas prices.

Figure 1 - United States - Oil & Gas Prices

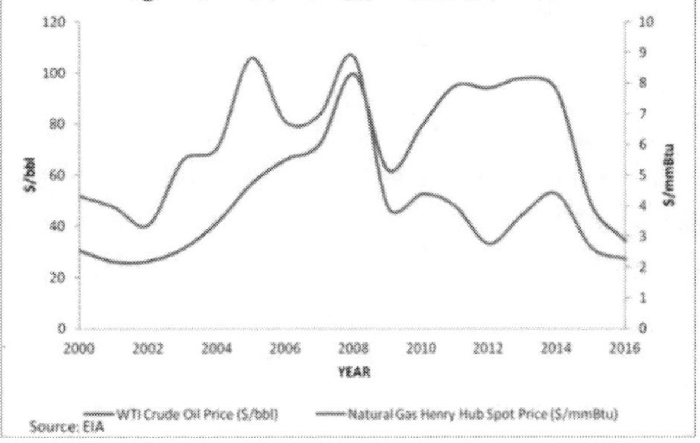

Figure 2 - U.S. dry shale gas production

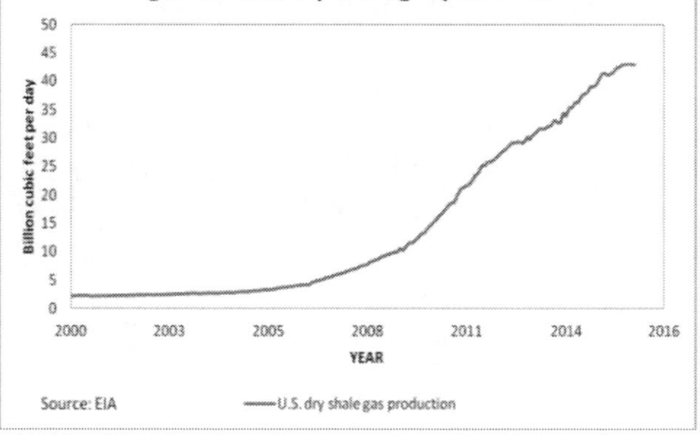

The locations of the active shale plays are important to understand the drastic shift that has occurred in the North American production. Among the largest active plays, the Barnett and Eagle Ford plays are located close to regions with a history of oil and gas production and enjoy a more developed infrastructure compared to the Bakken, Marcellus and Utica plays. This is especially important with regards to ethane recovery since pipelines or cryogenic transports are needed to transport ethane and ethane is only used as petrochemical feedstock. This infrastructure takes time to develop. The first transportation projects to move ethane from the Bakken to Canada's existing ethylene crackers and from the Marcellus and Utica to the east coast for shipment in refrigerated gas carriers to existing European crackers are just coming online now. The new ethane supply has spurred a wave of proposed ethylene cracker projects positioned to take advantage of the new ethane supplies; however, most of these projects are still in the engineering phase and are not expected to be online until 2017 or later.

Figure 3 – North American Shale Plays

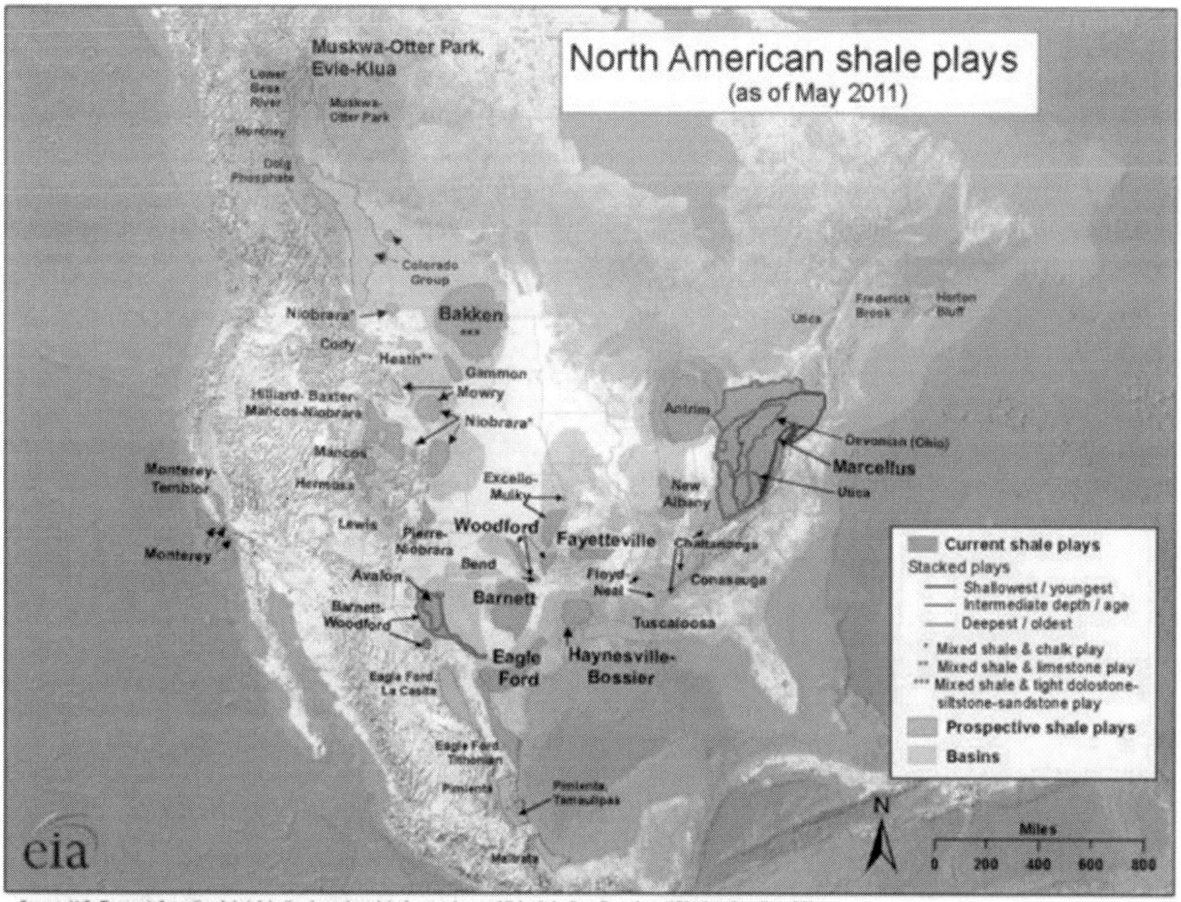

Source: U.S. Energy Information Administration based on data from various published studies. Canada and Mexico plays from ARI.
Updated: May 9, 2011

3. NGL Recovery Plants – The Impact of Shale Gas Revolution

As discussed in the background section, cryogenic turbo-expander processes have become the dominate plant type for NGL recovery. Among the many processes that have been developed, the Gas Subcooled Process (GSP) originally patented by the Ortloff Corporation in the late 1970's has become the dominate process for ethane recovery in North America[1]. The advantages of the GSP process are good ethane recoveries of 90+%, good tolerance to inlet CO2 which can freeze at cryogenic temperatures and the flexibility to reject ethane when required. Although there have been many processes that have been developed which have better performance than GSP process, the fact that GSP is now an open art process that provides good overall performance, relatively simple process design and simple operation has led to its continued use as a standard design.

The impact of the rapid growth of shale gas production has been the move towards "standard plants". Although the concept of modularized standard plants is not new to the industry, the high ethane recovery cryogenic plants had remained mostly a custom designed plant where the turbo-expander and plate-fin exchanger were designed specifically for the inlet gas stream and then put on order. The industry was also moving to increasing larger plants to take advantage of economies of scale. The penalty for custom designing these large plants is time. Several vendors realized that for the shale revolution, speed was more important that absolute efficiency. They designed "standard plants" that would work for a range of inlet gas compositions in several "rated" sizes such as 60 MMscfd, 120 MMscfd and 200 MMscfd. The long lead items such as the turbo-expander and plate-fin exchangers were pre-ordered and in some cases the module fabrication was already started before the customer committed to an order. The maximum plant size of 200 MMscfd was based on the ability to modularize the plant to the greatest degree possible while allowing easy transportation of the modules and quick installation. These vendors that had foresight to develop their "standard plants" early captured the majority of the market for new cryogenic plants installed in North America.

Figure 4 – GSP Process

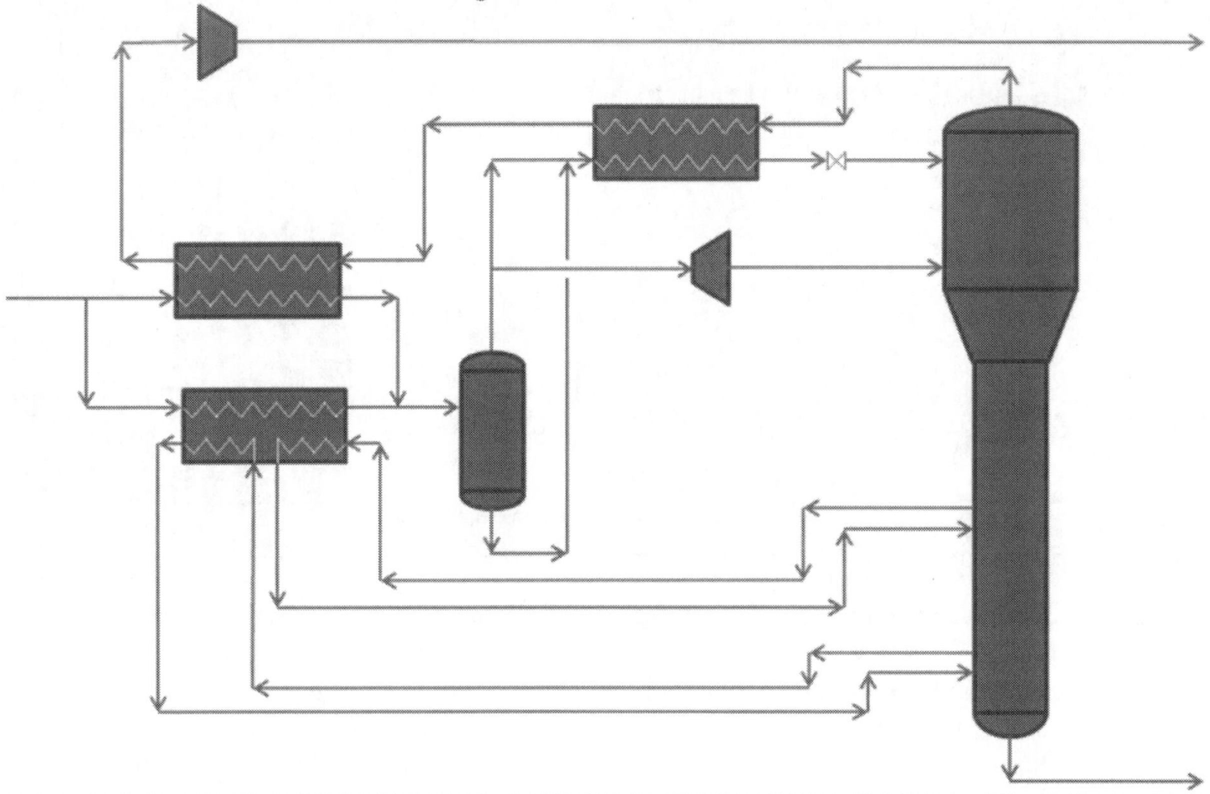

4. NGL Recovery Plants – Post Revolution

Although production from shale gas is not expected to stop, the sharp drop in oil prices at the end of 2014 has tempered the wild growth that was typical during the "Revolution". The industry has now entered into the evolution stage where only the strongest and fittest will survive. In this environment optimization of processing assets, both new and existing, is of prime importance.

The GSP standard plant is good overall, but has it weaknesses. Specifically, although it has the flexibility to reject ethane, high levels of ethane rejection causes the propane recovery to fall as well, possibly dropping into the 80% range in full rejection mode. The lost propane can limit the amount of rejection possible in cases where the high heating value of propane pushes the gas to its specification limit. The lost propane is also lost value as propane typically maintains a positive value for recovery.

There are a number of processes that have improved on the GSP design that allow higher propane recovery during rejection. A typical enhancement is the addition of recycle vapor stream that is used as a top reflux, shown in Figure 5 is Ortloff's Recycle Split Vapor (RSV) process[2]. This addition increases the maximum ethane recovery and maintains higher propane recovery during ethane rejection, but requires a higher initial capital investment. Several other technologies are also available that effectively enhance the reflux compared to the GSP process.

Figure 5 – Recycle Split Vapor Process

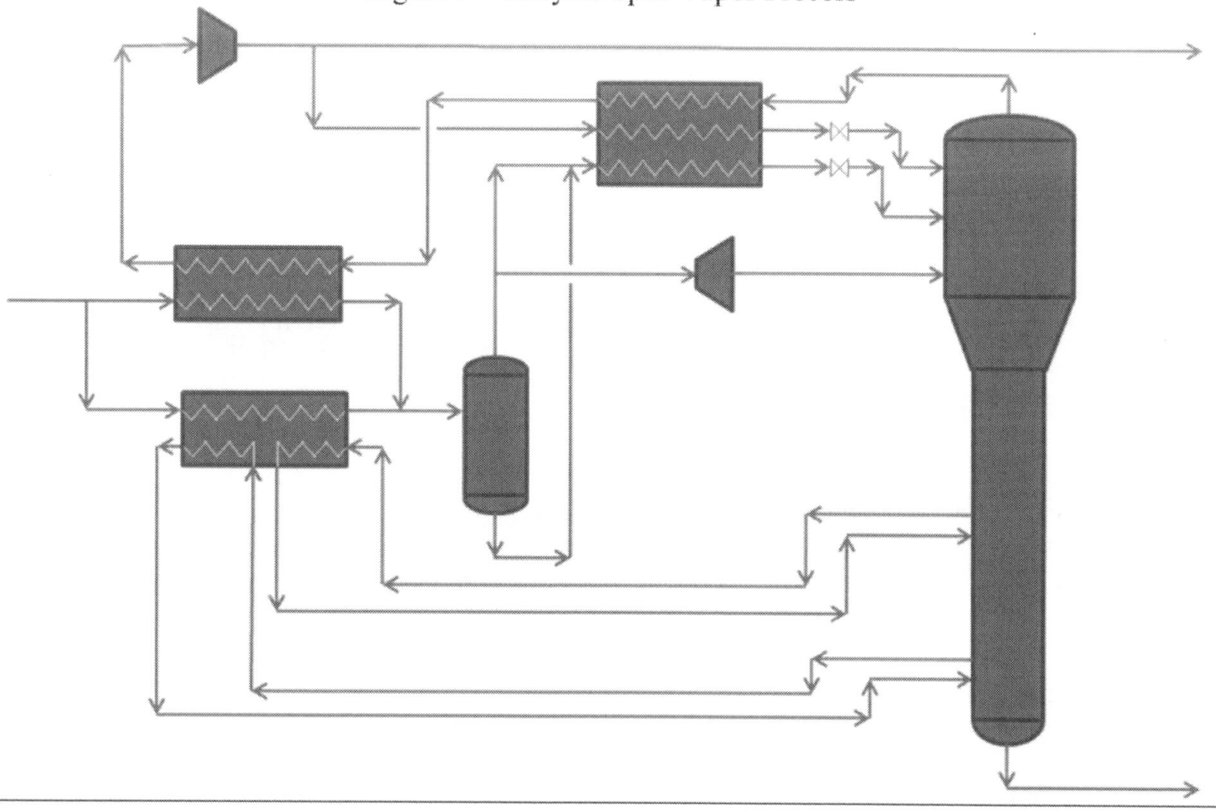

Processes that specifically focus on high propane recovery can typically achieve propane recoveries of 98+%, with nearly complete ethane rejection, with greater efficiency than an ethane recovery process being operated in ethane rejection mode. The designs using turbo-expanders typically use two fractionation towers; however, some designs have combined towers into a single column. Several mixed refrigerant designs that avoid the use of a turbo-expander are also available. If lower levels of propane recovery are acceptable, then a simplified refrigerated plant with a lower capital investment could be considered.

Revamping or retrofitting an existing unit or an uninstalled "standard plant" may be a better option than a new custom plant or simply installing a "standard plant" as-is. As an example Ortfloff Engineers is now specifically marketing their Retro-Flex™ package for this purpose[3]. This package is proposed as a standard add-on to a "standard plant" and bolts into the overhead system of a standard GSP plant to improve the reflux especially under ethane rejection conditions.

Making the decision whether to install a "standard plant" or consider one of the alternatives is currently not as clear. For each location a variety of factors should be considered:

- Constraints – product specifications, regulations, permit requirements
- Market Conditions – expected production rate, expected transportation costs, expected product
- Market Variably – development of sensitivity cases for the market conditions over the expected life of the plant
- Capital Cost – capital cost of the new plant or revamp of a existing unit should consider the possibility of obtaining distressed assets in the current market environment as well as current labor conditions for construction
- Operation Cost – the operating cost of the plant gains more importance as the product margins decrease
- Schedule – the economic advantages of a shorter construction schedule cannot be ignored

By using these factors the economics of each of the options can be compared and the optimum process selected for a specific installation.

5 Conclusion

The shale revolution has brought a wave of change to the North American and world markets, rapid expansion has led to an oversupply condition causing large swings in product pricing. The ethane market has specifically suffered since the ability to transport purity ethane is limited compared to other NGL products and the demand is almost exclusively for ethylene crackers. The low price of ethane has spurred a wave of new ethylene crackers as well as transportation projects to move ethane to existing crackers. However, the drop in crude prices has left many questioning where the market will stabilize.

Purchasing and installing a "standard plant" had become almost a standard practice. The margins were high enough that getting the plant running as soon as possible was the primary concern. As the margins have decreased, spending the time to perform economic screening studies looking at the available options to increase capacity or optimize processing efficiency is increasingly important; the "standard" answer may not be the best answer for all situations.

6. References

[1] Campbell, R.E., Wilkinson, J.D., U.S. Patent 4,157,904

[2] Campbell, R.E., Wilkinson, J.D., and Hudson, H.M., U.S. Patent 5,568,737

[3] Website - http://www.ortloff.com/files/Retro-Flex-01.pdf

Additional references not cited:

[4] "GPSA Engineering Databook" 12th Edition

[5] "Hydrocarbon Gas Liquids (HGL): Recent Market Trends and Issues," U.S. Energy Information Adminstration, November 2014

444172 Conversion of Synthesis Gas into Ethanol Using Homogeneous Catalysts

Abstract:

Before production of fuel ethanol became dominated by fermentation routes, many of the large oil and chemical companies utilized synthesis gas produced from natural gas and patented methods for conversion of H_2 + CO into ethanol using *homogeneous* catalysis. They often employed metals long used in Fischer-Tropsch synthesis, because of the ability of these metals to catalyze CO-transfer and CO-insertion reactions. By a judicious choice of ligands and reactor residence time, chain lengths of the Fischer-Tropsch products could be limited to the lower molecular-weight alcohols. Ongoing work at TDA Research, Inc. demonstrates that a wide variety of transition-metal oxides are reduced by the H_2 + CO of high-pressure synthesis gas into carbonyl anions and anionic metal-carbonyl hydrides that remain stable at ambient pressure and temperature. In general, the TDA carbonyl anions catalyze CO-transfer reactions; hydrides transfer hydrogen for reduction reactions, and carbonyl hydrides transfer *both* CO and hydrogen. Rate to ethanol exceeds 0.26 g ethanol/g metal catalyst/h, which on a metals basis, is equivalent to the best heterogeneous catalysts reported, but with the homogeneous system showing much higher selectivity, especially in minimizing low-value CH_4.

Acknowledgement: U.S. Department of Energy, Office of Science, Grant Award: DE-SC0004376.

16AIChE Spring Meeting, Houston, TX

12TH GLOBAL CONGRESS ON PROCESS SAFETY

Exploring Options for Mercury Removal Process within an Existing Condensate Stabilization Unit

Mahdi Nouri & Eberhard Lucke, CH2M OG&C, Houston, TX
Seyi Odueyungbo, Chevron ETC, Richmond, CA

ch2m

Chevron

Agenda

16AIChE Spring Meeting, Houston, TX

12TH GLOBAL CONGRESS ON PROCESS SAFETY

- Overall Block Flow Diagram
- Adsorbent Process Fundamentals
- Design Basis
- Condensate Stabilization Overview
- MRU Location
- Study Methodology for Data Screening
- Economic Analysis
- Conclusions
- Q&A

ch2m

Chevron

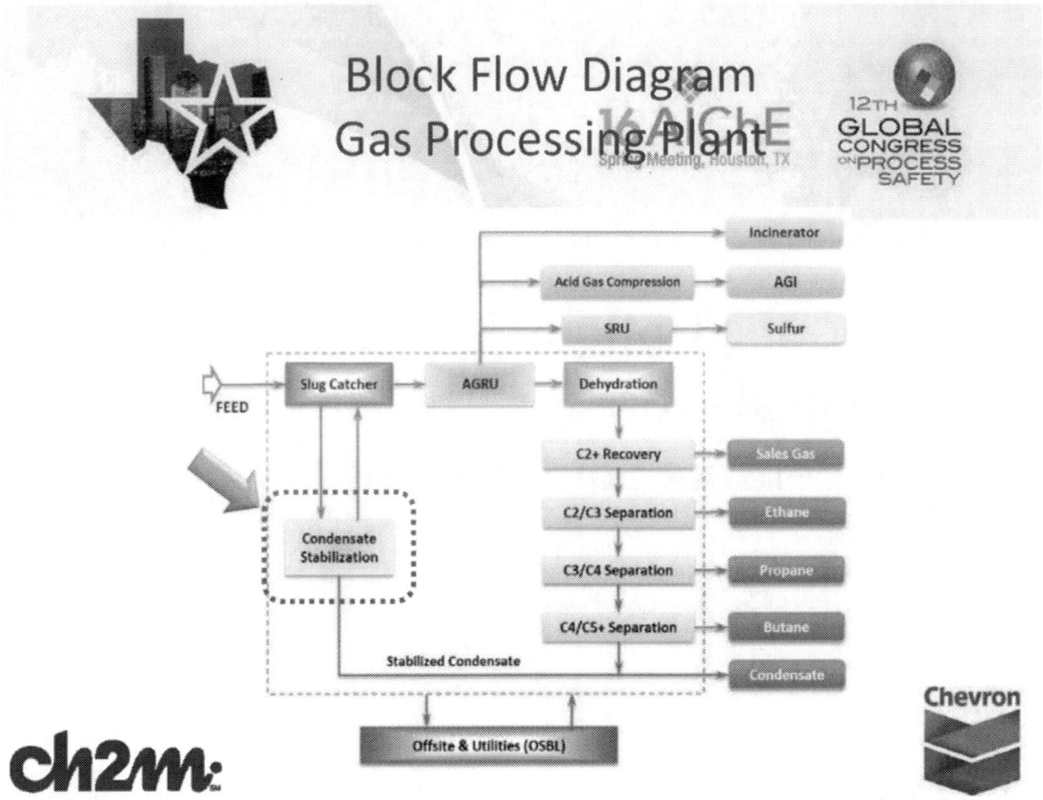
Block Flow Diagram
Gas Processing Plant
12TH GLOBAL CONGRESS ON PROCESS SAFETY
Incinerator
Acid Gas Compression
AGI
SRU
Sulfur
FEED
Slug Catcher
AGRU
Dehydration
C2+ Recovery
Sales Gas
C2/C3 Separation
Ethane
C3/C4 Separation
Propane
C4/C5+ Separation
Butane
Condensate Stabilization
Stabilized Condensate
Condensate
Offsite & Utilities (OSBL)
ch2m
Chevron

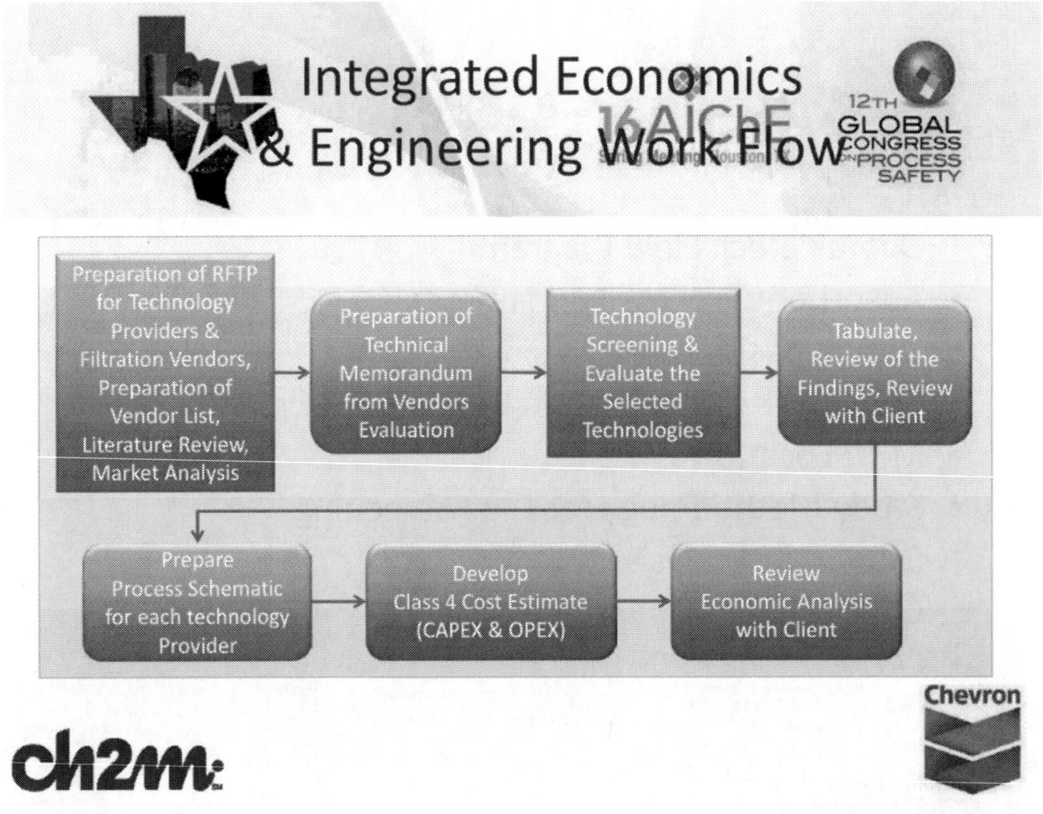
Integrated Economics
& Engineering Work Flow
12TH GLOBAL CONGRESS ON PROCESS SAFETY
Preparation of RFTP for Technology Providers & Filtration Vendors, Preparation of Vendor List, Literature Review, Market Analysis
Preparation of Technical Memorandum from Vendors Evaluation
Technology Screening & Evaluate the Selected Technologies
Tabulate, Review of the Findings, Review with Client
Prepare Process Schematic for each technology Provider
Develop Class 4 Cost Estimate (CAPEX & OPEX)
Review Economic Analysis with Client
ch2m
Chevron

Project Goals

16AIChE Spring Meeting, Houston, TX

12TH GLOBAL CONGRESS ON PROCESS SAFETY

- To review and evaluate mercury removal technologies for removal of elemental mercury from hydrocarbon condensate streams within an existing condensate stabilization unit
- To establish technology limitations and identify optimal location of the MRU and filtration package
- Develop and present optimal treating process through investigation into technical feasibility of various technologies and an economic evaluation of their application to liquid hydrocarbon stream.

ch2m: Chevron

Mercury in Natural Gas & Condensate

16AIChE Spring Meeting, Houston, TX

12TH GLOBAL CONGRESS ON PROCESS SAFETY

Location	Gas (mg/m3)	Liquids (mg/kg)
Europe	100 - 150	-
South America	50 - 120	50 - 100
Gulf of Thailand	100 - 400	400 - 1200
Africa	80 - 100	500 - 1000
Gulf of Mexico (USA)	0.02 - 0.4	-
Overthrust Belt (USA)	5 - 15	1 - 5
North Africa	50 - 80	20 - 50

ch2m: Chevron

Adsorbent Process Fundamentals

- Sulfur Impregnated Activated Carbon,

$$Hg + S \rightarrow HgS$$

- Metal Sulfides Dispersed in a Solid Carrier,

$$Hg + MSx \rightarrow MS(x-1) + HgS$$

- Molecular Sieve Technology

Design Basis

- Feed Actual Plant Data

Stream No.	<11>		<14>		<17>	<18>
Description	HC to Stabilizer Condensate Mixing Drum		HC from Stabilizer Condensate Mixing Drum		Condensate Stabilizer Bottom	Stabilized Condensate Product
Phase	V	L	V	L	L	L
Temperature (C)	18.1		17.2		200.1	41
Pressure (Psia)	145		101.5		89.92	88.03
Normal Flowrate (m3/hr)	202.8	75.21	46.14	74.56	88.69	71.26
Density (kg/m3)		783.9		784.7	619.6	771.1

- Product Specification: >10 ppbwt of Mercury Content

Condensate Stabilization Process Flow Diagram

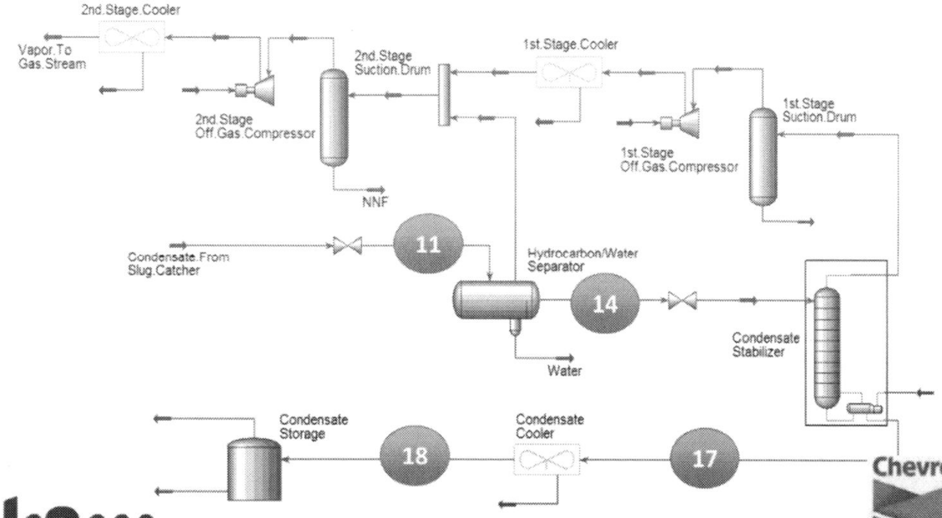

ch2m
Chevron

MRU Process Scheme and Proposed Location

F-100
Particulate Filter

V-100 A/B
Mercury Removal Beds

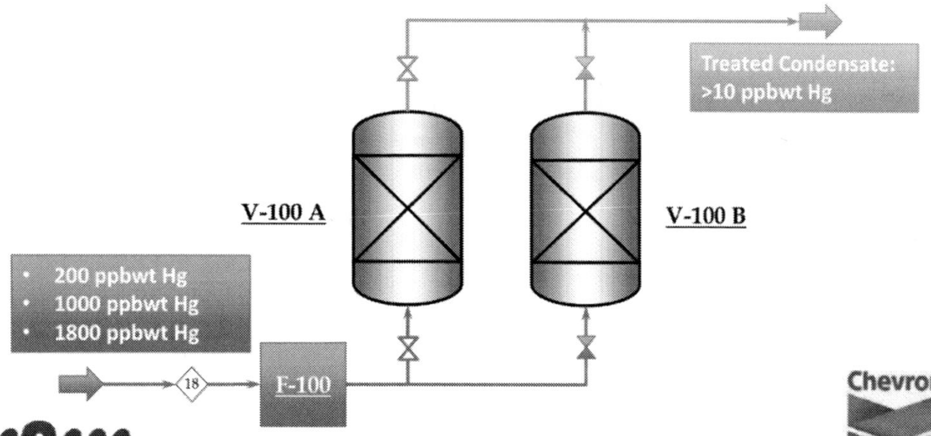

ch2m
Chevron

Study Methodology for Data Screening

DESCRIPTION	CASE 1	CASE 2	CASE 3	CASE 5
Hg inlet concentration (ppbwt)	1800	1000	200	1800
Hg outlet concentration (ppbwt)	< 10	< 10	< 10	< 10
Location of bed	--	--	--	--
Mercury removal bed				
- Flow direction	--	--	--	--
- Bed configuration	Single	Single	Single	Single
- Diameter (m)	--	--	--	--
- Height (m)	--	--	--	--
- Volume (m3) /bed	--	--	--	--
- Mass (kg) /bed	--	--	--	--
- Pressure drop (Psi)	--	--	--	--
- Design life of adsorbent	1yr	1yr	1yr	4yr

Mercury Removal Bed Mass of Adsorbent

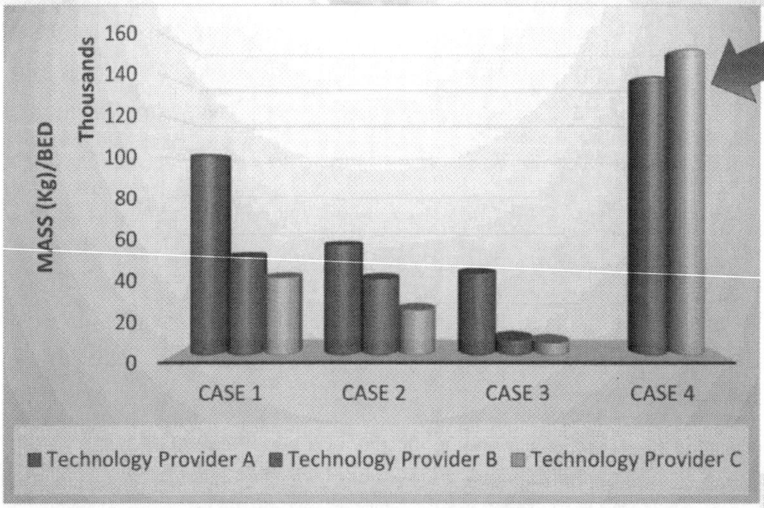

(1) Case 5 – Technology Provider A did not submit any design info (bed size limitations), others proposed 2xparallel beds.

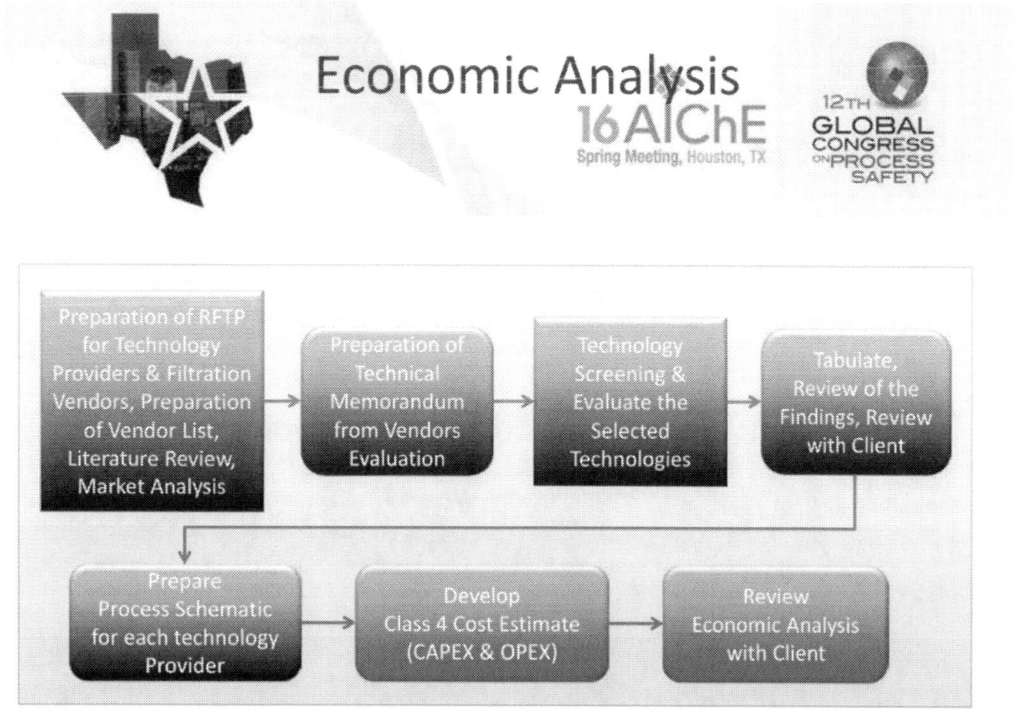
Economic Analysis
16 AIChE
Spring Meeting, Houston, TX
12TH GLOBAL CONGRESS ON PROCESS SAFETY
Preparation of RFTP for Technology Providers & Filtration Vendors, Preparation of Vendor List, Literature Review, Market Analysis
Preparation of Technical Memorandum from Vendors Evaluation
Technology Screening & Evaluate the Selected Technologies
Tabulate, Review of the Findings, Review with Client
Prepare Process Schematic for each technology Provider
Develop Class 4 Cost Estimate (CAPEX & OPEX)
Review Economic Analysis with Client

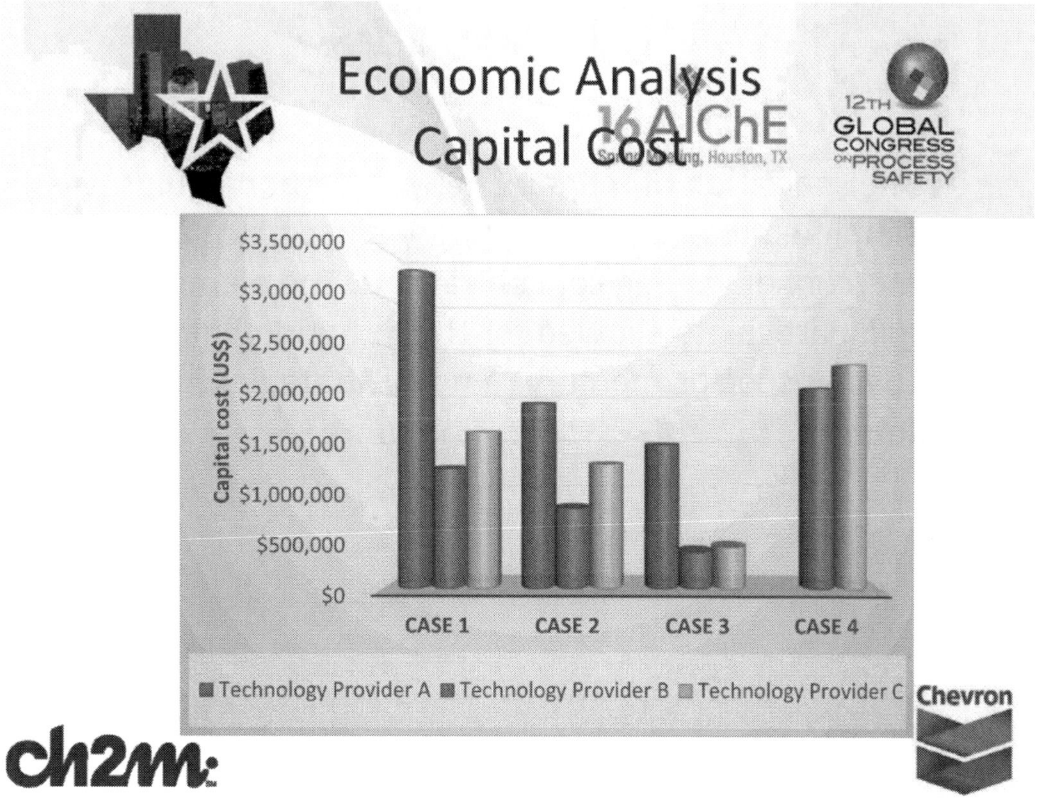
Economic Analysis
Capital Cost
16 AIChE
Spring Meeting, Houston, TX
12TH GLOBAL CONGRESS ON PROCESS SAFETY
Capital cost (US$)
$3,500,000
$3,000,000
$2,500,000
$2,000,000
$1,500,000
$1,000,000
$500,000
$0
CASE 1
CASE 2
CASE 3
CASE 4
Technology Provider A
Technology Provider B
Technology Provider C
Chevron
ch2m

Economic Analysis
Operating Cost

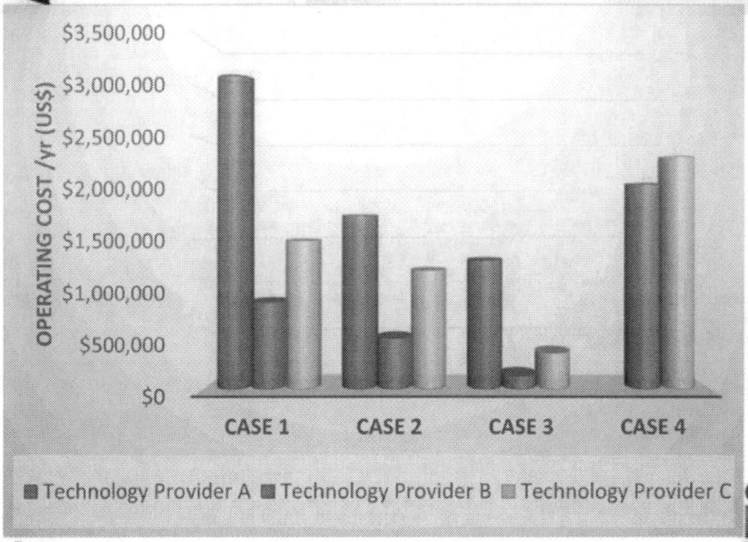

Conclusions

- The speciation of mercury in hydrocarbon liquids was simplified to elemental mercury and particulate mercury of varying concentrations. The elemental and particulate allowed for the development of species specific mercury removal strategies in the future.

Conclusion (Cont'd)

16AIChE Spring Meeting, Houston, TX — 12TH GLOBAL CONGRESS ON PROCESS SAFETY

- The distillation range of the hydrocarbon condensate stream indicates the presence of hydrocarbon molecules up to C12+ in size.
- Long paraffinic hydrocarbon molecules tend to plug the surface of the adsorbent and block access of mercury to the active sites of the adsorbent.
- three licensors claim that they have the capability to treat the specified liquid HC stream

ch2m: Chevron

Conclusion (Cont'd)

16AIChE Spring Meeting, Houston, TX — 12TH GLOBAL CONGRESS ON PROCESS SAFETY

- TP-A had operational experience with a heavy HC condensate, similar in composition and normal boiling point (NBP) to the basis for this study.
- TP-B shared a reference list; however, they did not provide their experience on C12+ compounds.
- TP-C submitted a technical proposal, they indicated a lack of confidence in their proposal and stated that they would be unable to provide standard performance guarantees for treating the specified condensate stream.

ch2m: Chevron

Conclusion (Cont'd) Location of MRU

- Several locations (stream 11, 14, 17, and 18) were evaluated within an existing condensate stabilization unit.
- Stream #18 (stabilized condensate product) was considered to be the most suitable location for the MRU package due the sub-cooled nature of the stream.

Conclusion (Cont'd) Safety Issues

- Most government agencies classify equipment and materials contaminated by mercury as hazardous waste that needs to be disposed of properly by local regulations.
- TP-A and TP-B indicated that they offer waste disposal programs to their customers to assist in managing the entire life cycle of the adsorbent.
- TP-C does not offer any waste management.

Conclusion (Cont'd) Summary

16 AIChE Spring Meeting, Houston, TX

12TH GLOBAL CONGRESS ON PROCESS SAFETY

Summary of economic analysis for MRU at worst operating case condition for achieving the product specification are:

- Mercury inlet concentration of 1800 ppbwt,
- Single bed configuration (2x100% beds),
- 1 year design life of adsorbent (Case 1)

ch2m

Chevron

Conclusion (Cont'd) CAPEX Selected Option

16 AIChE Spring Meeting, Houston, TX

12TH GLOBAL CONGRESS ON PROCESS SAFETY

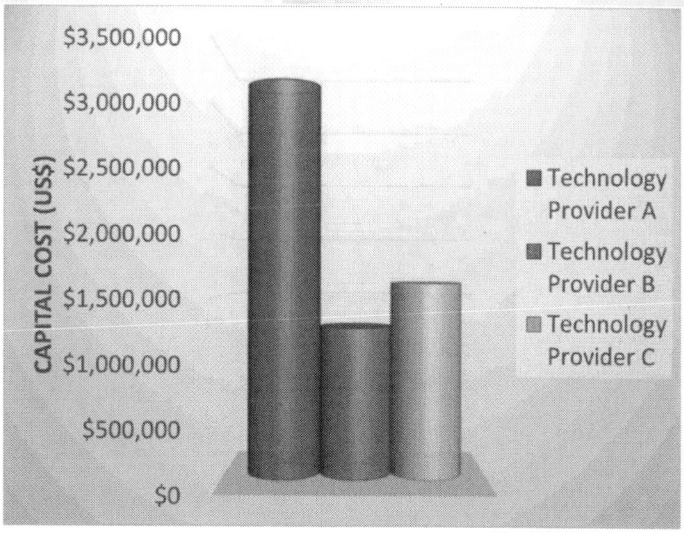

ch2m

Chevron

Conclusion (Cont'd) OPEX Selected Option

16 AIChE Spring Meeting, Houston, TX

12TH GLOBAL CONGRESS ON PROCESS SAFETY

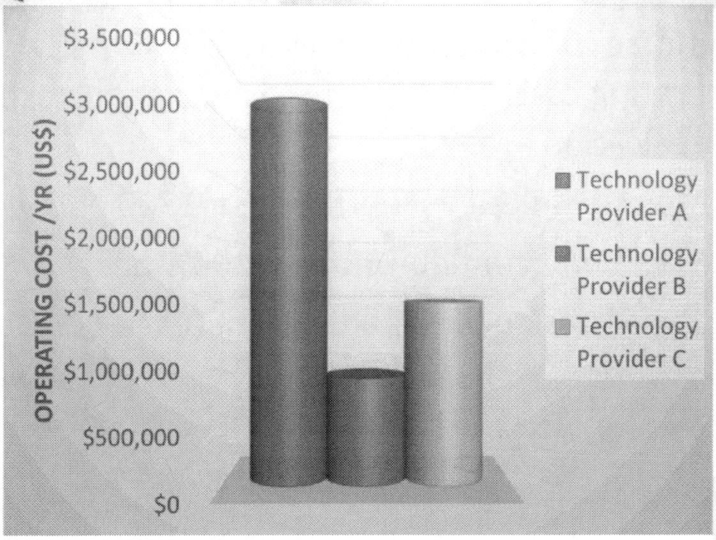

ch2m:

Chevron

16 AIChE Spring Meeting, Houston, TX

12TH GLOBAL CONGRESS ON PROCESS SAFETY

ch2m:

Chevron

16AIChE
Spring Meeting, Houston, TX

LNG Annual Production Rate Sensitivity Analysis

Christophe Duchateau
Bechtel OG&C
3000 Post Oak Blvd., MS42, Houston, TX 77056, USA
cduchate@bechtel.com

Philip Diwakar, Jaleel Valappil
Bechtel OG&C
3000 Post Oak Blvd., MS 42, Houston, TX 77056, USA
pmdiwaka@bechtel.com, jvalappi@bechtel.com

Prepared for Presentation at
American Institute of Chemical Engineers
2016 Spring Meeting
11th Global Congress on Process Safety
Houston, Texas
April 10-14, 2016

UNPUBLISHED

LNG Annual Production Rate Sensitivity Analysis

Christophe Duchateau
Bechtel OG&C
3000 Post Oak Blvd., MS42, Houston, TX 77056, USA
cduchate@bechtel.com

Philip Diwakar, Jaleel Valappil
Bechtel OG&C

Keywords: Liquefied Natural Gas, Computational Fluid Dynamics, Process simulation, Hot air recirculation, Meteorological impact

Abstract

Bechtel OG&C routinely performs hot air recirculation studies through its Advanced Simulations Group for its Liquefied Natural Gas (LNG) facility projects. These Computational Fluid Dynamics (CFD) studies aim at determining the effect of hot air recirculation from the process air coolers, gas turbines and inlet air chillers for an entire facility, based on local site meteorological conditions. The results provide input for process simulations, so as to determine the sensitivity of the LNG annual production rates based on wind direction simulations. We describe here the latest findings and best practices to obtain the most accurate simulation results and derive the most realistic production rate previsions out of them.

1. Introduction

Liquefied Natural Gas (LNG) production facilities work by cooling a natural gas feed down to cryogenic temperatures where it condenses into a liquid. The heat extracted from this gas is rejected by heat exchangers into the ambient air. As a result the air at the outlet of these process air coolers is much warmer.

The heat exchange capability of the different air coolers is a function of the ambient air temperature at their inlet. If the inlet ambient air is warmer, the heat exchange is lesser, thus reducing the capability of the equipment to correctly perform its task. Similarly, the drive turbines that power the entire process will have a reduced power output in case of a warmer air at their inlet. Both of these situations result in a reduction in the overall LNG production rate. Since the facility produces warm air and is sensitive to ambient temperature, it is critical to understand the air circulation patterns within a facility. Moreover, for a multi-train facility where the number of heat sources and of heat-sensitive equipment is multiplied. As a matter of fact, the predominant phenomenon leading to inlet air temperature rise is recirculation of hot exhaust within the plant. Two main reasons resulting in hot air recirculation are:

- Self-recirculation – Hot exhaust from the air coolers or gas turbines could be recirculated into their own air inlets.

- Cross contamination – Hot exhaust from some air coolers or gas turbines could be recirculated into the air inlet of other air coolers or gas turbines.

The extent of the impact of this air recirculation depends on various factors such as relative elevations of equipment, wind direction and velocity, exhaust temperatures and flow rates, topology and plant layout. Hot air recirculation studies are based on detailed Computational Fluid Dynamics (CFD) models. In order to keep reasonable computation times, the model has to make some simplifying assumptions, however, in order to keep a good level of accuracy the scope of the model has to include all major heat sources and equipment sensitive to air temperature, as well as all the major obstructions to wind (storage tanks, cold boxes, vapor barriers, topological features, equipment buildings and large groups of other equipment).

In order to account for the risk of having inlet air warmer than the ambient temperature, the different parts of a facility are slightly oversized by the inclusion of a temperature margin in their design calculations. This allows them to meet their nominal workload for an inlet air temperature that is warmer than expected. This design margin is generally sufficient; however, in some cases it might not be enough, making it necessary to include mitigations in the plant design.

We will cover here the required input data, recommended assumptions, general trends observed, and the best ways to analyze, interpret and use the results.

2. Air Recirculation Studies Required Input Data

In order to perform a hot air recirculation study, three types of data are required: local weather records, site geography, and plant data.

2.1 Weather Data

Among the input data needed to perform a hot air recirculation study, meteorological records is of utmost importance. The best possible case is to have detailed hourly wind velocity, wind direction and ambient temperature measurements over a period of twenty years or more. This allows the extraction of key parameters such as the average wind speed for a particular wind direction, the annual average percentage of time the wind blows from a specific direction, the annual average ambient temperature, and the annual average percentage of time there is concurrence of a specific wind direction and an ambient temperature above the project's specified high ambient temperature.

Such detailed weather records generally come from a nearby airport, sometimes from a non-airport weather station. In either case, it is unlikely that the recording has been done at the exact location of the LNG facility. As a result there can be a slight difference between the records used for the simulation and the usual weather at the plant site.

2.2 Site Data

Another important input, even if not as impacting as the other inputs, is topology. Any mountain, hill, valley, or slope, even as far as half a mile away from the plant, can result in local variations in wind speed and direction. It is consequently important to include these topological features in the CFD model, since they can modify the results of the study.

2.3 Plant Data

The other most important data is about the plant itself. The exact location, elevation and size of each air inlet and exhaust, the air temperature and velocity at every exhaust, the air velocity at the inlets, the location and dimensions of the main obstructions (LNG storage tanks, vapor barrier, cold boxes and other groups of smaller equipment) are all required since all of them can either modify the flow of air or its temperature.

3. CFD Modeling Assumptions

In order to reduce the computation time, some simplifying assumptions have to be made. For instance the ambient temperature is assumed constant, and the wind remains steady in terms of direction and velocity for each case. The effect of heat conduction and radiation is ignored, and so are most of the small topological features and isolated small pieces of passive equipment. The different air inlets and outlets are modeled as "velocity inlets", with dimensions and velocity – along with exhaust temperature – as per mechanical data. Note that the exhaust temperatures might need to be linearly adjusted when the mechanical calculations have been based on an ambient air temperature different from that used for the hot air recirculation study.

It is also important to maintain enough details in the model to ensure the reliability of the results. For that, the wind conditions (temperature, velocity and frequency) are determined from the weather records. All the major heat sources shall be included, as well as all the large obstructions to air flow. The model should also extend far enough around the plant – half a mile of modeled surroundings is advised. Significant groups of small equipment can be represented as a porous medium. Finally, inlet air temperature is calculated as an integral (mass-weighted average) over the inlet area.

4. General Observations

Once the CFD model is correctly setup and the calculations have successfully converged towards a solution, we can draw several conclusions that are valid for most of the LNG plants. We present the most important ones, along with some recommendations.

Because buoyancy effects carry the heat upwards and there are no recirculation vortices created by the wake of individual components, calm condition yields the lowest temperature rise for all equipment inlets. For this reason calm condition is the one leading to the best LNG production rate at a given ambient temperature.

When the wind direction is almost parallel to the longer dimension of most of the equipment, the temperature rise at the inlet is very limited for the main units. This alignment of the plant and the wind direction is in fact the second best scenario (after calm condition), and the reason why it is recommended to align as much as possible the plant with the prevailing wind.

When the wind comes at a 10 to 45 degree angle with the equipment length, the inlet air temperature is well above that of the ambient air. This is caused by the prevalence of self-recirculation. In other words, hot air from the equipment exhausts is re-entrained to the inlets of the same equipment.

When the wind is perpendicular to the equipment length, the cross contamination is predominant. For example, hot air from one bank of air coolers goes to the inlet of the units located downwind of it. Depending on the exhaust temperature and the distance covered by the hot air before it reaches the next inlet, the impact of this cross contamination can be very important.

5. Results Analysis

Compared to the ambient temperature, the mass-weighted average inlet temperature rise for each wind condition and each piece of equipment can be tabulated. Because the ambient temperature does not remain constant in real terms, the temperature rise above ambient is more representative than the temperature itself. Comparison with on-site measurements have shown the CFD-predicted temperatures to be within 0.3°C of the “real” values.

For most of the projects, the hot air recirculation study ends with the table summarizing the values of the inlet air temperature rise above ambient for the different units under the different wind conditions.

When the annual average temperature rise (weighted by the percentage of occurrence of the wind conditions) is greater than the temperature margin used for the design of a unit, when a key unit operates outside of its design margin for a very significant part of the year, or any similar concern, then a more detailed study can be performed. This detailed stage of the study consists in using the CFD-predicted temperature rise values as inputs for the process simulator so as to determine whether or not the guaranteed LNG production rate can be met under the most stringent set of climatic conditions.

Depending on the results of this advanced study, it becomes clear whether or not there is a need to implement mitigations, or to modify the plant layout during the design phase, in order to bring the expected production rate back above the guaranteed value. Mitigations can include increasing the design margin, adding skirts to a process air cooler, covering the gap between two bays of air coolers, increasing exhaust stack height, but are not limited to this list. The CFD-predicted hot air streamlines allow the identification of the impacting heat sources, and of the most effective type of mitigation.

6. Conclusion

Hot air recirculation in and around an LNG plant has a consequent effect on the overall production rate. Through the use of CFD, it is possible to predict the circulation patterns of hot air in and around the facility, and the resulting thermal impact on the different units, for the different wind conditions. The resulting temperature rise values can then be used as inputs in a process simulator to determine the impact of the weather conditions on the LNG production rate. If the simulated production rate does not meet the guaranteed one, any required mitigation can be investigated and included early enough in the design phase of the plant, so as to avoid more costly unplanned modifications.

7. References

[1] Diwakar, Philip, et al., "Plot Layout and Design for Air Recirculation in LNG Plants", *Bechtel Technology Journal*, Volume 2, NO. 1, 2009

[2] Shaari, Siti Farhana, and Kubota, Kei, *Understanding of Hot Air Recirculation Phenomena in Air-Cooled Base Load LNG Plant.* In Proceedings of the LNG17. Houston, TX. 2013

[3] Berkoe, Jon, "Fluid Dynamics Visualization Solves LNG Plant Recirculation Problem", *Oil & Gas Journal*, Volume 97, NO. 13, 1999

442935 Integrating Reliability, Availability & Maintainability into Process Design

Abstract:

Integrating Reliability, Availability & Maintainability (RAM) analysis into process design makes it a powerful tool for equipment selection during the design phase of the project. The traditional method for making equipment choices in the conceptual phase is experience based. The plant availability is generally assumed (usually 80-95%) based on similar installations at the beginning of the design phase. A RAM study is generally performed only to validate the assumptions from the beginning of the project.

This paper presents a unique approach of using simplified RAM models combined with economic analysis of a range of potential market factors to make process engineering choices at the design phase of the project. The RAM analysis results can be straightforward, but become more complex when combined with economic analysis of input parameters that are not discrete values and are driven by market factors. This paper presents a simple case study for an LNG plant with waste heat recovery vs. fired heaters and how RAM analysis can be used to make an engineering choice. Several methods for analyzing RAM are available. In a simulation method, a probability distribution is assigned to the "mean time to failure" and the "mean time to repair" and a Monte Carlo simulator is used to dynamically analyze the behavior of the system. A more simplified analytical method uses reliability flow diagrams and simple equations to analyze the RAM. Both methods will be discussed in the case study. While this paper focuses on a single case study, the concepts can be of value across applications.

444998 Process Optimization for Small Scale LNG Plants

Abstract:

LNG plants have recently been more sensitive to CAPEX due to plunging oil prices. An increasing trend is being noticed in reducing CAPEX at the expense of higher OPEX. Companies are willing to accrue more operating costs to justify lower CAPEX and to kick-start projects. This philosophy requires being acclimatized to lower revenues, but helps smaller operators to reduce debts. Implementation of this approach however requires significant optimization and cost benefit studies. The intent of this paper is to illustrate optimization case studies.

Small Scale LNG projects require integration of several licensed/proprietary and open-art process units, as listed below:

- Liquefaction Unit
- Gas Pre-Treatment Unit, which may include some or all of the following:
    - Dehydration (Water Removal)
    - CO_2 Removal
    - Sulfur Removal
    - Heavy Hydrocarbon Recovery
    - Mercury Removal
    - Removal of other impurities (oxygen, glycol, etc.)

Depending on whether a liquefaction project is greenfield or brownfield, peak-shaver or transportation fuel, etc., various open art units may be required in the facility. Auxiliary units may include some or all of the following:

- Boil Off Gas System
- LNG Storage
- Truck Loading or Marine Loading
- Regasification Facilities
- Feed Gas Compression

Process optimization of these units is critical for the success of a Liquefaction project. A host of supporting case studies will be presented to demonstrate various technically and economically effective solutions.

443635 Gas/Liquid Scrubber Performance Testing at Field-like Conditions

Abstract:

The Separation Technology Research (STAR) Program joint industry project was initiated in June 2014. The mission of the STAR Program is to benefit STAR Program participants by providing accurate, applicable performance data for design, characterization, and qualification of separation equipment using a multi-company, collaborative model. As part of the STAR Program JIP, gas/liquid scrubber performance testing has been conducted on a baseline case (with only inlet vanes) and sets of internals (with inlet vanes, mesh pads, and cyclones) from seven equipment manufacturers. The scope of this presentation will include an overview of the test procedure and results, including the impact of pressure and liquid volume fraction (LVF) on hydrocarbon gas/oil separation in gas/liquid scrubbers.

Based on input from the STAR Program participants, a testing project was defined, developed, and conducted with the focus of:

- developing standardized procedures for performance testing of scrubber internals
- collecting performance data on gas/liquid scrubbers at field-like conditions

The project plan was developed around an existing scrubber at Southwest Research Institute (SwRI) and the capabilities of SwRI's Multiphase Flow Facility. The objective was to conduct testing under field-like pressures and conditions while using hydrocarbon fluids. Variables included in the test program were gas flow rate, liquid flow rate, and pressure. The test gas was methane and the test liquid was Exxsol D110 (model oil). When measured in air at ambient conditions, the nominal Exxsol D110 properties were a density of 0.80 gm/cc, a viscosity of 2.5 cP, and a surface tension of 25.5 dyne/cm.

Standardized procedures were developed for testing gas/liquid scrubber internals. The test section used for the testing, especially the method used for determining scrubber performance, will also be discussed.

The baseline tests have been completed. The baseline tests showed that the separation efficiency was impacted greatly by test pressure. In addition, the results for all but one of the LVFs tested correlated in such a way that they almost fell on top of one another. This was not true for one of the LVFs, where the separation efficiencies were much lower than for the rest of the LVFs, but there was still a strong effect of pressure on scrubber performance at this LVF.

The testing of the seven sets of scrubber internals is currently underway and will be completed in early 2016. The paper will include a discussion of all of the results, which are not available at the time of this abstract submission. To date, we see strong effects of pressure and LVF on the performance of the internals tested.

A standardized procedure for measuring scrubber performance has been developed. In addition, accurate, applicable performance data have been collected that demonstrate the necessity of performance testing at field-like conditions. These data can be utilized by end users to qualify separation equipment for their field applications.

Solutions to challenges in hydrotreating sour fuel gas (Paper #137b)

Henrik Junge Mortensen, Chemicals R&D, Haldor Topsøe A/S, Lyngby, Denmark

Haldor Topsøe in brief

- Established in 1940 by Dr. Haldor Topsøe. Private 100% family owned company.
- Global market leader in heterogeneous catalysis and surface science for more than 75 years.
- More than 2,500 employees in 11 countries across five continents.
- HQ in Lyngby, Denmark. Production in Denmark, USA and China. New factory in Brazil 2016
- Three key business units: Chemicals, Environmental and Refinery. A fourth unit, New Business Unit, focused on developing new technologies
- Revenue DKK 5,685 million (~USD 1b) in 2014. Operating profit DKK 563 million (~USD 100m) in 2014
- We have engineered over 150 hydroprocessing units globally, and more than 500 units use our hydroprocessing catalysts

HALDOR TOPSØE

Fuel Gas Hydrotreatment – what is it?

In short:

Robust, efficient, simple fuel gas purification in a once-through process with a single downstream amine wash

✓ Full-scale commercially proven for over 2 years

3

HALDOR TOPSOE

Refinery Fuel Gas – Plentiful and problematic

- Refinery fuel gas is a mixture of off-gases from a number of processing units (including the Coker unit, the FCC unit, and various hydrotreating units)
- The fuel gas will thus typically contain (besides various hydrocarbons):
  - diolefins (dienes) = high coking potential
  - olefins (alkenes) = coking potential
  - CO_2, CO
  - H_2S
  - **COS and organic sulfur species which are not easily removed in an amine wash**

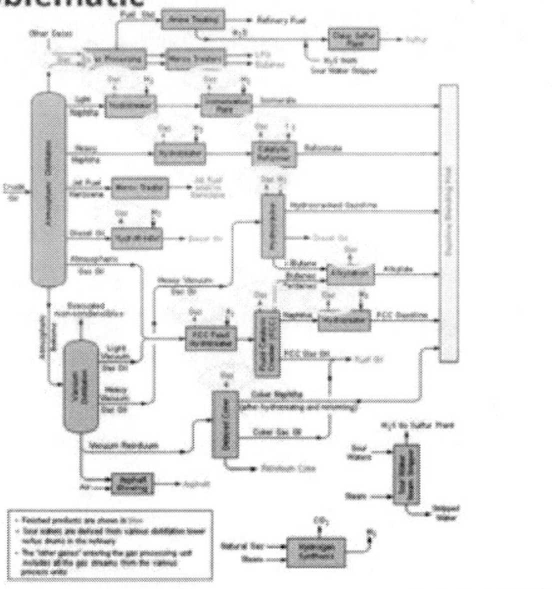

HALDOR TOPSOE

A generic diagram of an oil refinery clearly indicates the variety of refinery processes that generate off-gases. These gases are typically used as a fuel within the refinery, but they often contain compounds such as olefins and diolefins that can cause coking or otherwise deposit in the equipment, as well as sulfur compounds that should not be emitted after combustion.

The environmental driver for Fuel Gas Hydrotreatment

- Increasing restrictions on total sulfur emissions
- Fuel gas is burned in the many fired furnaces throughout the refineries, and the organic sulfur contributes to the SO_x emission from the refinery
- Topsøe's licensed fuel gas hydrotreating technology can reduce the SO_x emissions from the fuel gas by more than 99%

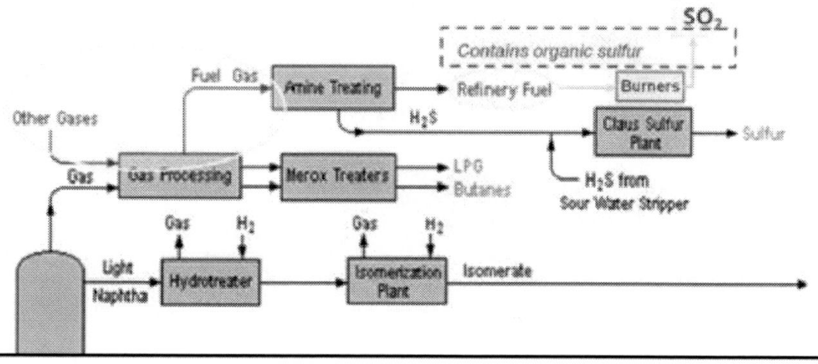

HALDOR TOPSOE

The primary reason to hydrotreat refinery fuel gas is restrictions on sulfur emissions. Whether due to tightened environmental regulations or a desire to increase production within the current environmental approval, a desulfurization beyond the capabilities of a standard amine treatment can be required. A central treatment prior to combustion is more efficient than would be cleaning the flue gas from the various consumers of fuel gas distributed around the plant. A very significant reduction of the sulfur emissions can be achieved using Topsøe's specialized hydrotreating technology.

The value-adding driver for Fuel Gas Hydrotreatment

- US lighter, paraffinic refinery feeds ⇒ more fuel gases, sometimes exceeding demand for burners
- Fuel gas can be purified and utilized as feed for instance in hydrogen generation (steam reformers) to replace more expensive Natural Gas feed
  - For every 1 MMBTU of fuel gas that is used as feed to the hydrogen plant, the refinery saves $2-3 in the US and as much as $6-12 in other parts of the world (LNG)

The second driver is a desire to improve refinery economics. This is possible whenever there is an excess of fuel gas.

Example: Refinery fuel gas

Properties:

- Methane rich
- High sulfur concentration
- High alkene concentration
- Other impurities
- Varying composition and availability

Sample gas composition:

CH_4	50-60 %	CO	0.3-0.7 %
C2	10-20 %	CO_2	0.1-0.2 %
C3	0.5-5 %	H_2	8-10 %
C4	0.1-1 %	N_2	0.2 %
C5	0-0.5 %	H_2O	1.8 %
C6+	0-0.2 %	H_2S	9-14 %
C2=	1.5-3 %	COS	35-50 ppm
C3=	1.5-3 %	Org.S	160-450 ppm
C4=	0.4-1.2 %	HCN	100 ppm
C4==	100 ppm	BTX	800 ppm

Flow rate: 22,000 - 56,000 Nm^3/h
Pressure: 11 - 15 bar g

Objective: (Hydro-)treat the gas to
1. minimize the alkene (olefin) content and
2. reduce the organic sulfur + COS content to < 20 mole-ppm
- with minimal consumption of hydrogen and energy!

7 HALDOR TOPSOE

Fuel gases have certain characteristics in common, although, due to their origin, the compositions of the gases may vary greatly among refineries and with time. Furthermore the flow rate of the gas can be highly variable. In this presentation we discuss pilot test results for a particularly challenging fuel gas composition.

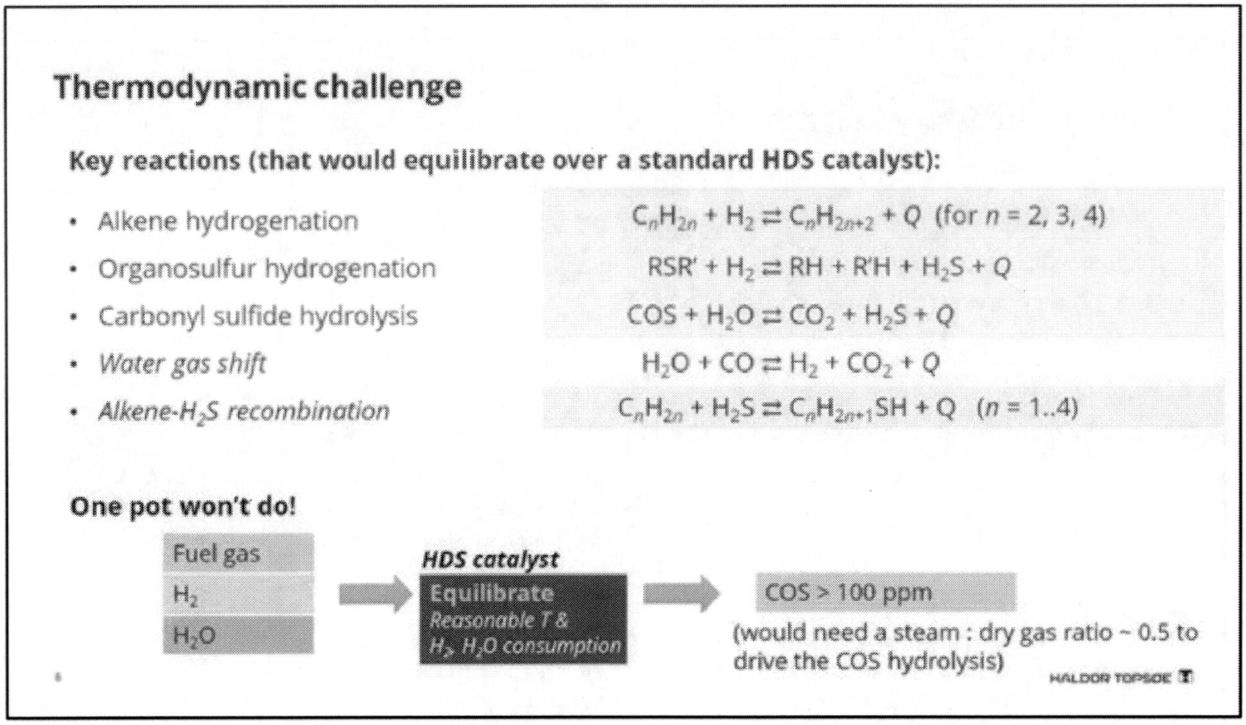

The five key reactions that would usually equilibrate over a standard hydrodesulfurization catalyst are shown here. As written, they are all exothermic. The two hydrogenation reactions and the hydrolysis are the desired reactions, whereas the two side reactions are undesirable in their forward direction. In particular, the water gas shift will tend to increase the slip of COS, and the recombination counteracts the organosulfur hydrogenation. Unless a very large addition of steam to the fuel gas can be accepted, equilibration of all of these reactions at one temperature will not lower the amount of unwanted compounds sufficiently.

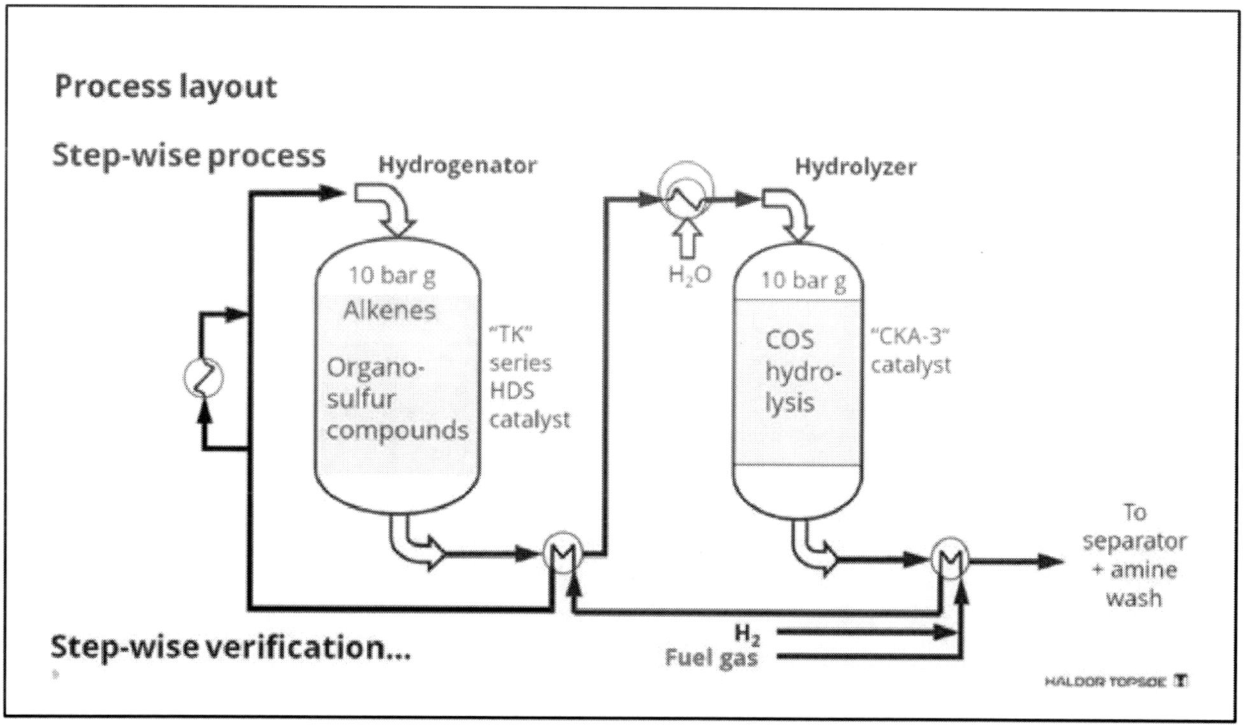

The simplest practical treatment is a three-step process: hydrogenation of alkenes and subsequently organosulfur compounds are carried out over our "TK" type catalyst in the hydrogenator followed by the hydrolysis of carbonyl sulfide by our "CKA-3" catalyst in the hydrolyzer. In the case of a high H_2S content, these must be followed by the usual amine wash. If the fuel gas contains insufficient hydrogen for the hydrogenation, more must be added.

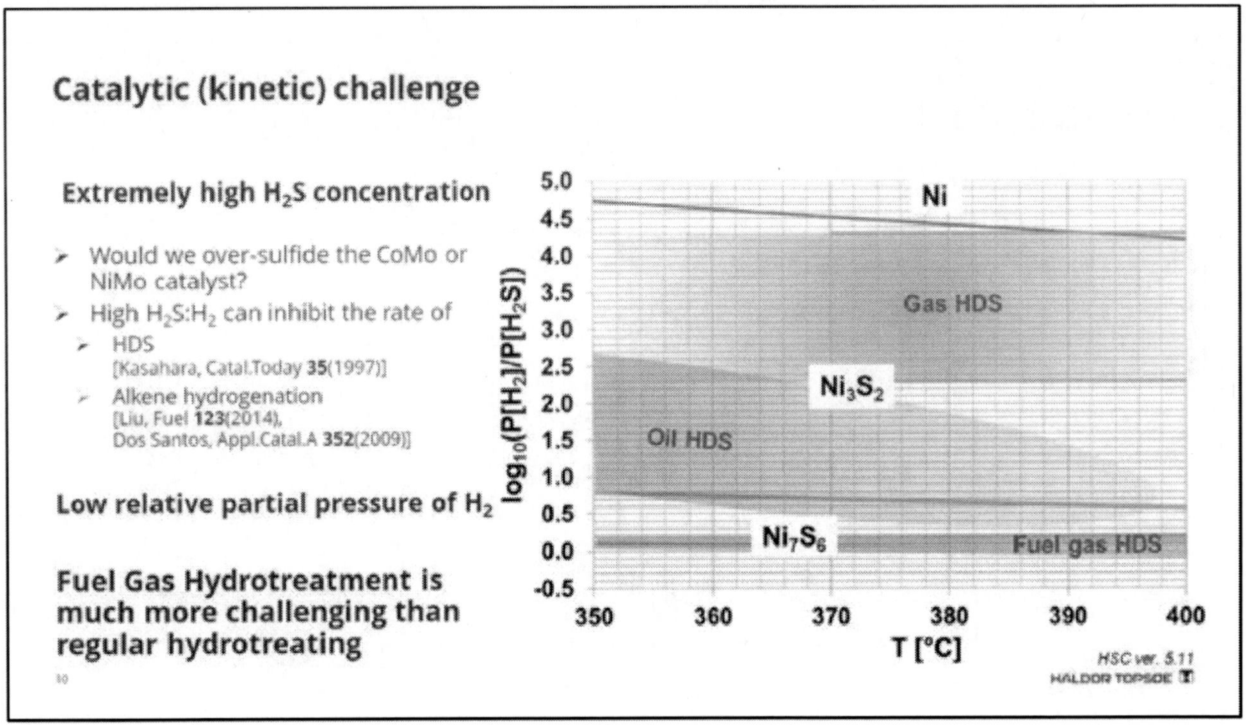

The extremely high ratio of H_2S to H_2 makes fuel gas hydrotreatment particularly challenging. Looking, e.g., at a NiMo HDS catalyst the nickel sulfide phase stable under such conditions differs from the one that is stable during regular hydrotreating of both natural gas and oil. The literature contains examples of an inhibition of both of the desired hydrogenation reactions by a high $H_2S:H_2$ ratio. E.g., Dos Santos found a 73-% reduction of the 1-hexene hydrogenation rate for 0.15 % H_2S/H_2, 20 bar total pressure, 11.5 bar H_2 pressure, 4 % 1-hexene in heptane.
Hence an experimental verification was necessary.

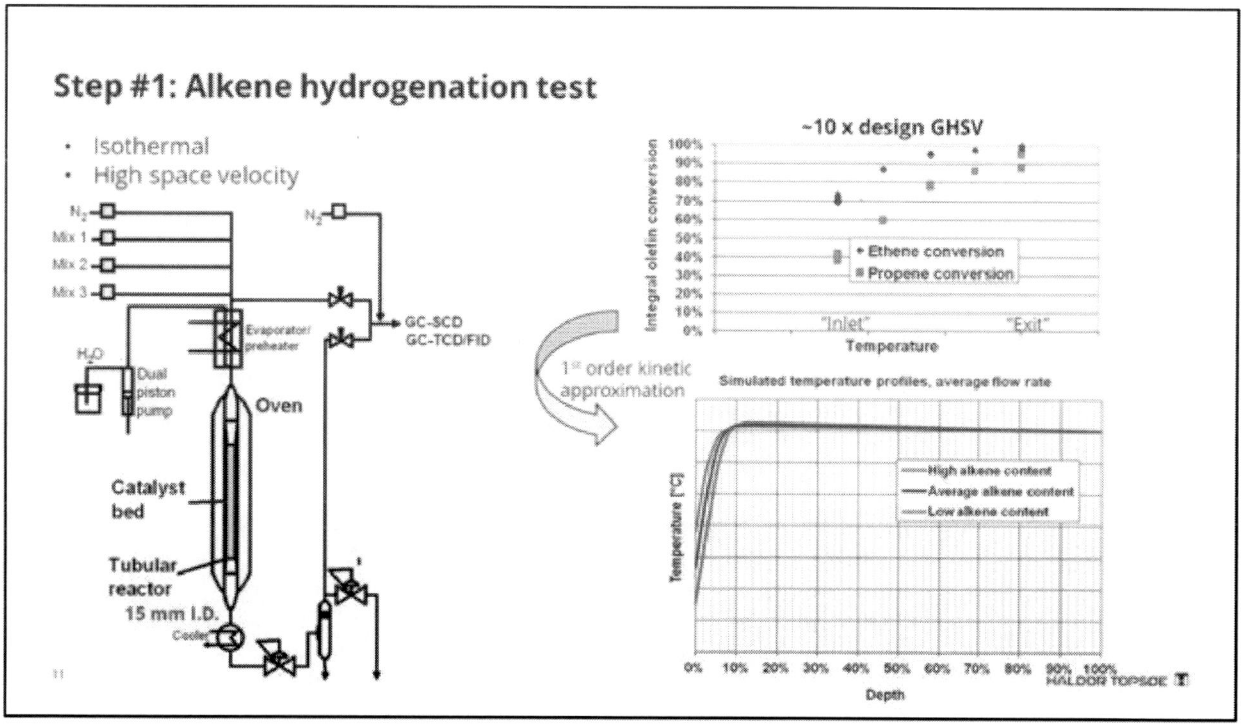

The alkene hydrogenation is usually fast, so it was tested at a high space velocity under isothermal conditions. The resulting partial conversions were used to derive a 1st order kinetic expression which confirmed that the hydrogenation rate over our catalyst would be sufficiently high to equilibrate the reaction within the upper 10 % of the reactor - even with this challenging gas composition.

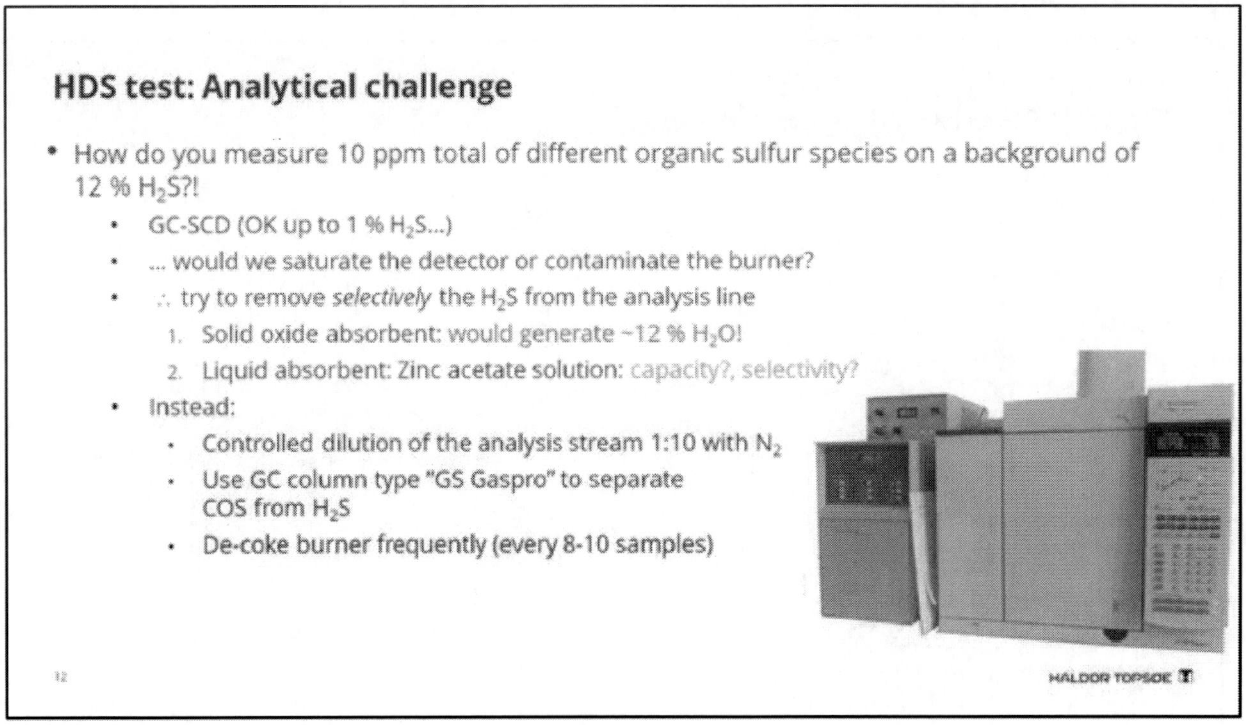

In order to verify the organosulfur hydrogenation it was necessary to measure ~ 1-ppm levels of organic sulfur species in a gas containing 12 % hydrogen sulfide. The method of choice (ASTM standard) for quantitative sulfur speciation at low concentrations is gas chromatography with a sulfur chemiluminescence detector. Measurements in gases with such high H_2S concentration, however, is not standard. After considering ways to remove selectively the H_2S we opted for a controlled dilution of the analysis stream to protect our detector. Even then, only 8-10 injections could be analyzed before the sensitivity decreased notably and de-coking of the SCD burner was necessary.

Step #2: Sulfur hydrogenation test

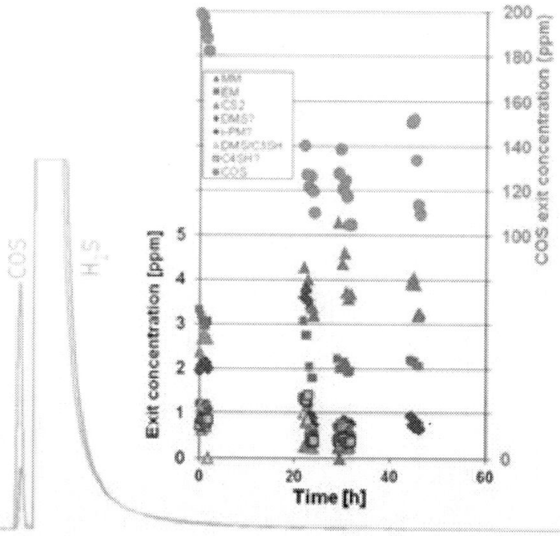

Species	*Feed [ppm]*	*Treated [ppm]*	*Equilibrium [ppm]*
Ethene	12,900	15	11
Propene	14,100	42	41
Butenes	5,600	80	67
COS	0	125	115
Methyl mercaptan	351	4	2
Ethyl mercaptan	0	2	4
Propyl mercaptans	0	1.3	2
Butyl mercaptans	0	0.4	0.4
Dimethyl sulphide	10.5	0.8	0
Tetrahydrothiophene	31.6	0	0
Carbon disulphide	0	0.3	0.4
Other S compounds	0	< 1	
Total non-COS	394	< 10	

HALDOR TOPSOE

A sulfur hydrogenation test with a worst-case gas composition and space velocity was carried out. The exit gas analyses showed a variety of sulfur species all at low-ppm concentrations. By comparison with calculated equilibrium concentrations it is clear that the catalyst is doing a great job.

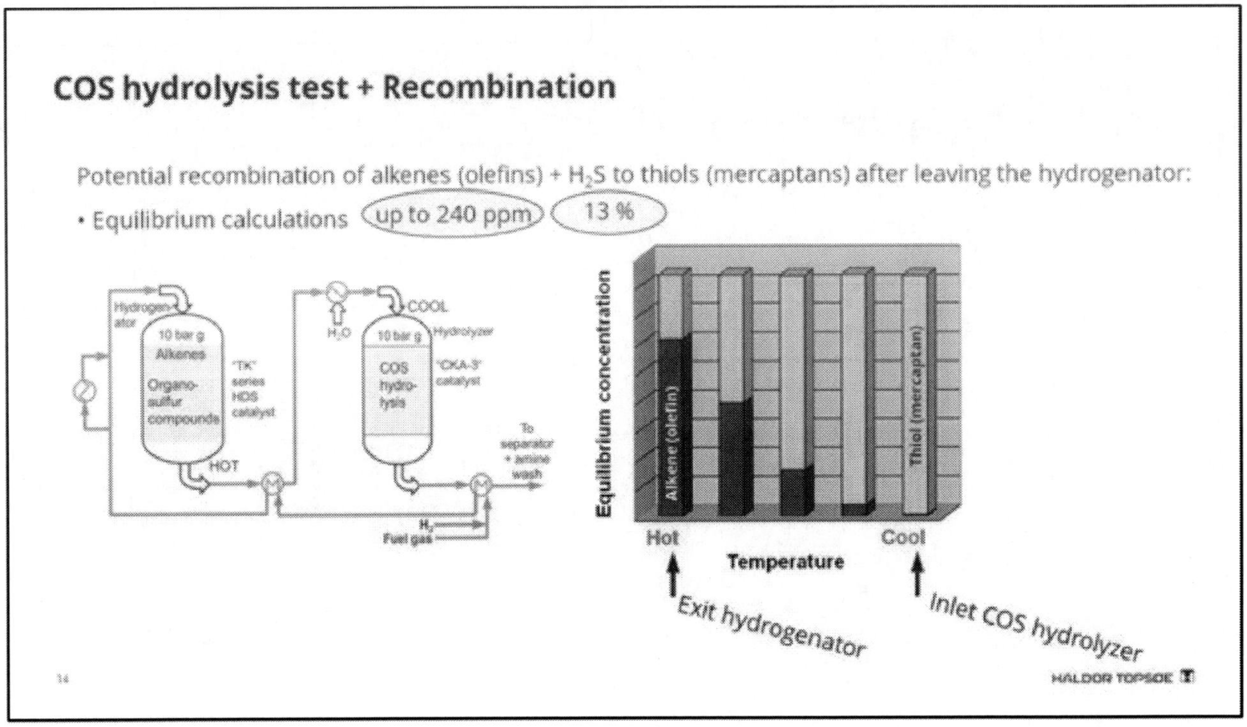

In order to minimize the COS slip, the temperature in the hydrolyzer must be lower than in the hydrogenator. With the equilibrium slip of alkenes and the extremely high concentration of H_2S, the thermodynamic potential for the (exothermic) recombination of these species to form thiols (mercaptans) becomes extreme as the gas is cooled.

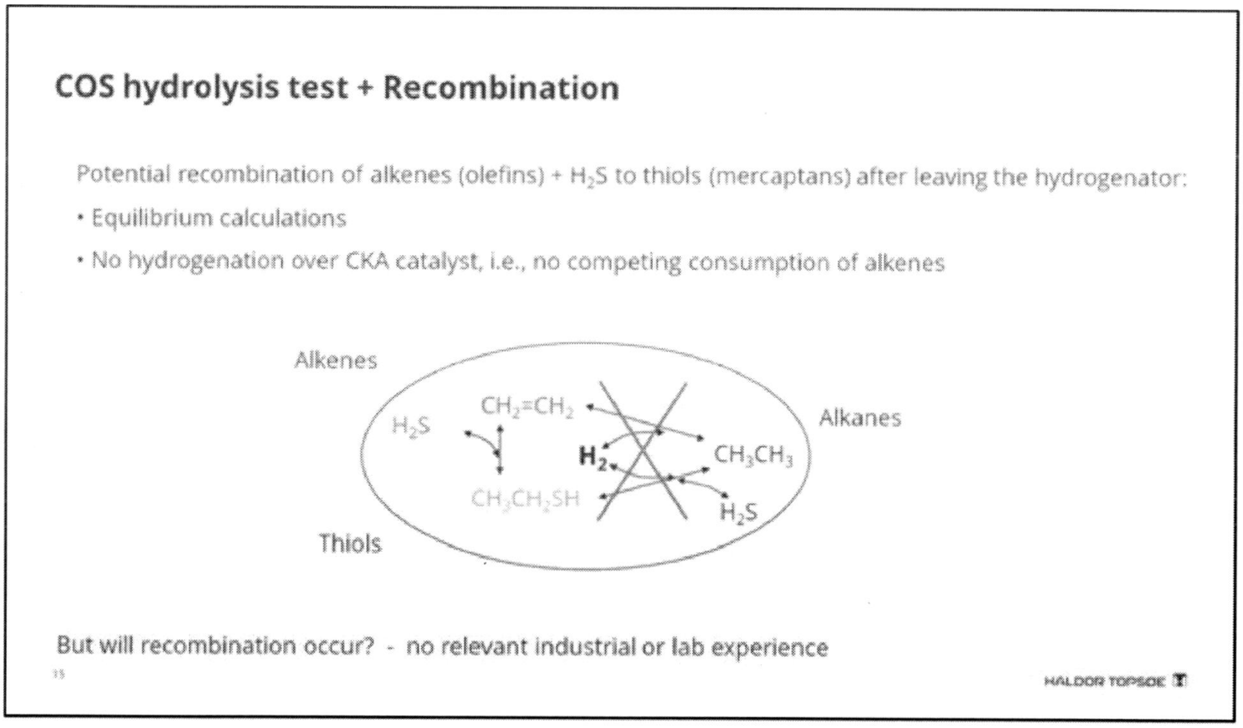

The CKA-3 COS hydrolysis catalyst does not activate the hydrogenation of alkenes or thiols, so although these reactions are exothermic, too, and hence could act as alkene sinks, they do not occur in the hydrolyzer. The outstanding question was whether recombination would.

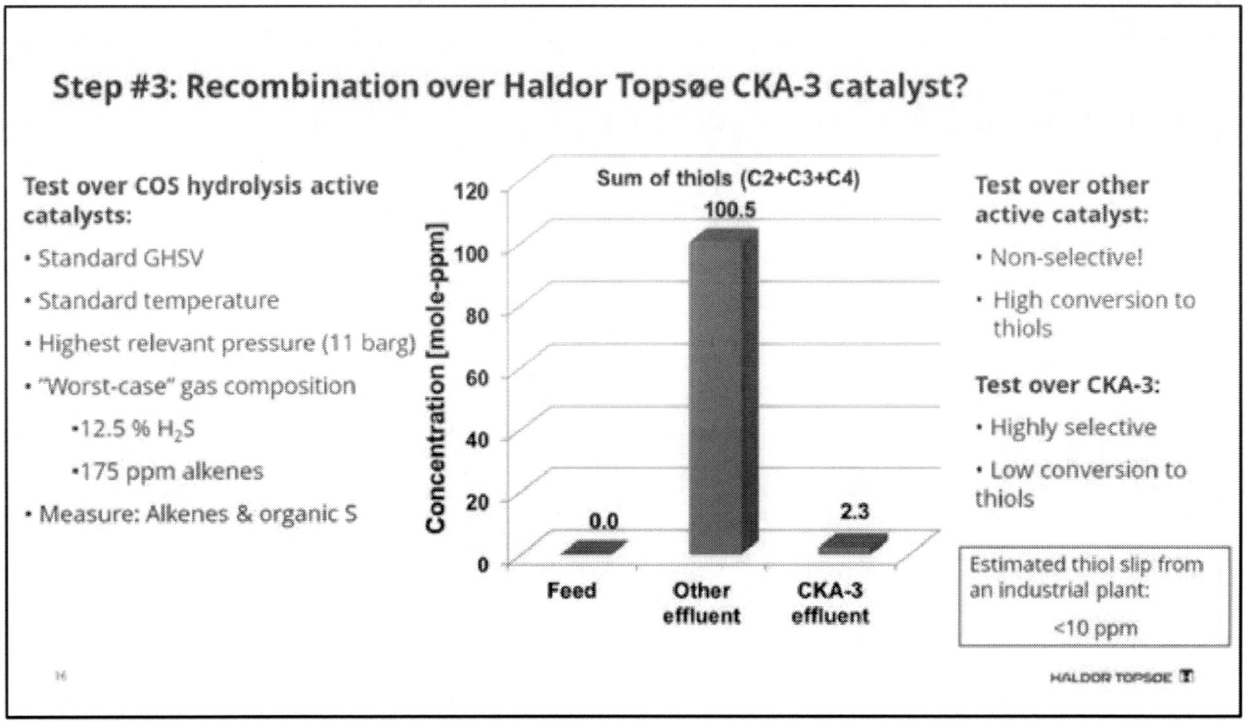

We took the opportunity to test an alternative catalyst known to be able to catalyze the COS hydrolysis. This provided a text-book example of the importance of both activity and selectivity in catalysis. Clearly, the recombination over a non-selective catalyst would completely ruin the work done by the hydrogenator. On the other hand, our selective CKA-3 catalyst displayed a very low recombination rate as required by the process. Therefore, a total slip of organic sulfur species below 10 ppm is achievable using this technology.

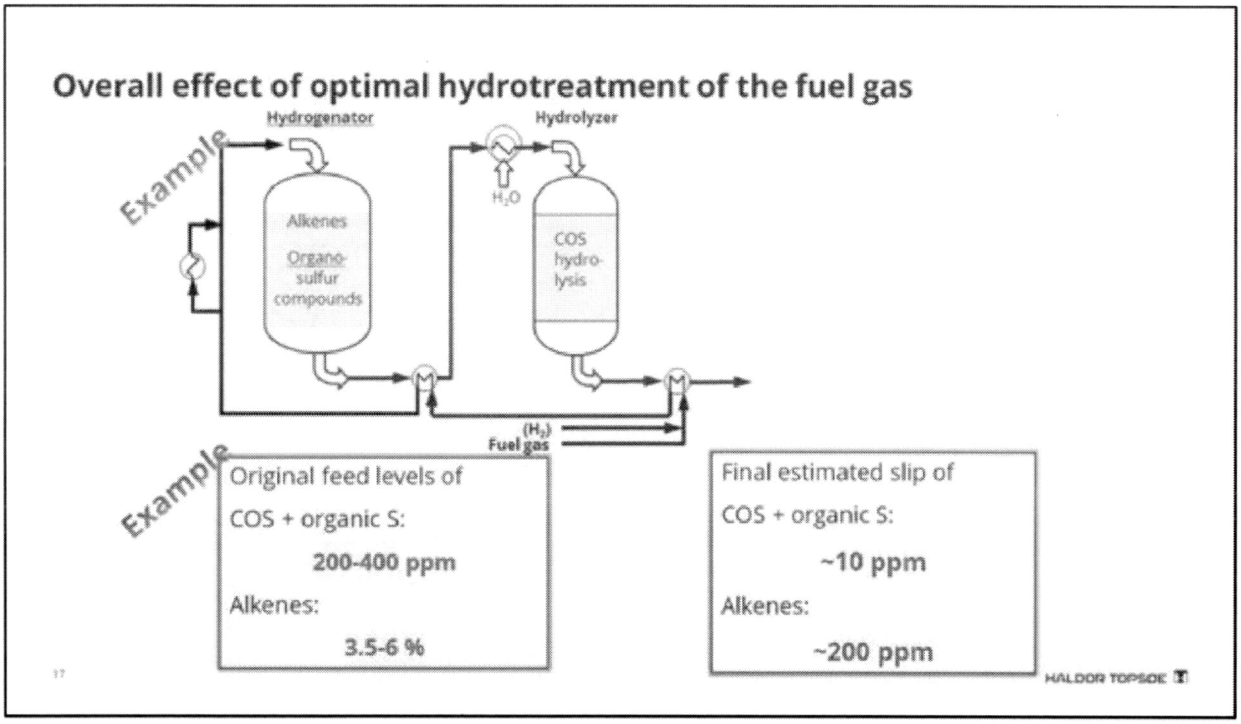

Using our three-step process organosulfur levels in the fuel gas can be reduced from hundreds of ppm to <10 ppm and percentage levels of alkenes can be reduced to a few hundred ppm.

Fuel Gas is converted to stream of alkanes with amine-removable H_2S, a clean feed gas with many uses, incl. hydrogen production

- Indicative compositions of feed and product:

Component	Feed, mol %	Product*, mol %
H_2S	15%	Same+
di-Olefins	5000 ppm	ppm-level
mono-Olefins	10%	ppm-level
COS	500 ppm	<5 ppm
organic Sulfur	5000 ppm	<5 ppm
CH_4, C2+, H_2	Balance	Balance

** Topsoe tailors each technology application to meet the requirements of various fuel gas compositions and purification.*

18

HALDOR TOPSOE

Summing it up, also looking at other gas components, it is evident that this process can turn a dirty fuel gas into a quite clean gas, also suitable for feed with specifications similar to natural gas with a content of hydrogen – the ideal feed to the purification section of a hydrogen plant.

Topsoe's Fuel Gas Hydrotreatment is commercially proven

- Large US refinery operational for >2,5 years
- Two more large US refineries have licensed – under engineering/construction

19

HALDOR TOPSOE

For more information or a quotation, please visit the Haldor Topsøe Fuel Gas Hydrotreater website.
Thank you.

16AIChE
Spring Meeting, Houston, TX

Dynamic Analysis of Flare System for Series of Multiple Failures

Haribabu Chittibabu
Bechtel Oil, Gas & Chemicals, Inc.
3000 Post Oak Blvd., Houston, TX
hchittib@bechtel.com

Subhash Patil
Bechtel Oil, Gas & Chemicals, Inc.
3000 Post Oak Blvd., Houston, TX
sspatil@bechtel.com

Jaleel Valappil
Bechtel Oil, Gas & Chemicals, Inc.
3000 Post Oak Blvd., Houston, TX
jvalappi@bechtel.com

Prepared for Presentation at
American Institute of Chemical Engineers
2016 Spring Meeting
Houston, Texas
April 10-14, 2016

UNPUBLISHED

Dynamic Analysis of Flare System for Series of Multiple Failures

AIChE 2016 Spring Meeting

Topical 6: 16th Topical Conference on Gas Utilization (T6A)
T6A00 LNG Simulation, Operation and Control

Key words
LNG, Dynamic simulation, Depressuring systems, Uninterrupted Power Supply (UPS).

1.0 Abstract

Proper design and verification of depressurization systems are critical in ensuring overall safety of the LNG facility. During an emergency, equipment failure or a planned maintenance event, pressure vessels or process systems are normally depressurized. The primary function of depressuring is to provide equipment with a controlled means of pressure reduction to prevent the possibility of over pressuring and rupturing the equipment or to reduce release rates if a loss of containment occurs.

Depressuring system consists of multiple components such as

- Source of overpressure - Equipment to be protected (vessels and pipe)
- Relieving device - Depressuring valve
- Disposal or flare system - Primarily flare header and flare stack.

Depressuring systems are generally designed for a single major relieving event and to depressurize certain section of the plant with one or minimum number of depressuring valves. The cause of the relieving event can be fire, major leak, power failure or an instrument air failure.

For process interruptions such as power failure and loss of instrument air, the facilities are expected to shutdown, hydrocarbon inventories are isolated and contained, and overpressure does not occur. In other words, the design is such that we will migrate to a "safe state" where depressurization is not required. During an extended loss of utilities the operator may opt to depressurize in a controlled manner.

General practice is not to design the flare system for a series of multiple failures. For a recent LNG project, Bechtel has undertaken an analysis of the flare systems to determine the effect of opening all depressuring valves simultaneously during a plant wide power outrage. The assumptions for a series of multiple failures during power outage are the inability to start the backup generator followed by complete discharge of Uninterrupted Power Supply (UPS) serving the depressuring valves.

Steady state simulation to analyze the flare system for this scenario is not appropriate as the relieving rate is not continuous and the flare system volume and accumulation are not accounted for. In recent times, dynamic simulation has found significant importance in carrying out such analysis of flare system due to several reasons.

- It takes into account initial accumulation and packing due to volume in flare network
- Peak flow out is typically delayed and at reduced rate relative to peak flow in
- More realistic and less conservative results are obtained

This article discusses the dynamic simulation analysis of an existing flare system during an extended general power outage scenario. The software used and the method employed in modeling the flare system are presented. The key results and the other design impacts to enhance the safety are also identified and discussed.

2.0 Introduction

Depressuring is a practice of lowering pressure in a system for safety or maintenance reasons. The primary function of depressuring is to provide equipment with a means of pressure reduction to reduce release rates when a leak occurs or to prevent the possibility of over pressuring and rupturing the equipment.

For example, in the case of fire or runaway exothermic reaction, the design temperature may reach a level at which stress rupture could occur. Under such circumstances the vessel will rupture.

Depressuring is used when pressure reduction is required, as opposed to relief devices which limit the pressure to the accumulated relief device pressure. Figure 1 illustrates the pressure profile inside a vessel as it depressure over time. Depressuring is common in hydrocarbon systems such as LNG, gas plants, refining, and gasification.

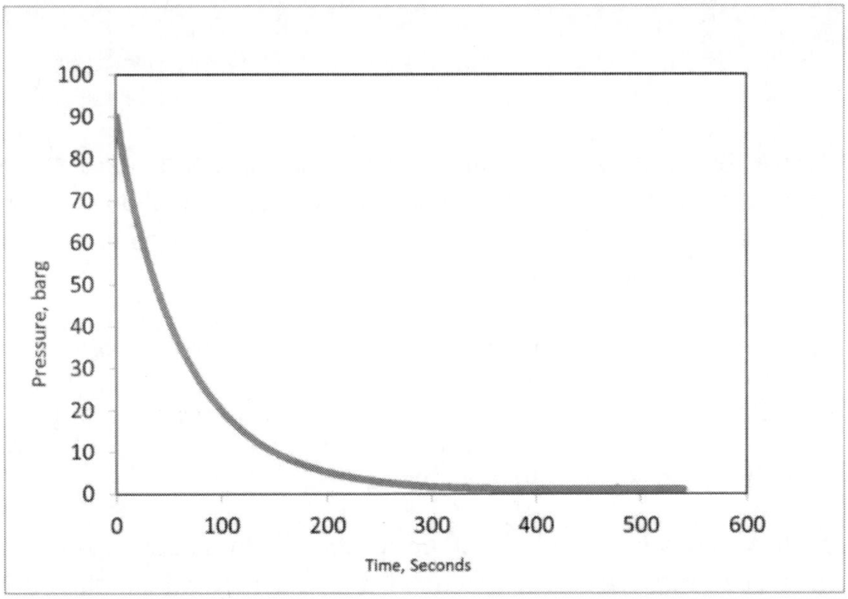

Figure 1: Depressuring – Pressure Profile vs Time.

3.0 Depressuring System Components

A depressuring system consists of multiple components such as

- Source of overpressure - Equipment to be protected (vessels and pipe)

  Isolated System - Blocked in section of a plant to limit the volume that is exposed to leaks and fire thereby making depressuring more effective.

- Relieving device - Depressuring valve (DP valve)

  Depressuring Valve– Manual or automatically operated valve used to vent the pressure from an isolated system to safe disposal

- Disposal or flare system - Primarily flare header and flare stack.

4.0 Series of Multiple Failures

There have been incidents in operating LNG plants where several depressuring valves opened simultaneously. Therefore Bechtel has undertaken an analysis as a part of Quality Risk Audit (QRA) to determine the effect of opening all depressuring valves simultaneously during a plant wide power outrage on an existing installation. The assumptions for a series of multiple failures during power outage are the inability to start the backup generator followed by complete discharge of UPS serving the depressuring valves.

5.0 Advantage of Dynamic Analysis

Steady state simulation to analyze the flare system for this scenario is not appropriate as the relieving rates are not continuous and the flare system volume and accumulation are not accounted for. In recent times, dynamic simulation has found significant importance in carrying out such analysis of flare system due to several reasons.

- It takes into account initial accumulation and packing due to volume in flare network.
- Peak flow out is typically delayed and at reduced rate relative to peak flow in.
- More realistic and less conservative results are obtained.
- Depressuring by incorporating multiple vessels, with flows between vessels simulated by a rigorous pipe segments. This avoids the necessity of combining all volumes and components into a pseudo-single volume which may misrepresent the amount and composition of the fluid.
- Piping network which allows the user to analyze complex single and multiphase systems in flare header networks.
- Dynamic heat, mass and momentum transfer that successfully models the changing pressure drop, temperature, and phases / flow regimes as the fluid flows through the flare network.
- The ability to dynamically calculate multiphase choke flow

6.0 Dynamic Modeling Approach and Input to Simulation

The modeling software utilized is VMGSim developed by Virtual Material Group.

6.1 Isolated system

Isolated systems can be modeled by incorporating multiple vessels, with flows between vessels simulated by a rigorous pipe segments. Figure 2 shows an isolated system of one section of plant consisting of

- Isolation valves (XV11, XV12 & XV13)
- Depressuring volume (Vessel - V1 and Process line – 36")
- DP Valve (HV1)
- Inlet piping to DP valve (16")
- Outlet piping from DP valve (20").

Instead of using a detailed model of all equipment and piping in the isolated system a simplified approach has been followed to lump the volume to a single vessel. The DP valve upstream and downstream piping has been modeled in detail.

The inputs as shown in figure 3 are

- Vessel and Piping volume
- Initial Pressure
- Temperature
- Compositions
- Liquid Levels

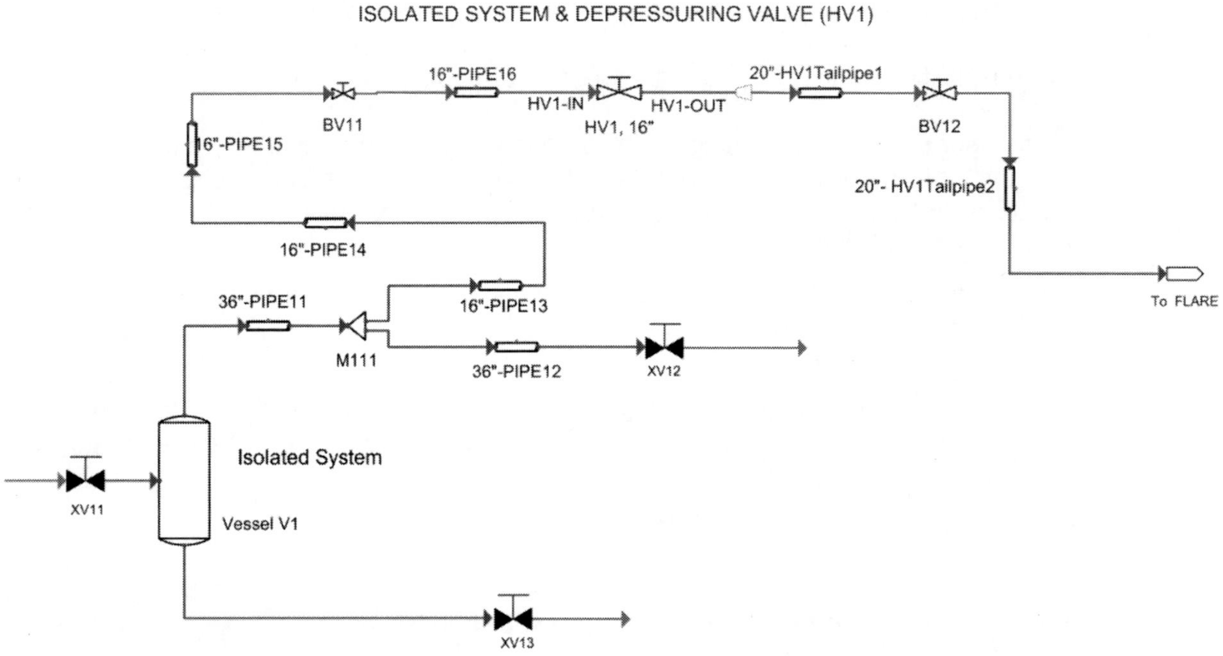

Figure 2: Isolated System V1

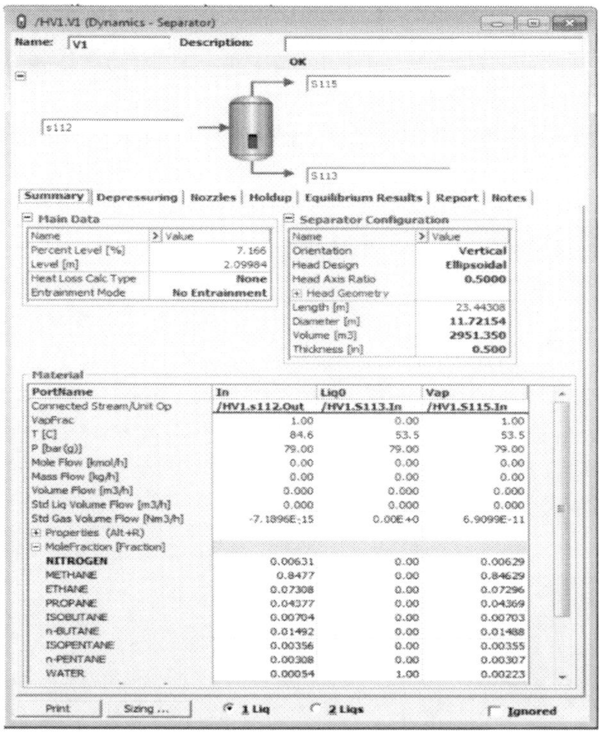

Figure 3: VMGSim – Vessel V1 inputs

Pipe pressure drop incorporating pipe segments as unit operations and rigorously calculating heat transfer and pressure drop. The pipe operation features many correlations for single and two phase flow.

VMGSim input to flare subheader as a pipe is shown in figure 4.

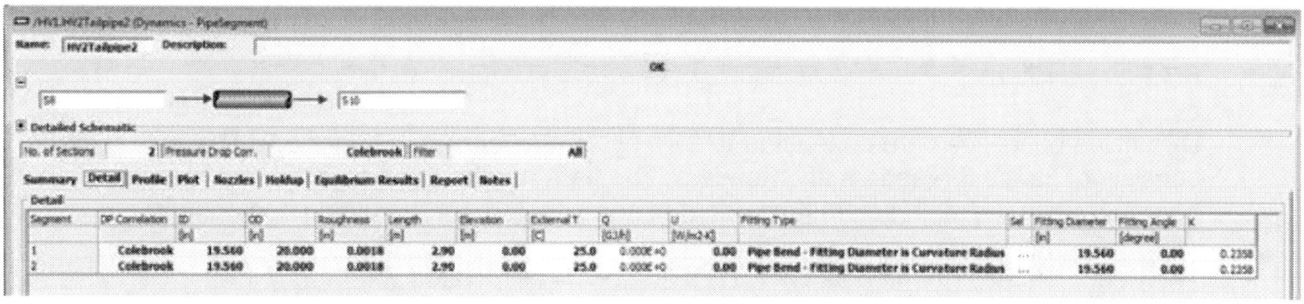

Figure 4: VMGSim – Pipe inputs

6.2 Depressuring valve

The inputs are

- Inlet line to DP Valve – Size, length and fittings.
- DP Valve – Valve Cv, characteristic, stroke time, XT factor, Heat capacity ratio (CPCV Ratio).
- Outlet line from DP Valve – Size, length & fittings

VMGSim input to DP valve is shown below in figure 5.0

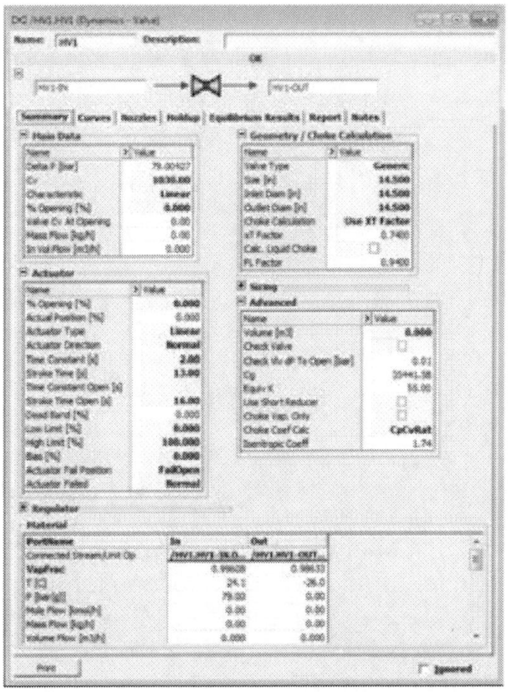

Figure 5.0: VMGSim – DP Valve (HV1) inputs

6.3 Flare system

DP valves outlet lines are connected to flare system as shown in figure 6.0. The flare system consists of flare sub headers and headers. The flare header piping system was modeled rigorously with the piping diameters, lengths and k factors for the piping components.

The dead volumes (marked as DV1, DV2 and DV3) associated with non-relieving DP valves and PSV outlet lines in the flare system have been lumped and modeled accordingly.

The inputs are

- Flare header piping Layouts
- Sub header / header sizes
- Dead Volume for non-relieving DP valves and PSV's.

- KO Drum (Knock Out Drum) dimensions
- Flare tip resistance

The flare tip at the flare stack is modeled as a valve, which is sized to maintain a back pressure provided by vendor at the maximum relief load.

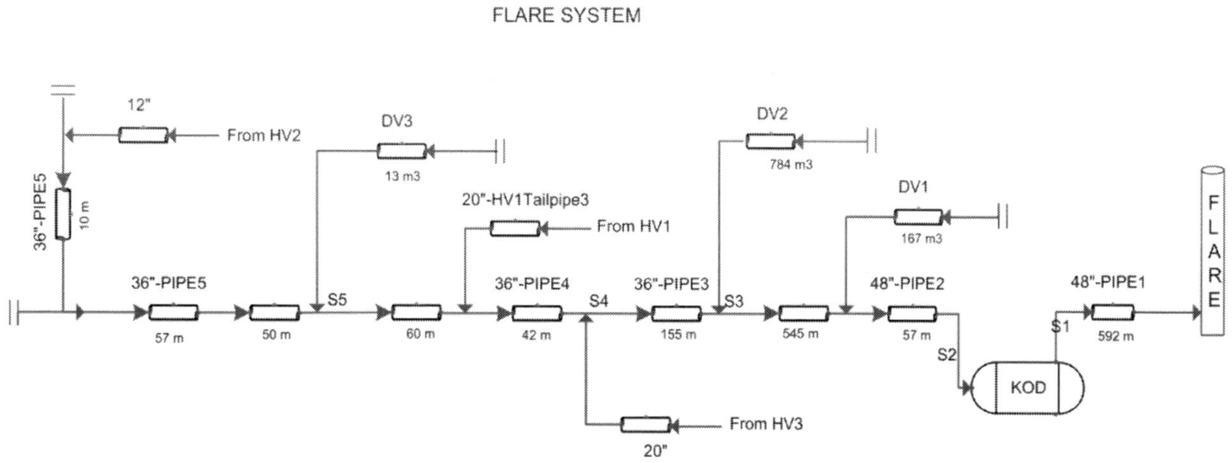

Figure 6: Section of Flare system

7.0 Results

Series of multiple failures are

- Plant wide power outage
- Inability to start the diesel generators
- Complete discharge of UPS

These consecutive failures will result in all of the DP valves opening simultaneously.

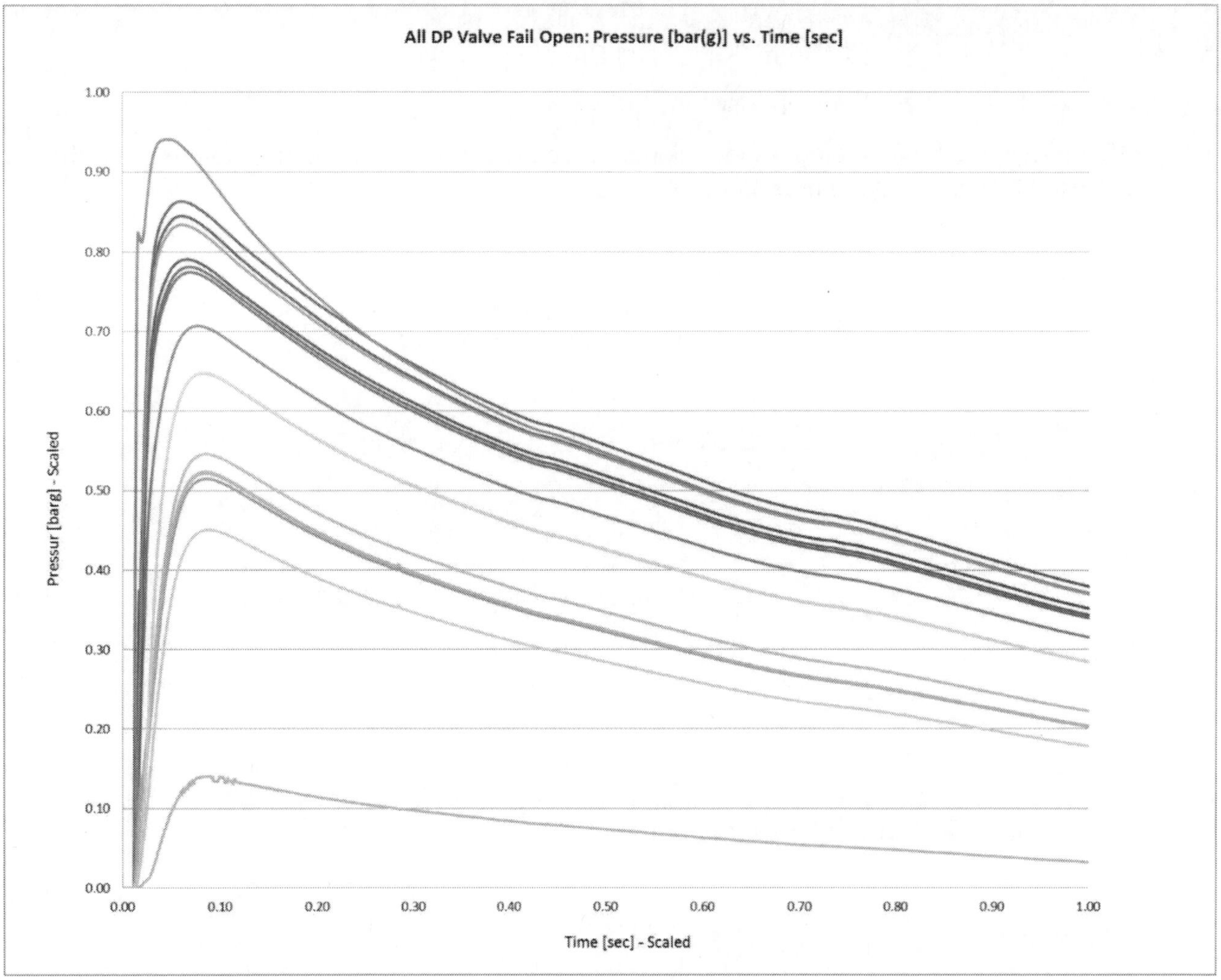

Figure 7: All depressuring valve Fail open – Flare system pressure vs Time

Figure 7 shows the increase in back pressure in the flare system when all the depressuring valves open.

- The maximum pressure in the flare system is less than 0.97 barg (scaled)
- The generated backpressure in KO Drum is 0.46 barg (scaled).

The flare header and sub-header system are 150# piping and tested pneumatically to 1.23 barg (scaled).

In this scenario, the simultaneous opening of all the DP valves will result in overpressure of KO Drum above the Maximum Allowable Working Pressure (MAWP) of the equipment, but the overpressure is below the KO Drum shop test pressures as shown in figure 8.

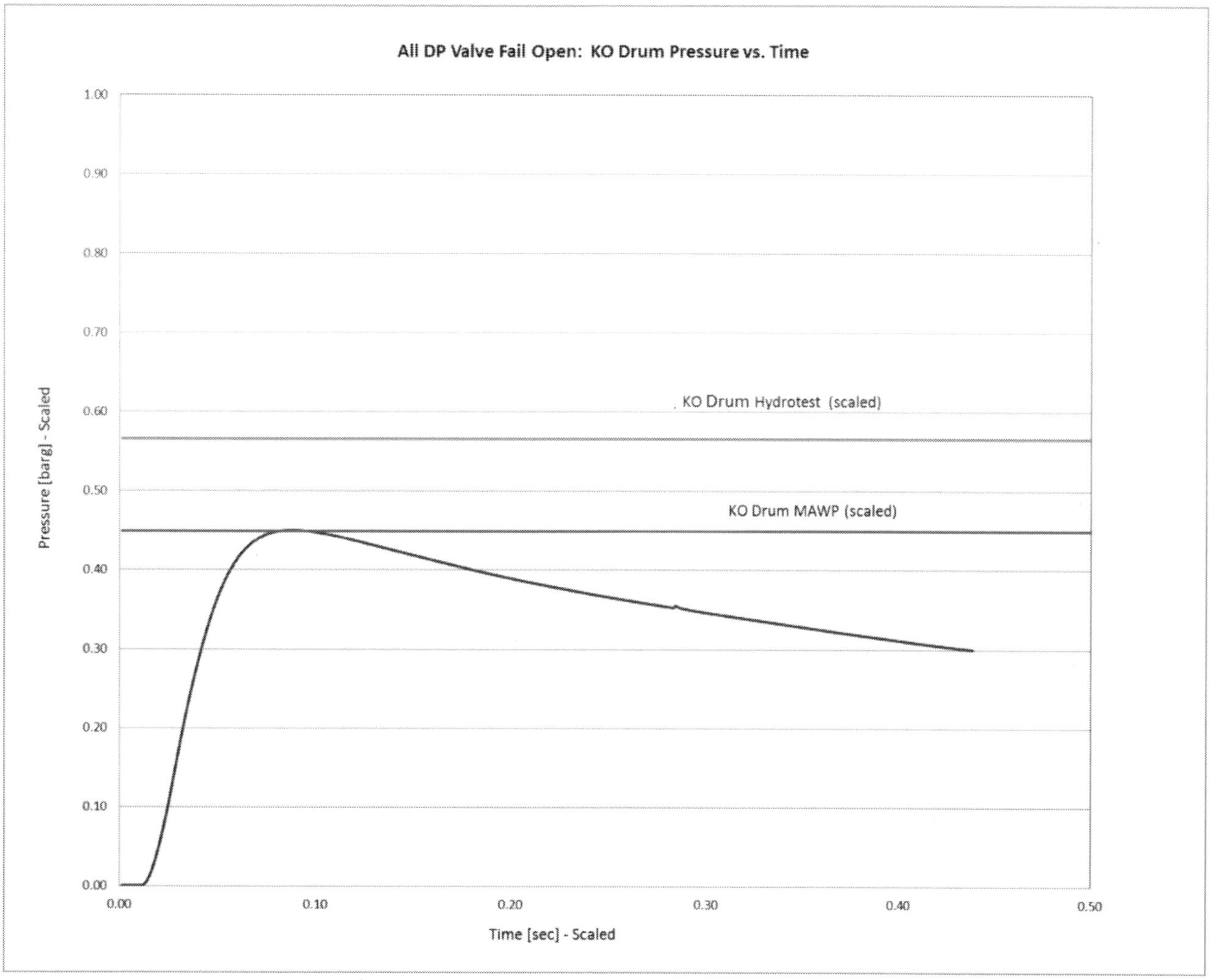

Figure 8: All depressuring valve Fail open – KO Drum pressure vs Time

The increase in pressure in the flare system has other potential impacts as mentioned below:

- **Overpressure of PSV balanced bellows**

The bellows on some of the relief valve services were identified as being vulnerable to damage due to back pressure that exceeds the design pressure of the bellows.

This event will shorten the life of the bellows and has the potential of rupturing the bellows over time resulting in a small leak to the atmosphere. If the bellows are damaged, the balance bellows relief valve will behave like a conventional relief valve. In order for these valves to not sustain potential damage during such an event, the bellows have to be replaced.

- **Overpressure of the LP Fuel Gas system**

If the check valves at the low pressure (LP) fuel gas connections to the flare leak, the built-up back pressure may exceed the design pressure of the LP fuel gas system. Check valve leakage rate calculations were made. However, the volume of the LP fuel gas system and distribution header is very large and the fuel gas system will not be overpressured by the reverse flow from the flare header during this event.

- **Reverse flow thru EDP valves**

During this event there was reverse flow through few DP valves as shown in figure 9. However the built-up back pressure is below the design pressure of the equipment protected by these DP valves. The equipment will not be overpressured but there is a potential for contamination. Over time, once normal operation is resumed the non-condensables will be vented from the system.

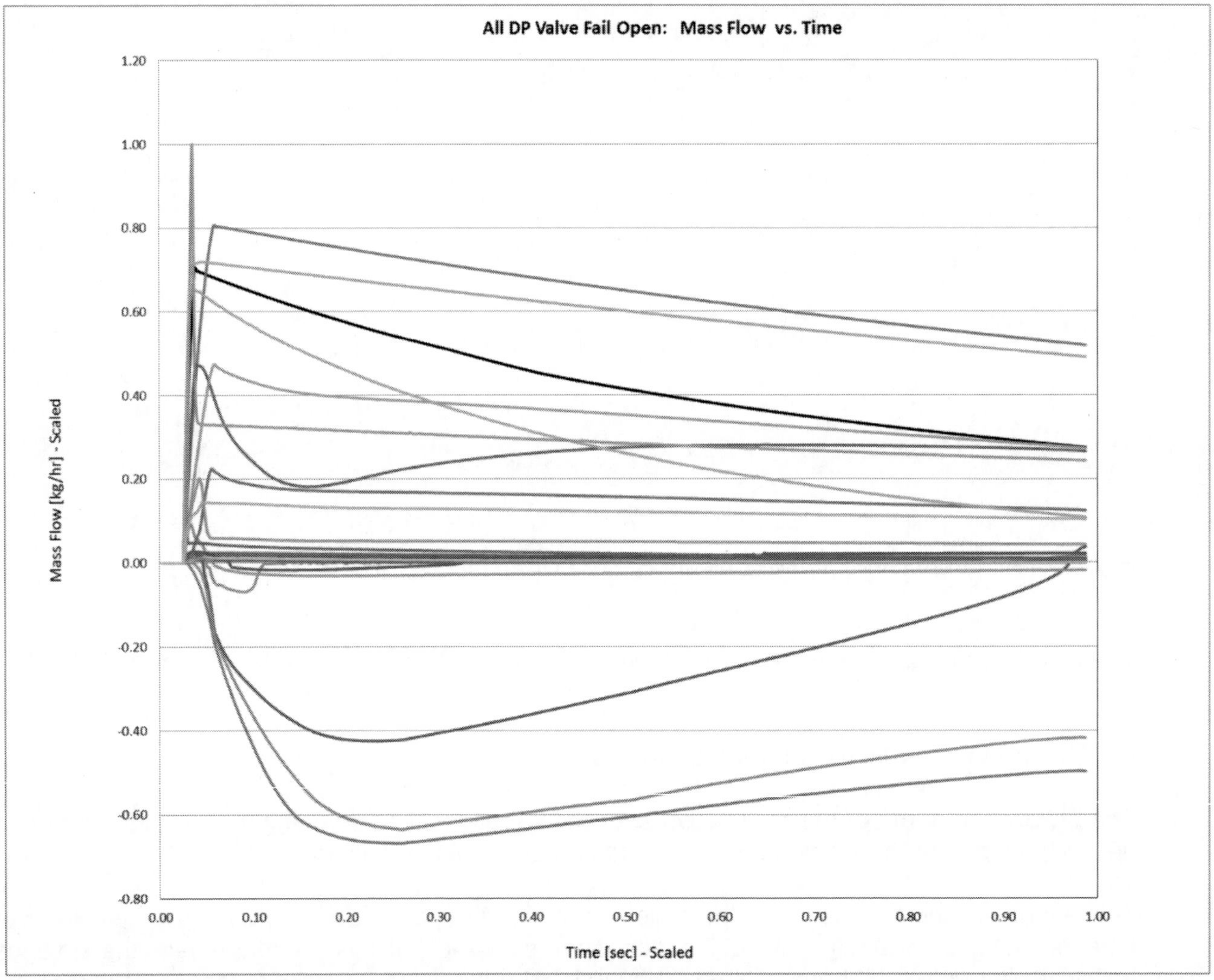

Figure 9: All depressuring valve Fail open – Mass Flow vs Time

- **High Mach numbers**

A number of lines were identified as exceeding a Mach number of 0.90. The duration of the choking in these lines was between 1 and 10 seconds. With such a short choking duration there is no risk for the development of acoustically induced vibration (AIV) due to the choking.

8.0 Conclusion

In recent times, dynamic simulation has found significant importance in carrying out analysis of flare system due to its advantage over steady state simulation.

The dynamic modeling approach combines both modeling the isolated system to be depressurised and the flare headers. It helps us to evaluate the flares system design for these types of multiple failures. It also helps us to determine the available flare capacity for blowdown of future tie-ins and clarifies the need of sequential blowdown on the existing system. For a new flare system dynamic simulation helps in reduction in header size resulting in significant capital savings.

440895 Using Dynamic Simulation to Improve a Prico LNG Liquefaction Plant Operation

Abstract:

The CHALLENGE

Meeting product specifications while optimizing power and fuel consumption can be a significant challenge as a plant operates at points away from the design conditions. Understanding how variables like feed flow, composition, weather, mechanical limitations, and others affect each other and the total system performance can be complex. Steady state process modeling struggles to provide meaningful insight as to what site operations should do or what is currently happening for a given set of conditions. Dynamic simulation is a more useful tool for optimizing plant operations and training plant operators. The benefits of dynamic simulation are explored through the lens of a PRICO LNG liquefaction plant. The system consists of a main cryogenic heat exchanger (MCHE), three (3) drums, one (1) compressor with two sections, two (2) air coolers and two (2) pumps each with operating variables that can be manipulated.

The SOLUTION

Dynamic simulation first validates the control strategy of the PRICO LNG plant. There are two major control loops for the mixed refrigerant: one is the mixed refrigerant flow and the other one is the compressor suction pressure. It is very intuitive to use the VFD on the compressor to control the mixed refrigerant flow and ultimately the LNG temperature. From a steady state simulation perspective, the mixed refrigerant control valve will be used to control the compressor suction pressure. However, dynamic simulation proves this is a bad strategy. From time to time, due to the wrong set point of the suction pressure, the compressor is in the risk of liquid at the suction. The correct way is to use the control valve to control the suction superheat, instead of the pressure. With this new strategy, the control is stable and there is very little chance the compressor is flooded with liquid.

Then dynamic simulation is used to optimize the plant operation. Since LNG temperature is specified, the compressor speed will be adjusted to accommodate the requirement. However, the superheat set point is not directly related to any hard product specification, thus it can be optimized to reduce the total power consumption of the compressor. Thanks to the dynamic simulation, users can optimize the virtual plant just like operators sitting in front of the HMI of the real plant. The full optimization results will be presented during the conference.

443482 Transient Natural Gas Liquefaction Process Comparison- Dynamic Heat Exchanger Under Transient Changes in Flow for Energy Storage with Cryogenic Carbon Capture (CCC-ES)

Abstract:

Transient modeling and optimization of two natural gas liquefaction processes have been done using Aspen HYSYS. In case of energy consumption, the optimized model has been chosen for transient modeling. These results pertain to Liquified Natural Gas (LNG) generally and to an energy storage process associated with cryogenic carbon capture (CCC) in which the LNG process plays a prominent role. The energy storage CCC process influences the time constants and magnitudes of the flow rate characteristics. These flowrate variations affect all units, especially compressors and heat exchangers. New types of heat exchanger design and optimization method; Model predicted control theory (MPCT) have been presented and transient responses have been compared with traditional design. The proposed process controls temperatures, pressures and other operating parameters. K-value- and U-value-techniques guide flowrate and heat exchanger stream variations. Transient responses to both ramping and step-changes in flow rates indicate process responses, including summary effects represented in transient efficiency graphs.Finally natural gas (NG) consumption and LNG production have been compared to show the optimality of new design.

444662 Simplifying Hydrogen Production

Abstract:

Steam reforming based hydrogen plants involve complex and energy intensive, multi-stage operations to convert hydrocarbons into high purity hydrogen. In a complex refinery, these plants are relied on to make hydrogen for hydroprocessing units in which the hydrogen is becoming of greater value to the overall operation. The barriers to achieve this reliability are increasing for the operator. For feed purification, high level sulfur specs on natural gas pipelines, varying sulfur speciation and levels in natural gas sources, and use of high-hydrogen refinery off-gases with heavier tails can result in condition changes, spikes, and transients in operation that challenge this reliability. For the steam methane reformer (SMR), heat transfer limitations across the reformer tubes, carbon formation and poisoning, rate changes in response to hydrogen demand, and increase in pressure drop in response to shutdowns and operation also challenge this reliability.

New purification and reforming technologies that deal with and remove these barriers for the hydrogen plant operator will be presented; these technologies can be implemented in existing plants with little to no capital expenditure. Experiences, support work and case studies will demonstrate how these barriers to reliability are being removed. For new plant designs, these technologies can miminize cost by removing vessels or decreasing the size of the plant, providing process simplification and intensification to hydrogen production as ever increasing demand for hydrogen and pressure on costs in the refinery make these technologies an attractive option.

445327 Hy-(drogen) Hopes in Clean Energy Landscape

Abstract:

Manifesting the challenge of future energy demand against global strife and mandates for curbing GHG induced climate change; hydrogen is poised to be a strong contender in the future energy spectrum. Strategic scenarios and roadmaps have been stipulated for addressing the future energy needs against the backdrop of strong environmental focus and clean energy targets. Its appreciable leverage and share is put on the power and transport sectors with gas, syngas and eventually hydrogen pathways rendering a progressive impetus in the longer run.

Hydrogen's role as a unique carrier of ultra clean energy is imminent despite any barriers or roadblocks put by skeptics debating its claimed viability. Based on its unparallel versatility and flexibility in culminating the different contemplated energy sources, whether fossil, renewable or nuclear into 'one clean energy platform', the hydrogen landscape is destined to be in the offing.

Deliberations over fossil-derived syngas with carcon capture and storage or sequestration (against current IGCC and NGCC) and gradual deployment of hydrogen from fuel substitution to stationary and mobile fuel cells is expected to take off exponentially in the coming decades. The availability of abundant low-cost (unconventional) gas is presently boosting as a cost-effective low-carbon feedstock for generating large scale syngas and hydrogen while well proven state-of-the-art technology.

With renewable energy as well as nuclear power expected to gain more grounds in future defying current setbacks, concerted roadmaps and future trajectories are being forecasted on syngas and hydrogen as dominant clean energy routes in the longer term. The clean energy drivers will be well-to-wheel efficiency in conjunction with mandatory reduction in the CO2 footprint expanding to use of renewable power to generate "green hydrogen". The economies of scale may also shift from large centralized power facilities to decentralized and distributed power generation, even digressing to household level.

The talk will provide an overview on the imminent role of hydrogen in the realms of future clean energy landscape.

444264 Modelling and Integration of Diesel Hydrotreater Kinetics with Refinery Hydrogen Networks

Abstract:

The current global trends to refine heavier crude oils and the stricter environmental regulations have enhanced the importance of hydroprocessing. Hydrogen, the essential feed of hydroprocesses, has a direct impact on the process performance and the final economic profits. Therefore, kinetic modelling of hydroprocessing units would provide detailed insights into the process as well as would enable opportunities to perform process optimisation and better hydrogen network management.

Based on an industrial configuration, a diesel hydrotreater model has been developed using industrial operating data. Bulk properties of feedstock measured through laboratory analysis are transformed into 18 molecular lumps based on the database that contains both bulk properties and molecular information of pre-collected samples. A kinetic model with molecular lumps is used for the reaction considering the operating conditions, including LHSV, makeup H2, recycle H2, reactor inlet temperature, H2 quench, and catalyst life. The model also predicts the light hydrocarbon generation that is critical for accurately modeling H2 recycle composition (on process level) and H2 network optimisation (on site level). The difference in properties between product and feed, such as distillation profile, cetane number and flash point, is predicted from the change in molecular lumps. A multi-period process optimisation model has also been developed. The proposed methodology changes the optimisation objective from the traditional methodology of reducing operating cost to improving the overall economic performance.

The developed kinetic hydroprocessing model is then applied to overall H2 network optimisation. This integration enables accurate evaluation of several changes to the diesel hydrotreating unit, such as: increased throughput to make full use of hydrogen surplus, increased proportion of cheaper feedstocks, variable H2 purity in the makeup for maximum H2 cascade utilization, etc. A case study is carried out to show the benefits of the developed methodology.

155a

Full Range Natural Gas Processing

Heinz C. Bauer
Andreas Bub (speaker)

Linde Engineering, Pullach, Germany

Abstract

Large and untapped natural gas reservoirs offer the opportunity to design a versatile and efficient gas processing complex. An optimized concept for the integrated production of liquid helium, liquid nitrogen, LNG (optional), sales gas with on-spec nitrogen content, ethane, and C_{3plus} will be discussed for a plant with about 49 BCM (5.1 BCF) feed gas flow rate.

Introduction

Virtually every inspired chemical engineer dreams about an opportunity to design a complex green field natural gas processing plant from bottom up. This paper will talk about such a rare situation and the findings evolving from this process synthesis.

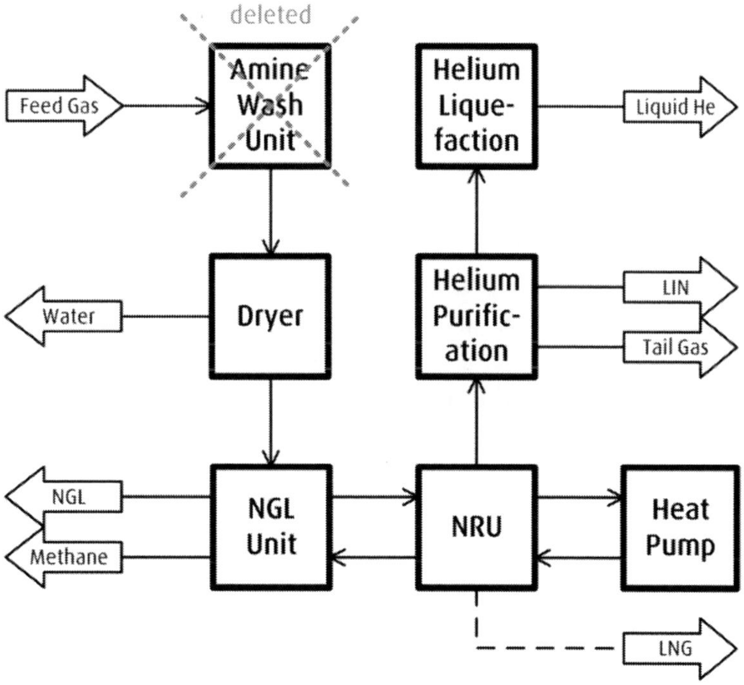

Figure 1 Block Diagram of an integrated Gas Processing Plant (iGPP)

Process arrangement

The overall task is full processing and separation of a large natural gas stream, which arrives at the plant site via a long distance pipeline in dew point controlled conditions. The main products include

- NGL separated into ethane and additional C_{3plus} fractions
- Sales gas with controlled nitrogen content below 2 mol%
- Liquid helium grade 5.5 with 99.9995 mol% purity

Other products comprise

- Liquid nitrogen as utility for the helium liquefier
- Nitrogen rich tail gas (partly used as utility, balance to atmosphere)
- LNG (optional)

The plant concept had to be flexible with respect to feed gas composition, easy to start-up and operate and last but not least energy and cost efficient.

Overall pressure profile

Compression systems are expensive in investment and operation. Thus, their use shall be minimized in a clever overall concept. Ideally, the process sequence should be operable with the net available pressure drop over the entire plant. If, however, the BL pressure of the sales gas product is higher than the highest possible operating pressure of any main process step, at least sales gas compression will be required. Now, the target was to identify an efficient arrangement, with which compression of the main gas streams except for sales gas can be avoided.

In addition to the path 'feed gas → sales gas' there is another main gas route: 'feed gas → nitrogen/helium fraction'. A compression system in this second line can be circumvented only, if nitrogen rejection is done at a pressure level, which is high enough for the subsequent helium clean-up and liquefaction. In a nutshell, the use of a so-called double column for nitrogen rejection would not be compatible with the desire to minimize compression steps. Eventually, the following overall concept was selected:

- NGL extraction and methane stripping relies on feed gas work expansion and sales gas compression, which is the only main stream compression step
- Nitrogen rejection (NRU) is operated slighted below the demethanizer pressure and is supported by a heat pump system
- Nitrogen/helium separation is operated slighted below the NRU pressure and does not require compression except for a small recycle gas compressor

- Helium liquefaction is achieved with an open loop helium cycle including work expansion, accordingly, there is no single compressor in the main path between feed gas inlet and liquid helium product

The details of the process arrangement will be described in the following section. For the sake of better understanding heat exchange systems, rectification column arrangements and compression trains are represented in a simplified manner.

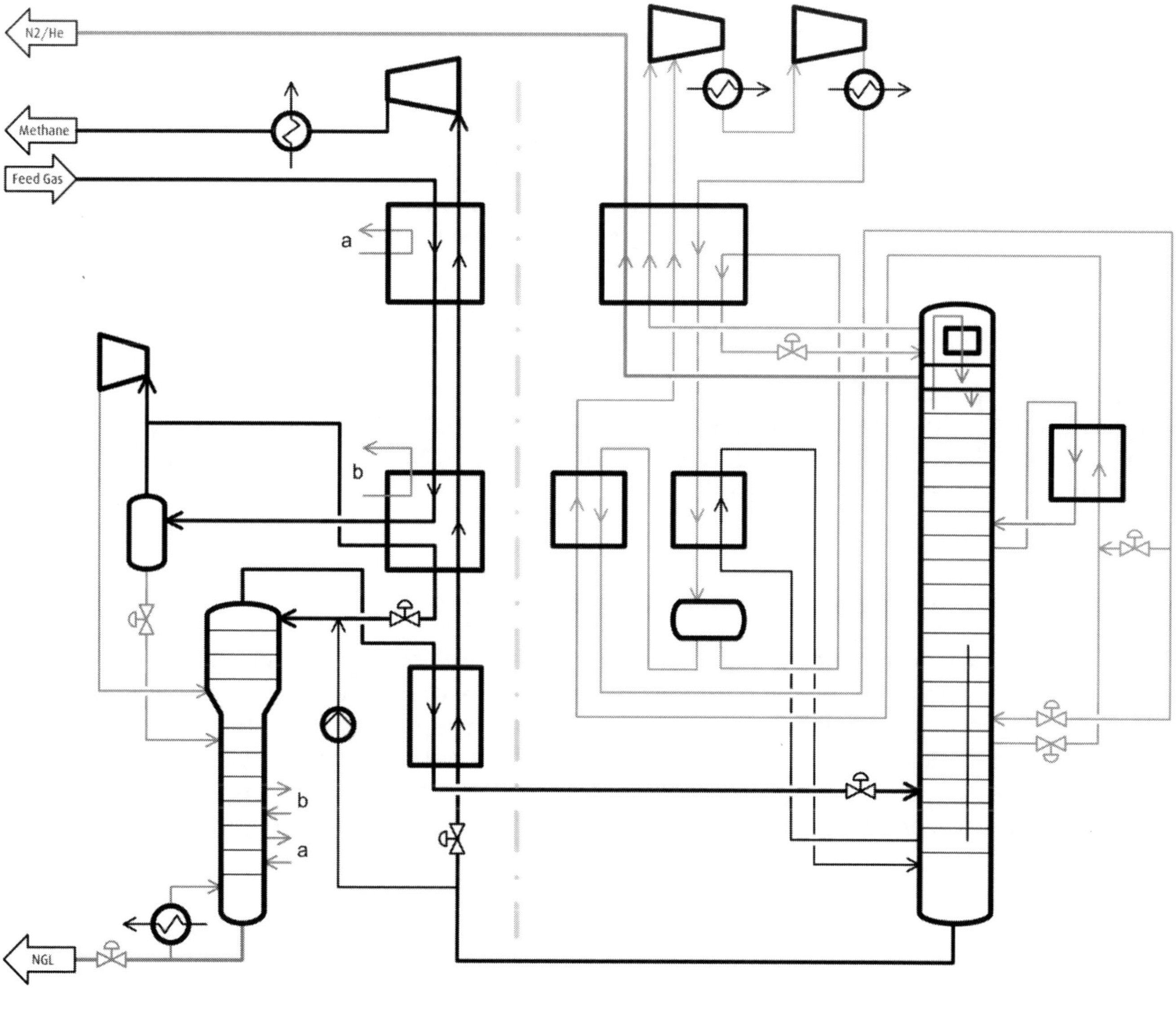

Figure 2 NGL and NRU section of the integrated gas processing plant

NGL recovery unit (left hand side in Figure 2)

NGL extraction and methane stripping is based on a proven GSP[1] concept with two side reboilers (streams a and b in Figure 2). High ethane recovery can be achieved using part of the NRU column bottoms as additional reflux. This feature compares in efficiency to a RSV[2] design, but avoids additional load to the sales gas compressor. The operating pressure of the demethanizer has to be kept at 30 bar (450 psia) or higher to establish the correct operating conditions in the subsequent NRU column.

Nitrogen rejection unit / NRU (right hand side in Figure 2)

As explained above a double column design is not desirable, as the nitrogen/helium fraction would require compression upstream of the helium purification and liquefaction step. Thus, a well proven and easy-to-operate single column concept has been chosen. Instead of taking part of the NRU column bottoms as overhead condenser refrigerant, a side draw behind a partition wall is selected as refrigerant source. This patented[3] concept minimizes the risk of refrigerant contamination with CO_2 or C_{2plus} components, which might cause operational problems in the cold end of the plant. If the CO_2 content in the feed gas is below a certain threshold an amine system can be avoided completely.

The perceived power consumption disadvantage of a single column versus a double column concept is offset by the integration of a side condenser for the NRU column, which unloads the recycle compressor significantly. The nitrogen/methane separation inside the NRU column is driven by a heat pump system (blue lines in Figure 2), in which the HP fluid is condensed in the NRU column reboiler and vaporized at two different pressure levels in the side condenser and the overhead condenser.

Helium purification and liquefaction (Figure 3)

The overhead product of the NRU column is still sufficiently pressured for purification and final liquefaction without need for compression in the main stream (red lines in Figure 3). The process sequence in this main stream is a proven arrangement of catalytic H_2 and CH_4 conversion, dehydration, cryogenic processing and final helium purification in a PSA. Only the tail gas of the PSA plus some minor recycle streams from the cold box need to be recompressed. Thus the pressure energy provided by the helium fraction is well utilized.

The driving force contained in the pressure of the feedstock's nitrogen fraction, is used in the flowing manner: after condensing the bulk of the nitrogen fraction the nitrogen rich liquid is stripped under pressure to recover dissolved helium. The resulting pure liquid nitrogen fraction (green lines in Figure 3) is vaporized partly at elevated pressure, so that its energy can be recovered by means of work expansion. The overall enthalpy balance allows for liquid nitrogen extraction from the cold box in sufficient quantities to satisfy the needs of the helium liquefier and the storage tank radiation shields.

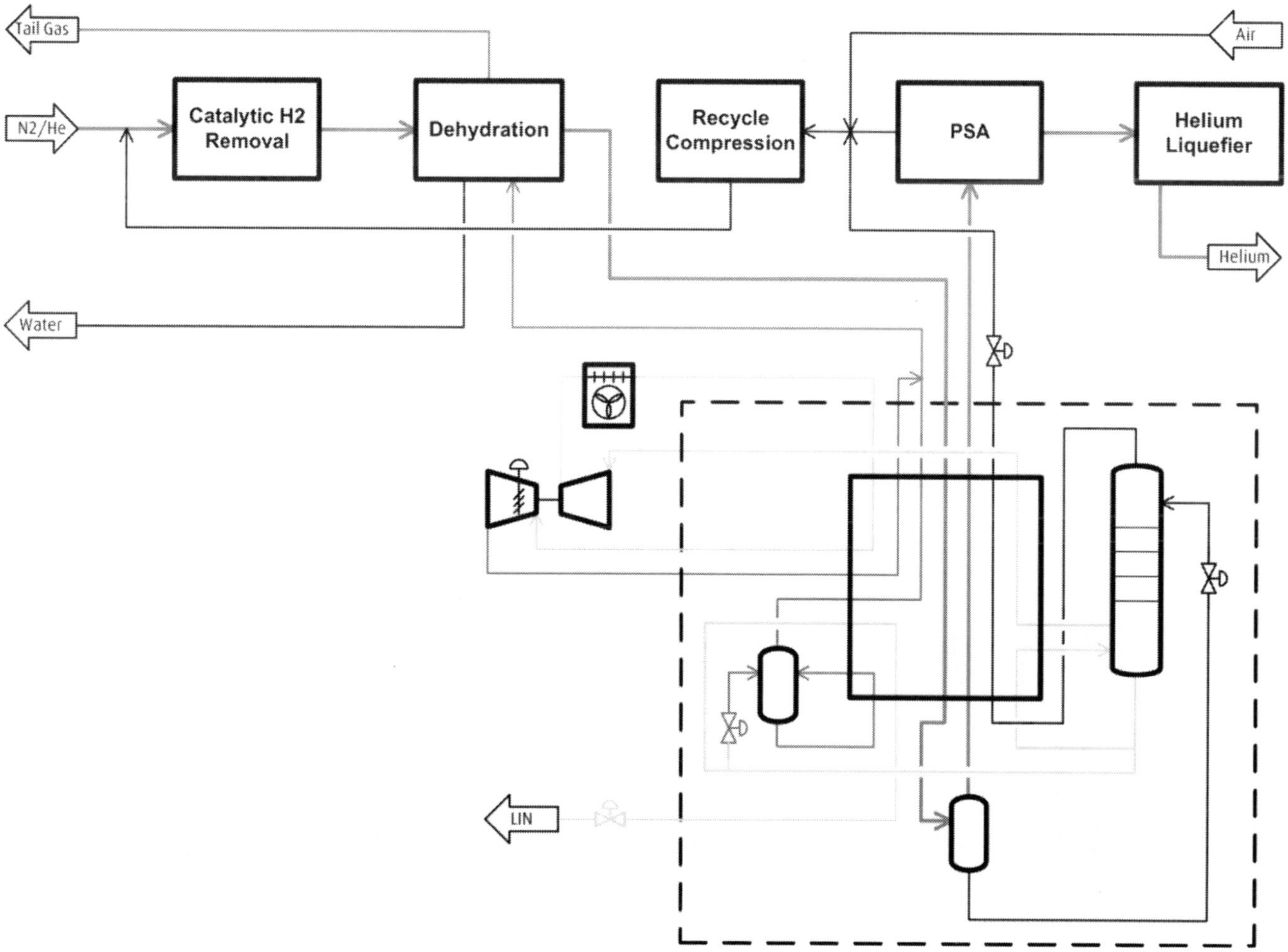

Figure 3 Helium section of the integrated gas processing plant

LNG production (optional)

Even though the main products of the integrated gas processing plant are NGLs, sales gas and helium there may be a demand for some quantities of LNG, if they can be produced at reasonable expense.

An elegant method for marginal LNG production is draining part of the LP refrigerant from the overhead condenser of the NRU column. This fluid (stream LNG a in Figure 4) is almost pure methane with less than 1% nitrogen. No additional equipment is required. The production is limited, however, to a few tons per hour, which is just good for make-up purposes.

In case more LNG (up to 10 % of the feed gas) can be marketed locally supplementary cooling systems are required. Figure 4 shows in green dotted lines tie-ins of two additional gas expanders, which provide refrigeration for LNG condensation and subcooling of part of the NRU column bottoms (stream LNG b). In this case the CO_2 concentration of the feed gas cannot exceed 100 vppm to avoid freezing problems in the subcooled LNG product.

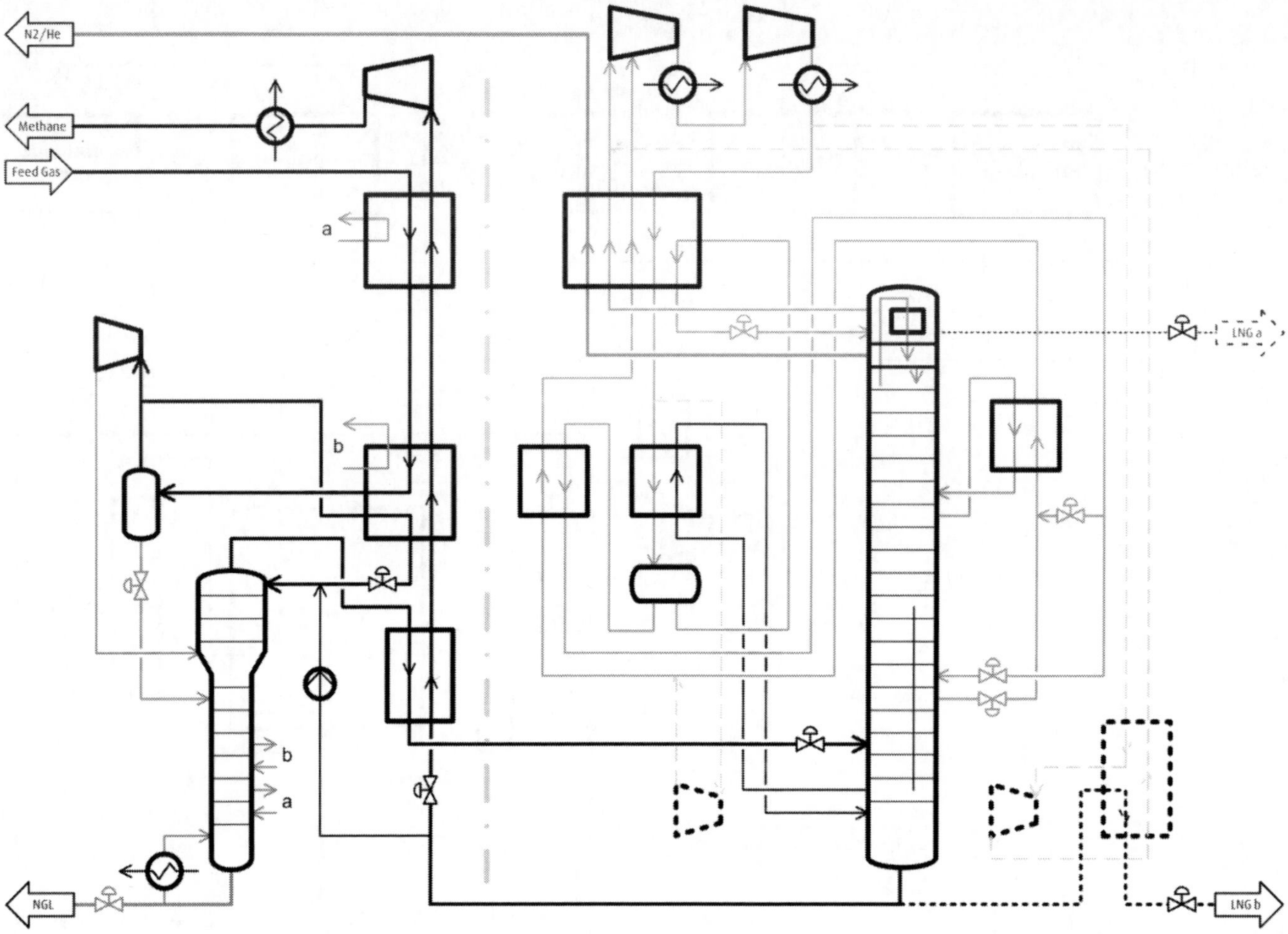

Figure 4 NGL and NRU section of the integrated gas processing plant with optional LNG extraction

AMUR GPP

Smart concepts need a proof of competitiveness in real projects. This proof was delivered end 2015, when Gazprom[4] (via its general contractor NIPIgas) selected Linde Engineering[5] for the design and supply of the main process units for the Amur GPP. The plant is part of Gazprom's project for the supply of Russian gas to China by the 'Power of Siberia' pipeline from Eastern Siberian gas fields and will be built in five phases up until 2024.

In late December 2015 Linde and NIPIgas entered into a binding engineering and supply contract in respect of the said units for all five construction phases of the Amur GPP. Phase one will consist of two ethane and NGL (propane, butane, pentane, hexane) extraction and nitrogen rejection units as well as one helium production unit. Related engineering works are in progress. When completed, the Amur GPP will be one of the largest gas processing plants in the world with a capacity of up to 49 billion standard cubic meters of natural gas per year. It will also include a liquefied helium production block with an annual output of up to 60 million cubic meters (2,100 MMSCFA) – the most extensive one around the globe.

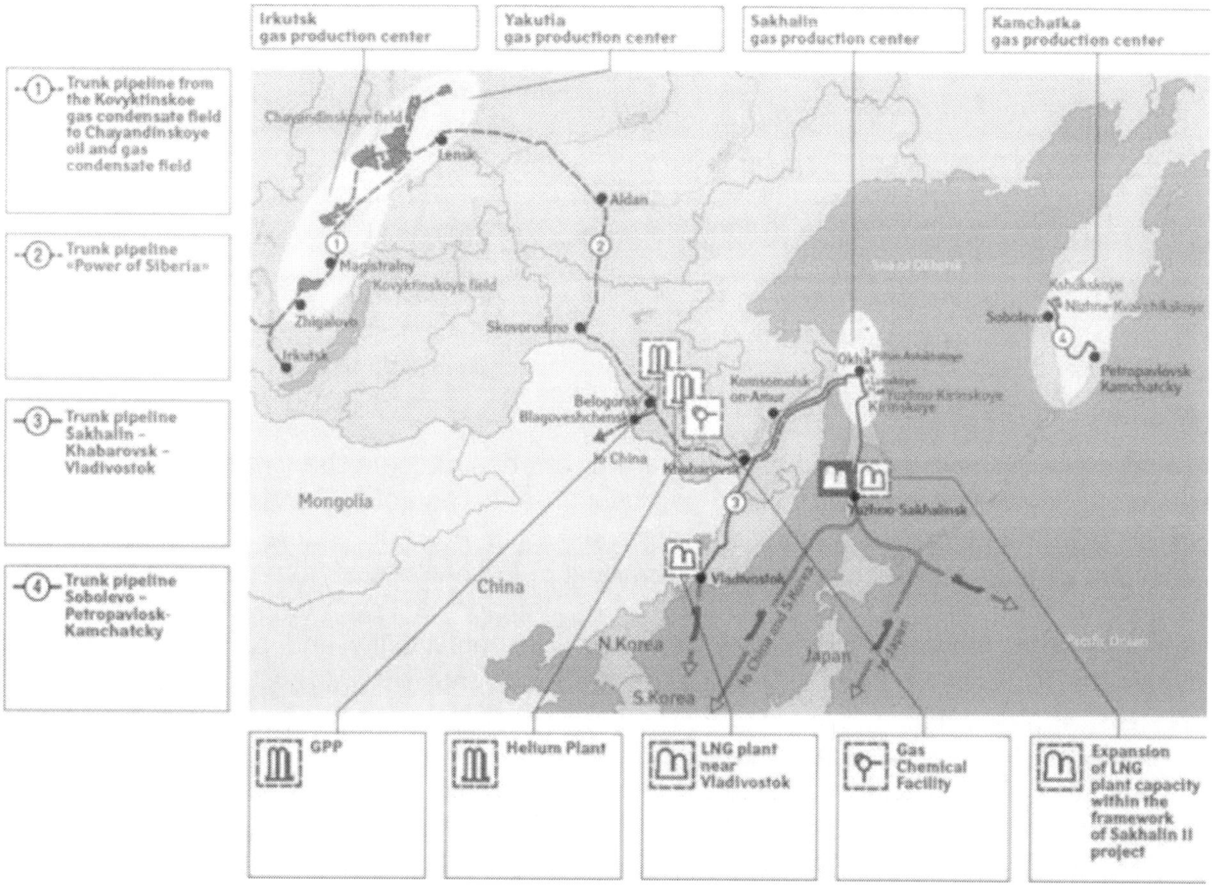

Figure 5 Map of Gazprom's Eastern Gas Program including Amur GPP and helium plant close to Belogorsk

Conclusions

This paper demonstrates the technical aspects of a successful synthesis of a full range gas processing plant. A highly attractive solution had been developed while avoiding technological step-outs.

References

[1] Gas Subcooled Process, US patent 4,157,904

[2] Recycle Split Vapor process, US patent 5,568,737

[3] US patent 9,003,829 and further patents in other countries

[4] http://www.gazprom.com/press/news/2015/october/article249177/

[5] http://www.the-linde-group.com/en/news_and_media/press_releases/news_20160121.html

Review of Incumbent and Emerging LNG Regasification Technologies Part 1 of 2: Onshore Terminals

Chevron, ETC, FE, LNG, MCP
J.C. Kuo, Senior Advisor, Ph.D, P.E.
Noma Ogbeifun, Process Engineer

Prepared for Presentation at the AIChE 2016 Spring Meeting
LNG Section
April 10-14, 2016
Houston, Texas

Abstract

Today's energy landscape is changing. Natural gas has become the fuel of choice for power generation due to its relatively clean combustion and ease of transportation. The selection of an LNG regasification technology depends on a host of factors including plant site location, climatic conditions and desired throughput capacities to name a few. LNG terminal operators are placing more emphasis on higher energy efficiency and minimal environmental footprint. This paper examines the applicability and suitability of different types of LNG regasification technologies that are currently available and new regasification designs proposed for future terminals. The paper is divided into two parts. Part 1 covers the LNG regasification technologies used in conventional onshore receiving terminal while Part 2 covers the LNG regasification technologies used in offshore receiving terminal.

Introduction

Today's energy landscape is changing. Natural gas has become the fuel of choice for power generation due to its relatively clean combustion and ease of transportation. When natural gas is cooled to cryogenic temperatures (-162°C at atmospheric pressure), it liquefies undergoing a 625: 1 volume reduction. The volume reduction allows large volumes of natural gas to be cost effectively transported over long distances where pipelines would be technically, geopolitically or economically constrained.

In 2014, 241.1 million tonnes of LNG was traded globally, which represents a 1.8% increase from 2013[1]. The International Energy Agency in its November 2014 report estimates a sustained annual growth rate of 1.6% in natural gas consumption over several years, positioning natural gas as the second largest energy source behind oil by 2040[2].

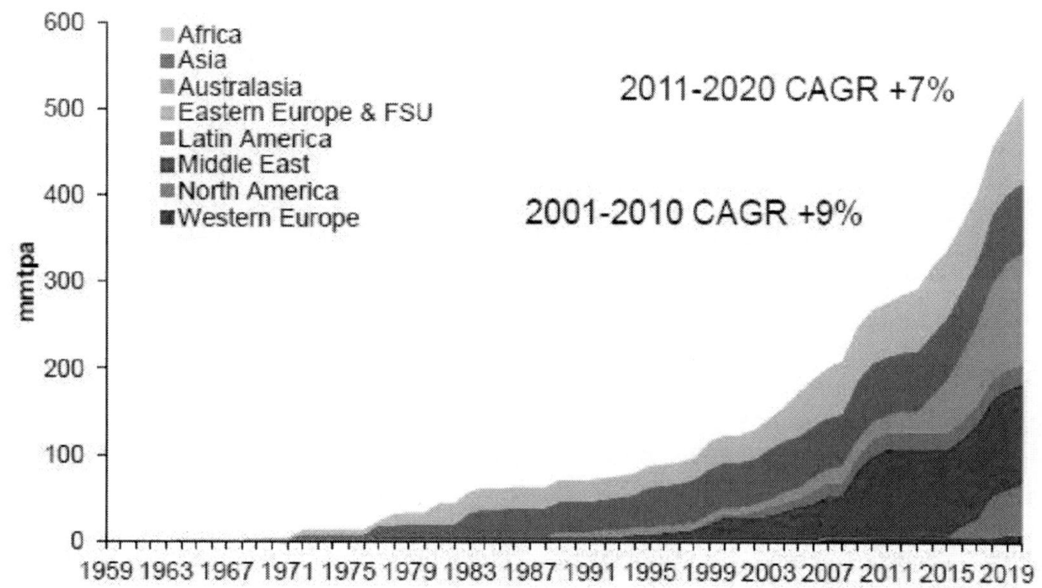

Figure 1: LNG Demand since 1970[3]

With the rise in natural gas consumption, many LNG importing countries are constructing LNG terminals to receive and regasify LNG for local consumption. Since 2000, Global LNG receiving capacity has increased by 260% to 724 MMTPA in 2014[1] (See Figure 2 below). This increase in global LNG receiving capacity has been driven by the emergence of floating regasification technologies, allowing new countries to become LNG importers, as well as access to a wide range of LNG supply options and flexible shipping strategies.

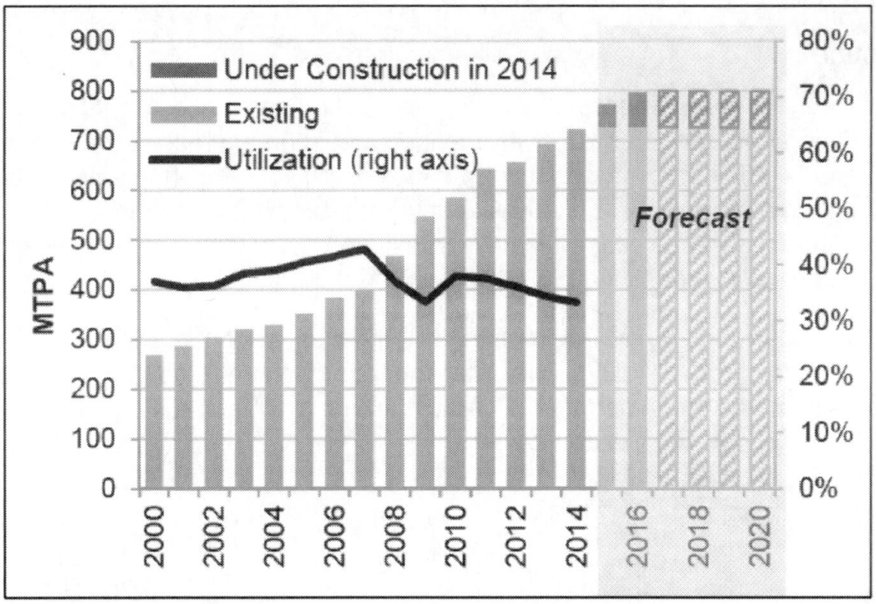

Figure 2:Growth in LNG Regasification Terminal Capacity[1]

In today's operating environment, LNG terminal operators are placing strong emphasis on higher energy efficiency and minimal environmental footprint. This paper examines the applicability and suitability of different types of LNG regasification technologies that are currently available and new regasification designs proposed for future large-scale baseload terminals. Cost comparisons are subjective and are not part of the scope of this paper. The paper is divided into two parts. Part 1 covers the LNG regasification technologies used in conventional onshore receiving terminal while Part 2 covers the LNG regasification technologies used in offshore receiving terminal. The onshore LNG regasification technologies in Figure 3 will be evaluated in this paper. Combustion based technologies are shown in the pink branch of the tree diagram and ambient sourced heat technologies are depicted in the green branch. Specific technologies highlighted in white will be covered in detail this paper.

Figure 3: Overview of Onshore Regasification Technologies

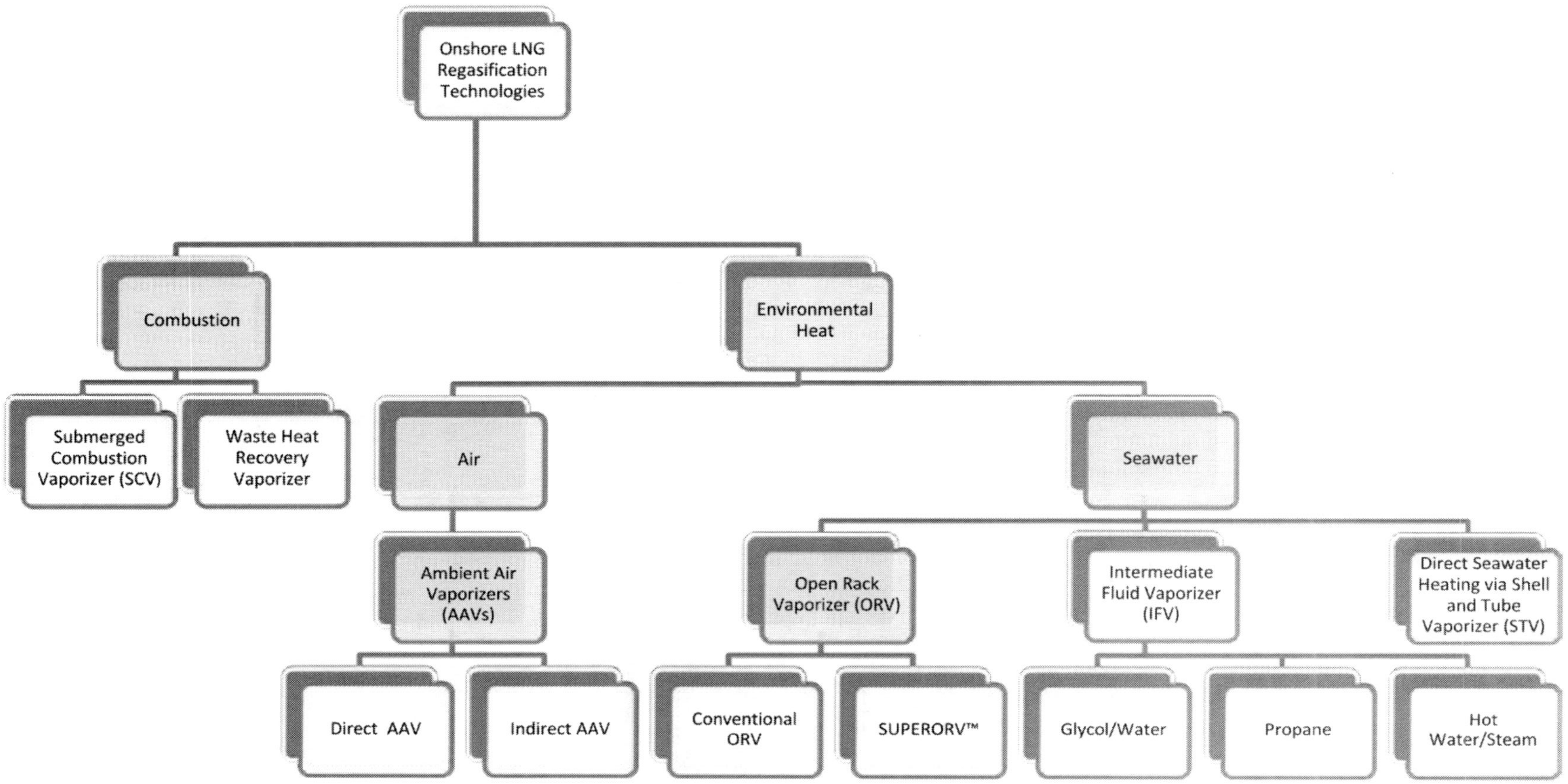

OVERVIEW OF ONSHORE LNG REGASIFICATION TECHNOLOGIES

Conventionally, LNG regasification terminals have been built coast side using thermal energy from ambient seawater or fuel combustion as heat sources to regasify LNG. Considering LNG regasification costs represent the largest terminal operating costs, the selection of a regasification system is a critical step in the design of a regasification terminal since it impacts terminal costs, throughput, operational flexibility, reliability, emissions, public perception and regulatory compliance[5]. For example, a one billion standard cubic feet per day (BSCFD) LNG regasification terminal requires approximately 600 million BTUs per hour (MMBTU/hr.) of heat for LNG regasification. Assuming a cost of US$ 5 per MMBTU, this amounts to around US$ 26 million per annum in operating expenses.

Onshore regasification terminals have generally used Open Rack Vaporizers (ORV) or Submerged Combustion Vaporizers (SCV) for LNG regasification purposes with a proven track record across Asia, Europe, and the United States. Use of SCVs creates environmental concerns due to CO_x and NO_x emissions. Use of ORVs could impact marine life due to the cold water effluent. Consequently, tighter environmental standards have led to the development of alternative regasification technologies such as intermediate fluid vaporizers, ambient air vaporizers that have been implemented across baseload LNG regasification service.

1.0 Environmental Heat

1.1 Ambient Seawater

1.1.1 Conventional Open Rack Vaporizer (ORV)

Open Rack Vaporizers (ORV) is a heat exchanger that uses seawater as the primary heat source. ORVs are well proven technology and have been widely used in Japanese, Korean and European LNG terminals. Around 70% of baseload LNG regasification terminals use ORV technology.

ORV technology uses naturally occurring heat in the form of warm seawater flowing over a series of panel coils to vaporize the LNG flowing countercurrent within the panels. Seawater lift pumps bring treated seawater to the distribution troughs at the top of the ORV panels. Warm seawater flows down the outside of the panels in a falling film. The seawater is cooled by exchanging heat with the LNG. The panels consist of multiple LNG tubes with integral fins. LNG flowing upward through the tubes is vaporized and warmed to approximately 40°F. A 4°F-6°F LNG approach temperature is possible depending on prevalent conditions i.e. summer or winter conditions. The preferred seawater temperature for ORV operation is above 42°F. The cooled seawater from the ORVs is collected in a concrete basin beneath the ORV and routed to the seawater outfall system.

Seawater is pumped into the ORV through intake screens to eliminate debris and marine life. The ORV sea water pumps consume approximately 12.38 BTU/lb_m(LNG)[4] for lifting and driving seawater. Sodium/Calcium hypochlorite is injected into each seawater pump suction continuously to control marine growth in the system. Typically, a water concentration of 0.2 ppm of Sodium/Calcium hypochlorite is desired. In order to shock the system, elevated concentrations of 2.0 ppm would be injected for 20 minutes every 8 hours, during ORV operation. It is expected that dechlorination will naturally occur as the treated water flows over the panels, through the trench, and to the outflow basin, such that no special treatment will be required to meet permitted discharge limits for residual chlorine.

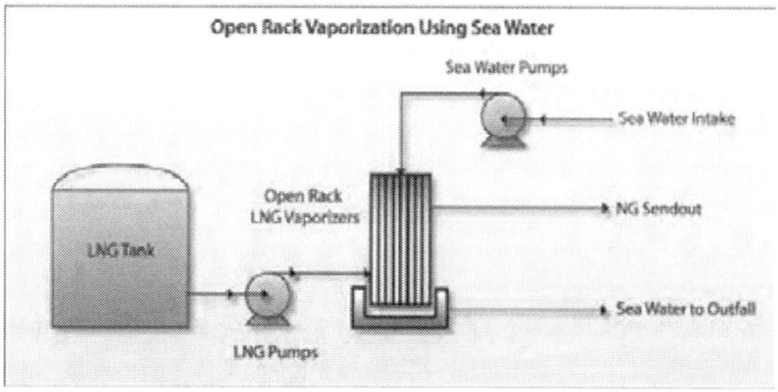

Figure 4: Simplified Process Flow Diagram for LNG Regasification using ORV[5]

ORV units are generally constructed of aluminum alloy for mechanical strength suitable to operate at the cryogenic temperature. The material has high thermal conductivity which is effective for heat transfer equipment. The tubes are arranged in panels, connected through the LNG inlet and the regasified product outlet piping manifolds and hung from a rack (Figure 5). The panels are coated externally with zinc alloy, providing corrosion protection against seawater. Regular maintenance is required to keep the finned tube surface clean.

The panel arrangement feature provides ease of access for maintenance. Depending on the design of the units, it is also possible to isolate sections of the panels and vary the load on the units. The unit can be turndown to accommodate fluctuations in gas demand, gas outlet temperature and seawater temperature.

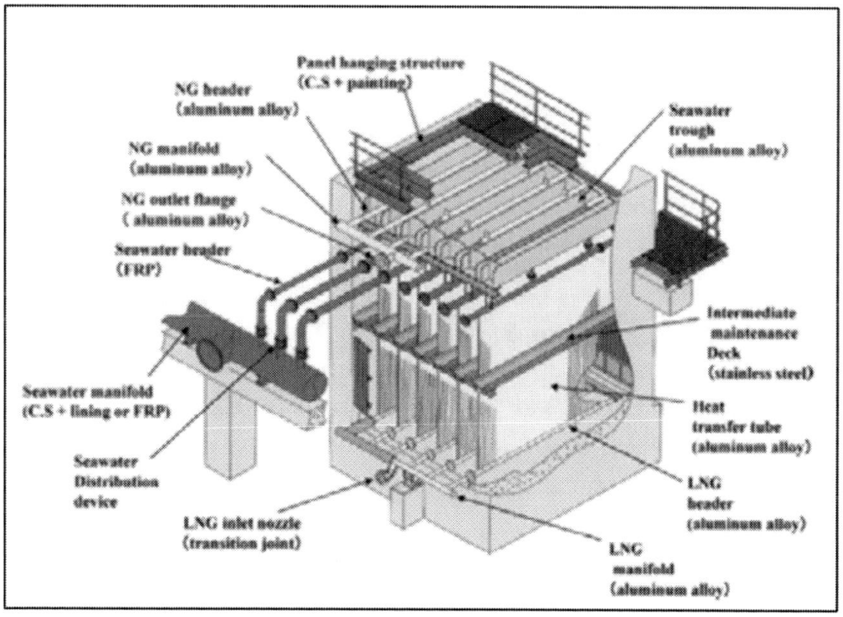

Figure 5: Schematic for Open Rack Vaporizer (ORV)[6]

1.1.2 SUPERORV™

The SUPERORV™ is a new type of open rack vaporizer that uses seawater as the primary heat source. It was developed through the collaborative efforts between OSAKA Gas and Kobe Steel. Seawater lift pumps bring warm seawater to the top of each SUPERORV™ to flow down the outside of the panels in a falling film. The seawater is cooled by exchanging heat with the LNG. The panels consist of multiple LNG tubes with integral fins. LNG flowing upward through the tubes is vaporized and warmed to approximately 40°F. The preferred seawater temperature for ORV operation is above 42 °F.

The key features of the SUPERORV™ are the double structure of the lower panel tubes which improve heat transfer performance by suppressing ice formation on the tube surface as shown in figure 6 below.

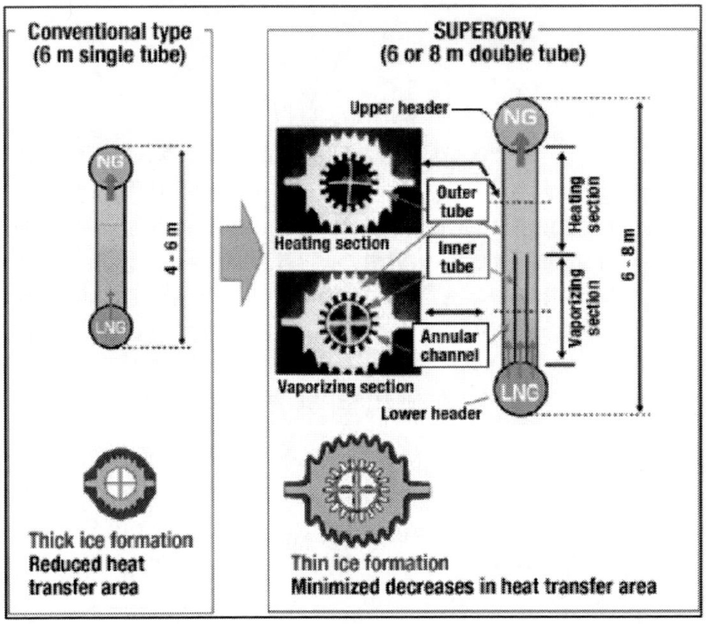

Figure 6: Icing on heat transfer tube surface [Courtesy of OSAKA Gas]

The twofold increase in inside diameter of the SUPERORV™ heat transfer tubes result in a three-five fold increase in throughput when compared to the conventional ORV. The consequent reduction in the required number of heat transfer tubes results in a 10% construction cost reduction, 40% reduction in required land area, and 15% reduction in seawater requirements.

The first SUPERORV™ started operation in November 2008 at Osaka Gas Himeji Terminal validating proof of concept. A case study of a regasification terminal using SUPERORV™ technology (table 1) highlights the aforementioned benefits as well as the reduced seawater demand.

Design Parameters	SUPERORV™	Conventional ORV
LNG vaporizing capacity	$331x10^3$ lb_m/hr LNG	$331x10^3$ lb_m/hr LNG
Design pressure	669 psia	669 psia

Design sea water temperature	50°F	50°F
Minimum outlet gas temperature	36°F	36°F
Maximum sea water flow rate	$9921x10^3$ lb_m/hr	$15432x10^3$ lb_m/hr
Number of panels	9	13
LNG vaporizing capacity per tube	660 lb_m/hr /tube [7]	220 lb_m/hr /tube
Installation Space	1938 ft^2 (54.1 ft x 35.4 ft)	2583 ft^2 (54.1 ft x 47.6 ft)

Table 1: Comparison between SUPERORV™ and Conventioal ORV

As of December 2013, a total of 102 SUPERORV™ units have been installed in LNG regasification terminals worldwide as shown in Figure 7 below.

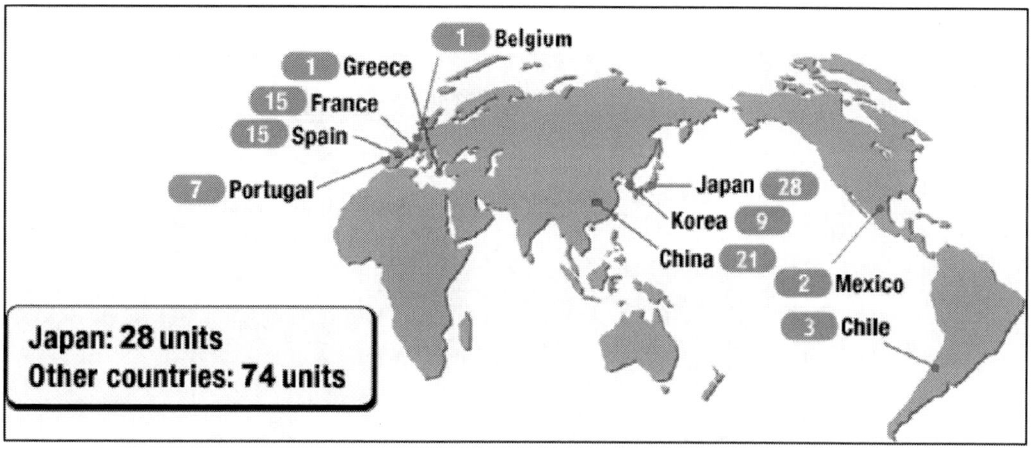

Figure 7: SUPERORV™ Installations (Courtesy of OSAKA Gas)

1.1.2.1 Environmental Impact

Environmental issues associated with open rack vaporizers include seawater intake, seawater outfall and air emissions. A primary environmental concern with seawater intake for ORVs is marine life in the water column. Screen depth, intake velocity and backwashing are key considerations in the design of the seawater intake screen system. The maximum through slot velocity of the intake screens should be around 0.5 ft/s to minimize impingement. An automated air backwash system should be installed to remove impinged organisms and debris on the screen surface.

Seawater from the ORVs is routed to the seawater outfall and returned to the ocean. The seawater discharge temperature needs to meet the World Bank guidelines (37°F within 330 feet from the point of discharge). Also the residual chlorine levels need to meet the regulation required (0.2mg/l at point of discharge per US EPA standard).

Sodium Hypochlorite is highly corrosive and relatively unstable with the potential to gasifiy, crystallize and foul water system components. Chlorination piping should be designed to maintain sodium hypochlorite flow velocity of 0.7 -2 ft/s to avoid gasification and crystallization within flowlines.

1.1.2.2 Safety

Open rack vaporizers are considered extremely safe. No moving parts are in contact with flammable fluids. Open rack vaporizers do not require combustion and pose no ignition risk.

1.1.3 Intermediate Fluid Vaporizers (IFV)

Currently, intermediate fluid LNG vaporizers account for around 5% of total LNG regasification markets worldwide[5]. In this technology, LNG is vaporized using a heat transfer fluid (HTF) in a closed intermediate loop to transfer heat to vaporize LNG as shown in Figure 8. Typically, the regasification process takes place in a shell and tube configurations in two stages. In the first stage, LNG is heated by the HTF in a heat exchanger where it is vaporized. In the second stage, the vaporized natural gas is circulated through a second shell and tube heat exchanger, with seawater as the heating medium used to bring the gas to the pipeline specified temperature for send out. This technology was originally developed by OSAKA GAS Co., Ltd. in the 1970s under the TRI-EX™ name[6].

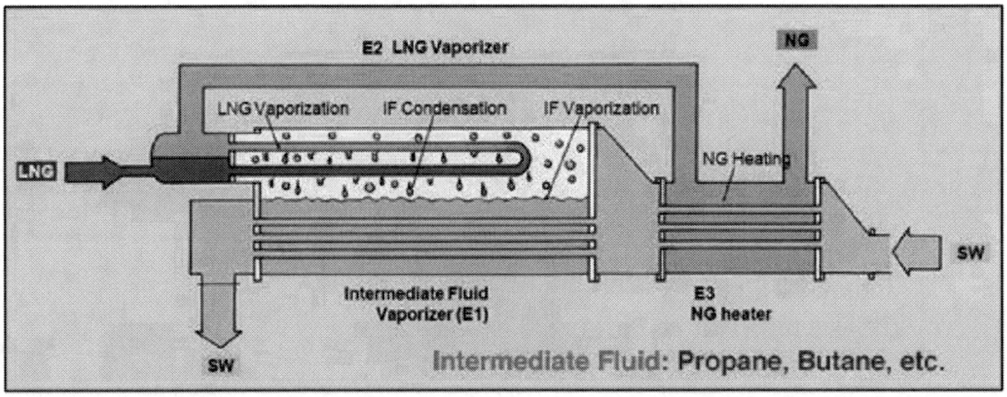

Figure 8: Schematic for IFV Regasification System with Seawater (Courtesy of Kobelco)

Several intermediate heat transfer fluids (such as Hydrocarbon Based HTF (Propane/Butane), Glycol-Water, and Steam/Hot water) have been utilized over the years. A summary of the comparative differences between these HTFs is provided in table 2 below:

Table 2: Comparative Differences between various IFV Heat Transfer Fluids

HTF Considerations	Hydrocarbon Based HTF (Propane/Butane)	Glycol-Water	Hot water/Steam
Thermal Energy Available	Latent Heat	Sensible Heat	Water: Sensible Heat Steam: Latent Heat

Cryogenic Temperature Properties	No freezing issues anticipated	Some freezing may occur with poor fluid distribution	Some freezing may occur with poor fluid distribution
Area Classification	May require hazard area classification	May require hazard area classification	Equipment may be located in non-hazardous
HTF Makeup	Refrigerant grade propane	Industry strength glycol or glycol-water pre-mixture	Water pretreatment may be required to meet boiler specifications

In comparison to water-glycol mixture and hot water/steam, propane generally provides better heat transfer properties as it can be easily vaporized using seawater and condensed by LNG thereby requiring a smaller heat transfer area and equipment size.

1.1.3.1 Environmental Impact

Environmental issues associated with IFVs involve seawater intake, seawater outfall, air emissions and handling of glycol. The same environmental concerns around seawater handling are equally applicable with the IFV vaporizer option as described in the ORV option. IFVs do not directly produce air emissions. They do indirectly generate air emissions due to the power generation requirement for seawater lift and circulation pumps.

Secondary containment will be required around all equipment and vessels. Leaking and spilled liquid will need to be collected in sealable containers, or vacuum (pump) into a storage vessel/container for disposal.

Ethylene glycol is classified as a combustible and will burn producing toxic decomposition products. Use of ethylene glycol will add another toxic, combustible and environmentally regulated chemical to the inventory on the facility. However, this is a controllable and acceptable risk if handled properly. Its use will require reporting to regulating agencies. Ethylene glycol is currently used offshore in closed systems, and therefore is not expected to be denied use by regulators, especially in a closed loop system.

1.1.3.2 Safety

The freezing of glycol/water solution on the shell side of the exchanger in the event of tube rupture is a concern as LNG will enter the shell side of the exchanger impeding the path of the expanding fluid to the relief valve inlet of the shell side of the exchanger.

1.1.4 Direct Heating Shell and Tube Vaporisers

Shell and tube vaporizers (STV) also use seawater as the source of thermal energy. STVs are usually of TEMA type NEN or NJN design and have been in service since 1974. Shell and tube vaporizers have been used onshore in regasification terminals in the USA (Freeport LNG), India (Dahej Terminal), and Egypt in a number of operating configurations.

Figure 9: Onshore LNG STV Application (Courtesy of Chicago Power & Process, Inc.)

In the open-loop, configuration (see Figure 10), LNG is pumped through the bottom of the STV, which is mounted vertically to optimize vaporization efficiency. Warm seawater is pumped to the top of the STV and flows down the outside of the tubes on the shell side. The seawater is cooled by exchanging heat with the LNG. LNG flowing upward through the tubes is vaporized and warmed to approximately 40°F. A 4-6°F approach temperature is possible depending on prevalent conditions i.e. summer or winter conditions. The preferred seawater temperature is above 42°F.

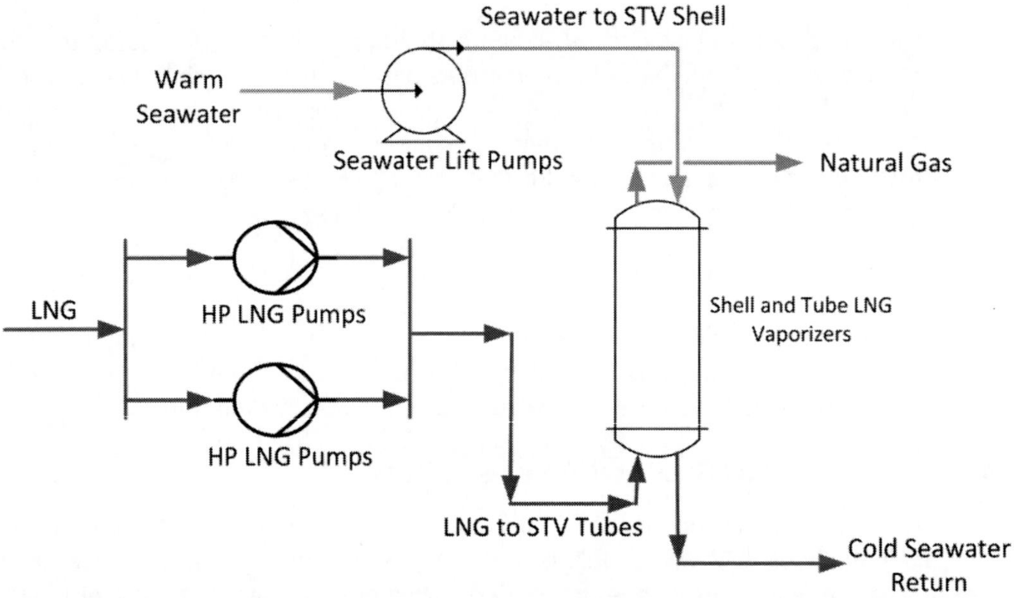

Figure 10: STV Open loop Configuration

In the closed-loop configuration (see Figure 11), a heat transfer fluid such as propane or glycol-water mixture is used to vaporize LNG. The intermediate fluid flows through tubes in separate heating equipment such as a fired heater or boiler for heating, then the HTF is passed through the STV unit to re-gasify the LNG. Since there are two heat exchangers, this requires a large amount of space.

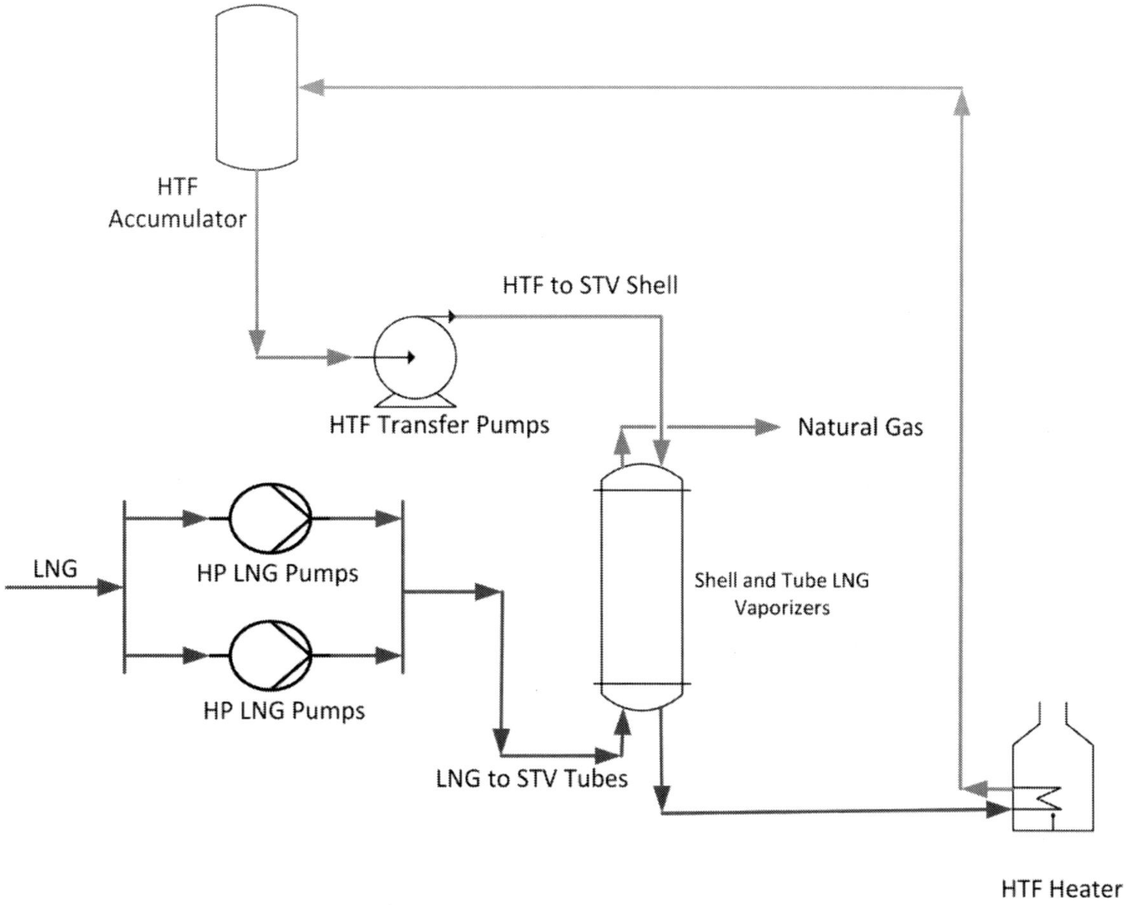

Figure 11: STV Closed loop Configuration

1.1.4.1 Environmental Impact

Environmental issues associated with STVs involve seawater intake, seawater outfall and air emissions. The open-loop technology reduces air emissions, since there is no combustion. A primary environmental concern with seawater intake for STVs is marine life in the water column. Features of the seawater intake screen system should be designed to minimize impingement and entrainment of marine organisms include screen depth, intake velocity and backwashing.

STVs do not directly produce air emissions. However, closed loop operation may indirectly generate air emissions due to the power generation requirement for seawater lift pumps and air circulation fans.

Also, secondary containment will be required around all HTF equipment and vessels. Leaking and spilled liquid will need to be collected in sealable containers, or vacuum (pump) into sealed storage vessel/containers for disposal.

1.1.4.2 Safety

Slosh formation and freezing of seawater on the shell side of the exchanger in the event of tube rupture is a concern as LNG will enter the shell side of the exchanger impeding the path of the expanding fluid to the relief valve inlet of the shell side of the exchanger.

2.0 Air

2.1.1 Direct Ambient Air Vaporizer (AAV)

Direct ambient air vaporizers use air to vaporize LNG. Air is introduced either by natural convection or forced draft to drive heat transfer. The design is such that heat from ambient air is used to supply majority or all of the LNG regasification load. In order to meet the specified send out temperature, supplimentary heating may be obtained from the exhaust of power generation prime movers such as gas engines or turbines or other avaialble heat sources.

Daily and seasonal ambient air temperature is an important factor, which affects the size and viability of ambient air vaporization technology. Depending on the degree of seasonal and daily temperature fluctations, the vaporizer surface area must be sized to accommodate the lowest possible ambient temperature while maintaining a 25-30°F differential approach temperature. Direct AAV operation requires multiple daily cycling of vaporizers to avoid freezing problems. Direct AAV have been used extensively in the regasification of cryogens and liquefied gases like liquid nitrogen. According to Cryoquip, the Sabine Pass LNG Terminal in Cameron Parish, Louisiana uses direct AAV (18 units) for supply a portion of the LNG regasification duty.

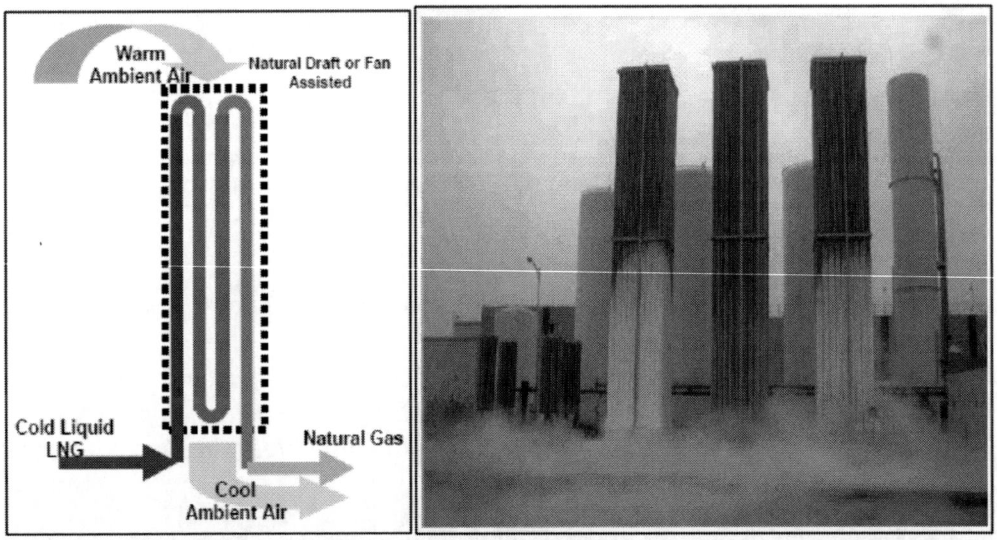

Figure 12: Schematic and Operation of Direct Ambient Air Vaporizer[9]

2.1.1.1 Environmental Impact

There are no emissions except the fog, water condensate, and icing created by cold air for AAV as shown in Figure 12 above.

2.1.1.2 Safety

The AAV is inherently safe.

2.1.2 Indirect Ambient Air Vaporizer

Indirect ambient air vaporizers operate by transferring heat from ambient air to a heat transfer fluid which in turn transfers heat to LNG for regasification as shown in Figure 13. The system consists of shell and tube LNG vaporizers, air-fin heaters and a closed HTF loop. Air draft fans drive air flow across the tube bundles maximizing heat transfer to the HTF and minimizing required heat transfer area. The hot HTF is sent to the shell and tube LNG vaporizers. The cooled HTF flows into the accumulator and is then pumped back to the air fin heaters for heating. The technology is installed in the Dahej LNG Terminal in India.

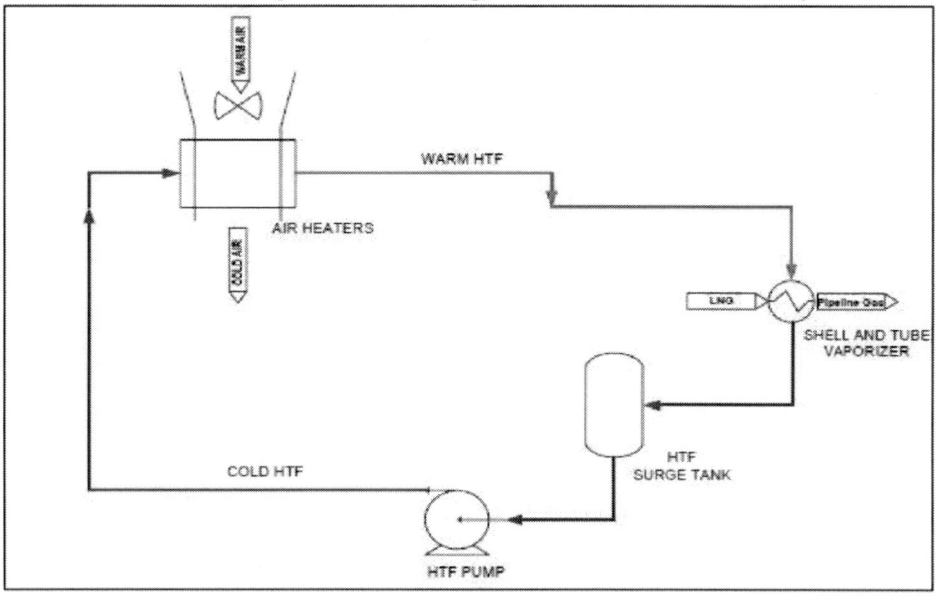

Figure 13: Schematic of Indirect Ambient Air LNG Regasification[9]

2.1.2.1 The Smart Air® Vaporization System by Wood Group Mustang

The Smart Air® vaporization system by Wood Group Mustang (see Figure 14) has a similar setup to the indirect ambient air regasification system except that the Smart Air® system uses a alkali metal formate/acetate (like Potassium Formate) and water mixture as the heat transfer fluid [10] . The system consists of Air-fin Heaters, accumulator, HTF Pumps, and Shell and Tube LNG vaporizers. The Smart Air® vaporization system Air-fin heaters are proprietary to Mustang Engineering. Each Air-fin heater contains layers of finned tubes with fans forcing ambient air from the surroundings to flow through the top of the unit downwards, over the tube bundles. The Air-fin heaters are elevated above grade to prevent recirculation of cold dense air from the bottom of the unit to the fan inlet. As with other ambient air vaporization systems, supplementary heat may be required when heating load cannot be met with design ambient conditions.

The alkali metal formate/acetate and water mixture, henceforth referred to as K-HTF, is a water based low temperature heat transfer fluid engineered to deliver excellent heat transfer performance within the temperature ranges of -58 °F to 424 °F[11]. K-HTF has been gaining popularity due to its good thermophysical properties, low toxicity, non-flammability, and lower pumpimg duty requirements when compared to other water based chemistries such as glycol /water mixtures, alkali chloride/water as well as hydrocarbon and silicone based fluids. However, K-HTFs are corrosive especially towards zinc, aluminium, and galvanized steel.

Trunkline LNG in Lake Charles, Louisiana, uses the LNG Smart® Air Vaporization process to vaporize 2.1 BCFD send out gas.

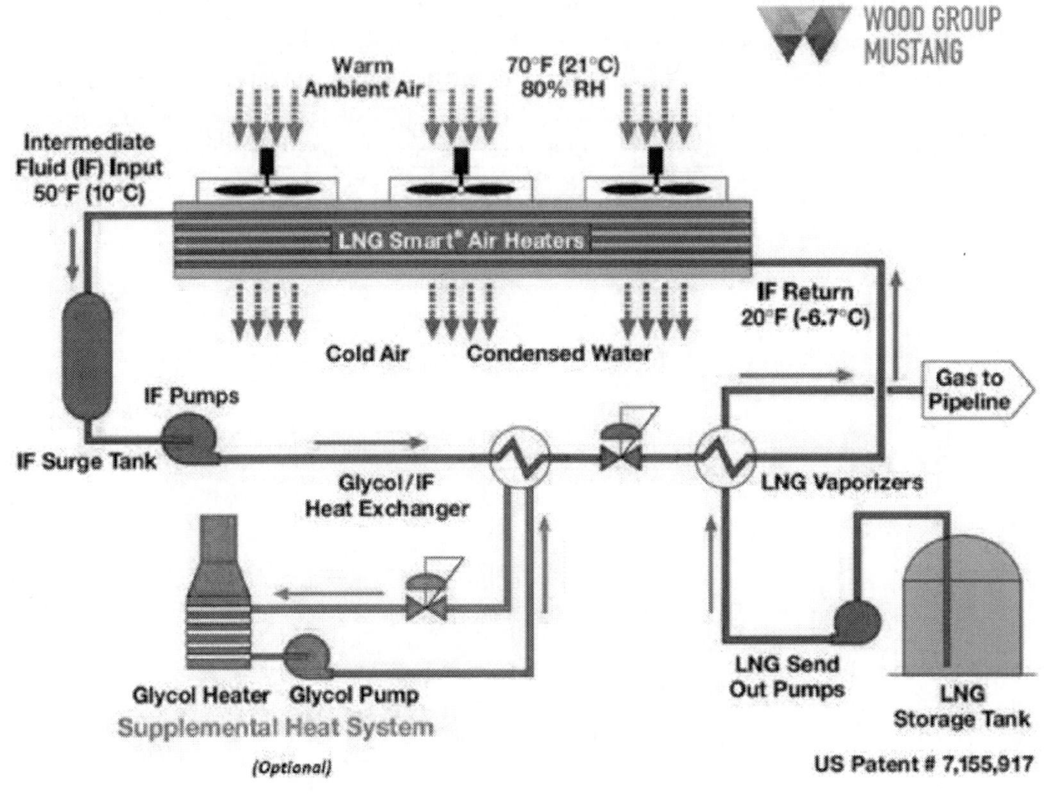

Figure 14: LNG Smart Air® Vaporization Technology (Courtesy of Wood Group Mustang)

2.1.2.2 The Reverse Cooling Tower

The reverse cooling tower design (Figures 15 and 16) extracts ambient heat (sensible and latent heat) from air and humidity by direct contact with cooling water. The design takes advantage of the thermal properties of cooling water to exchange heat with the cold heat transfer fluid via a heat exchange coil. Freeport LNG terminal plans to use the Reverse Cooling Tower upon startup in 2018.

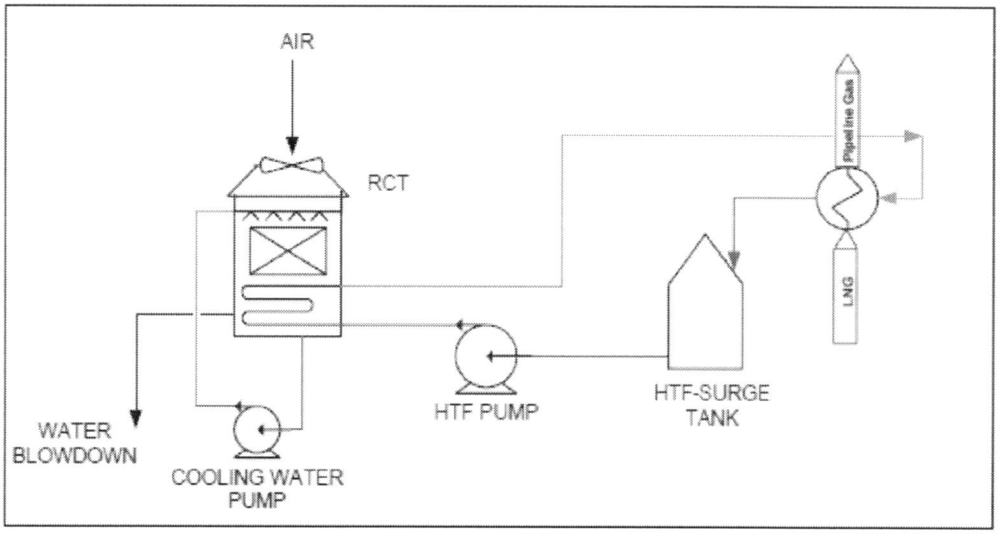

Figure 15: Schematic for IFV Regasification System with Reverse Cooling Towers

Figure 16: Freeport LNG IFV Reverse Cooling Towers

(Courtesy of Freeport LNG Development, L.P)

2.1.2.2.1. Environmental Impact

Environmental issues associated with LNG Smart® Air Vaporization System include air emissions from the HTF heater and handling of the HTF (glycol water mixture or K-HTF). The glycol water closed loop system need to meet the environmental/regulatory implication as listed in the IFV option. Water produced from condensation of water vapor from the surrounding air must be continuously discharged and safely discarded. For Reverse Cooling Tower systems, the circulating water stream may require chemical treatment and appropriate handling of blowdown.

2.1.2.2.2. Safety

The heat transfer fluid heater is source of ignition and the HTF loop has same concern as IFV. K-HTFs are corrosive especially towards zinc, aluminum, and galvanized steel so material selection is key consideration to minimize loss of containment events.

3.0 Combustion

3.1 Submerged Combustion Vaporizer (SCV)

Submerged Combustion Vaporizers (SCV) are typically used when there is insufficient water supply or when the water temperature is too low for regasification. Around 25% of baseload LNG regasification terminals use SCV technology. The primary energy source for vaporizing LNG, and raising the natural gas to the required send out temperature, is the combustion of vaporized LNG which consumes approximately 1.5-2% of the vaporized LNG as fuel thereby reducing natural gas throughput and increasing operating expenses. SCVs are compact and do not require much plot area when compared to the other baseload LNG regasification options.

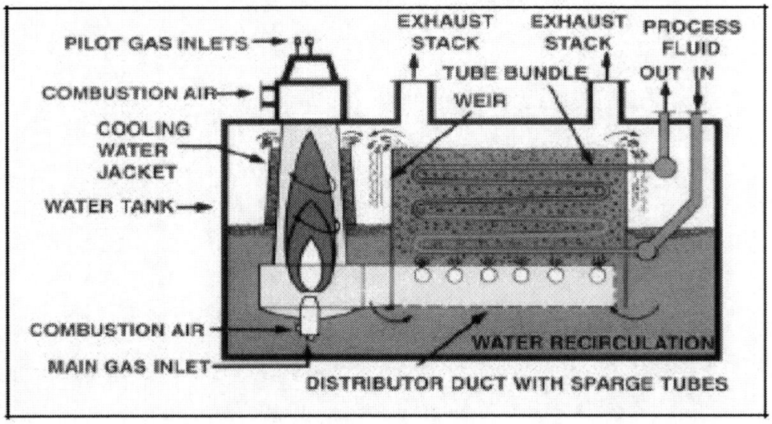

Figure 17: Schematic for Submerged Combustion Vaporizer (SCV)[9]

SCV is an indirect fired heat exchanger (see Figure 17) with the burner and LNG tube coil contained within a water bath. Since the water bath is always maintained at a constant temperature, the system copes very well with sudden load changes and can be quickly started up and shutdown. The burner combustion products (COx, NOx) are exhausted directly into a water bath below the surface, providing heat to the bath. Flue gases are sparged into the water using a distributor located under the heat transfer tubes.The water bath acts as the heat transfer media that vaporizes the LNG in the immersed tube coil. The sparging action promotes turbulence resulting in a high heat transfer rate and thermal efficiency (~98%). The turbulence also reduces deposits or scales that can build up on the heat transfer surface. Low chloride water is required for the initial fill of the bath, to avoid chloride stress cracking in the stainless steel tube coil.

3.1.1.1 Environmental Impact

Environmental issues associated with SCVs involve air emissions and water effluent. SCVs directly produce air emissions by burning approximately 1.5-2% of the LNG throughput as fuel. The turbine generators produce air emissions by providing power for the air blowers. Total air emissions for SCVs are much higher than the ORV emissions.

Approximately 22 gpm of water (per 200 MMSCFD of LNG vaporization capacity)[12] is produced per unit, and the pH must be adjusted with chemicals. The submerged combustion process creates nitrous and carbonic acid in the water bath. Therefore, pH monitoring and control of the water bath chemistry is required to minimize the effects of nitrous and carbonic acid tube corrosion and effulent disposal.

The typical stack temperature of an SCV is about 80°F. In order to minimize NOx emissions, low NOx burners can be used to meet the 40 ppm NOx emission limit. Selective Catalytic Reduction (SCR) may be used to further reduce NOx emission levels to ~5ppm but this comes at significant cost since SCR catalyst required high temperature for favorable reaction kinetics.

3.1.1.2 Safety

SCVs are a source of ignition. Minimizing ignition sources is preferred in any LNG regasification terminal. Safety concerns associated with a technology involving combustion offshore, where paths of egress are limited in the event of an emergency make SCV technology less suitable for an offshore location.

Additional vapor control would be required surrounding the SCV as opposed to ORV. This may mean additional combustible gas detection to indicate gas emissions and initiate mitigation measures to prevent ignition of the vapors.

SCVs require different spacing from ORV because of the combustion used in their process. Also, the SCV exhaust stacks may produce fog-like vapor plumes obstructing visibility during normal terminal operations.

3.2 Waste Heat Recovery and Integration

Where base-load LNG regasification terminals are located close to an industrial facility, such as a refinery or power station, there are opportunities for heat integration and improved regasification economics since waste heat from a power plant or refinery turbine exhaust/ waste heat recovery can be used to supply all or part of the LNG regasification duty via an integrated process between the industrial facility and regasification terminal as shown in Figure 18.

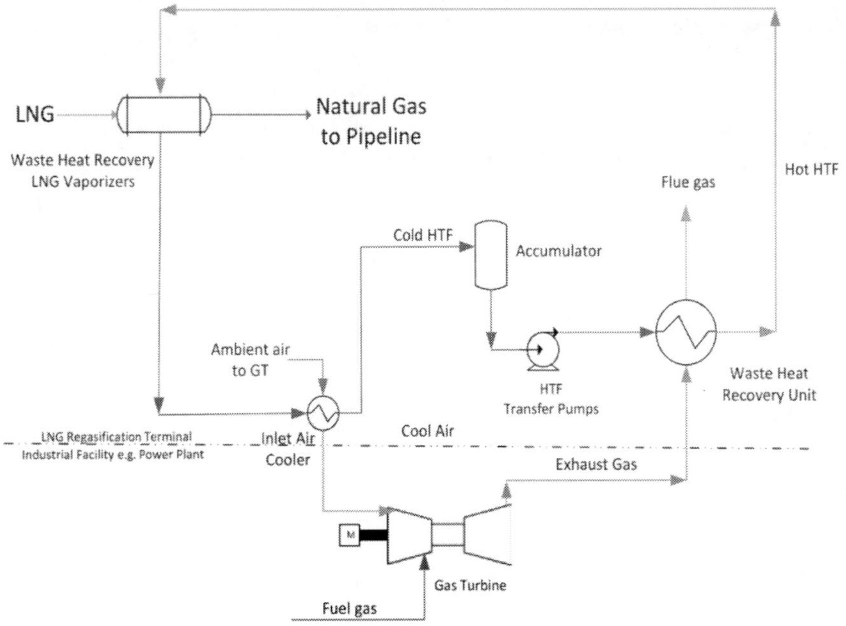

Figure 18: Waste Heat Recovery using integrated process between nearby industrial facility and regasification terminal.

In this process, cold HTF effluent from the Waste Heat Recovery LNG Vaporizers is used to provide inlet air cooling for the gas turbine before circulation to the waste heat recovery unit. The turbine inlet suction cooling will result in an increase in power and consequently, higher electricity generation. Typically, for each degree Celsius drop in air temperature, power output can be increased by about 1%.

The hot exhaust gas from the turbine is routed to a waste heat recovery heater and used to heat the HTF. The hot HTF is pumped back to the shell and tube Waste Heat Recovery LNG Vaporizers and used to vaporize LNG in a closed loop configuration.

3.2.1.1 Environmental Impact

Environmental issues associated with waste heat recovery regasification system involve indirect air emissions associated with electricity production to drive key equipment like seawater lift pumps. The turbine generators produce air emissions by providing power. Total air emissions could be lower than other sources when compared to SCV emissions.

3.2.1.2 Safety

In the event of a loss of containment event, gas turbines are a potential source of ignition. Minimizing ignition sources is preferred in any LNG regasification facility. Safety concerns associated with a technology involving combustion offshore, where paths of egress are limited in the event of an emergency.

Additional vapor control would be required surrounding the waste heat recovery regasification systems compared to ORV. This may mean additional combustible gas detection to indicate gas emissions and initiate mitigation measures to prevent ignition of the vapors.

4.0 COMPARISON OF ONSHORE LNG REGASIFICATION TECHNOLOGIES

The optimum choice of an LNG vaporization system is determined by the terminal's site selection, the environmental conditions, regulatory limitations and operability considerations. Table 3 compares ten LNG regasification options in term of their applications, operation and maintenance, utility and chemical requirements, environmental impacts and relative plot sizes for a LNG regasification terminal.

Table 3: LNG Regasification Type Qualitative Comparison

	ONSHORE									
							IFV			
LNG Regasification Technology	SCV	Waste Heat Recovery Vaporizer	Direct AAV	Indirect AAV	Conventional ORV	SUPERORV™	Glycol-Water	Propane	Hot Water/Steam	Direct Seawater Heating Shell and Tube Vaporizer
Feature	Indirect LNG vaporization by hot water which is heated by combustion of vaporized LNG	Indirect LNG vaporization by cooling HTF which is heated by hot exhaust gases	Direct LNG vaporization using air	Direct LNG vaporization using air via HTF	Direct LNG vaporization using sea water	Direct LNG vaporization using sea water	Indirect LNG vaporization by cooling hot glycol which is heated by any available heat medium	Indirect LNG vaporization by condensing propane which is heated by seawater	Indirect LNG vaporization by cooling steam which is heated in boilers	Direct LNG vaporization using sea water
Unit Max Train Throughput, Tons/hr	200	Design to desired throughput	Design to desired throughput	Design to desired throughput	300	900	Design to desired throughput	Design to desired throughput	Design to desired throughput	Design to desired throughput
Heating Medium	Vaporized LNG	Hot Exhaust Gas	Air	Air	Seawater	Seawater	Any available heat medium	Seawater	Hot steam condensate/Steam	Seawater
Fuel gas Consumption for LNG Vaporization, %	1.3-2	-	-	-	-	-	-	-	~2.5	-
Operations and Maintenace	High thermal efficiency with excellent turndown. More complex control. Need to balance waste heat and fuel gas to SCVs.	Seawater pumps and filtration system Maintenance of vaporizers and cleaning of exchangers	Max heating Duty is based on ambient conditions. Cyclic operation, requiring adjustment of the defrosting cycle according to ambient changes	Max heating Duty is based on ambient conditions. Watch out for HTF Corrosion	Seawater pumps and filtration system Maintenance of vaporizers and cleaning of exchangers	Seawater pumps and filtration system Maintenance of vaporizers and cleaning of exchangers	Similar to ORV with addition of a glycol loop system · More Complex, requiring coordination with thermal plant.	Similar to ORV with addition of a propane system	Similar to ORV with addition of a glycol loop system	Seawater pumps and filtration system Maintenance of vaporizers and cleaning of exchangers
Utilities Required	Fuel gas and electrical power	Fuel gas and electrical power	Electrical power only	Electrical power only	Seawater and electrical power	Seawater and electrical power	Seawater and electrical power	Seawater and electrical power	Seawater and electrical power	Seawater and electrical power
Chemicals	Neutralization required for pH control and NOx reduction by SCR	Neutralization required for pH control and NOx reduction by SCR	None	None	Chlorination for seawater treatment.	Chlorination for seawater treatment.	Sodium/Calcium Hypochlorite	Sodium/Calcium Hypochlorite	Sodium/Calcium Hypochlorite	Chlorination for seawater treatment.
Environmental Impacts	Flue gas (NOx, CO2, CO) emissions and 140 gpm acid water condensate discharge	Flue gas (NOx, CO2, CO) emissions and 140 gpm acid water condensate discharge	No significant impact on environment except dense fog Water Condensate, ice formation	No significant impact on environment except dense fog Water Condensate	Impacts on marine life from cold seawater and residual chloride content	Impacts on marine life from cold seawater and residual chloride content	Impacts on marine life from cold seawater and residual chloride content	Impacts on marine life from cold seawater and residual chloride content	Impacts on marine life from cold seawater and residual chloride content	Impacts on marine life from cold seawater and residual chloride content
Safety	Leakage of HC to water system which can be vented to safe location via the SCV stack and surge vessel	HTF Leakage	Leakage of HC from AAV to atmosphere at ground level	Leakage of HC from AAV to atmosphere at ground level	Leakage of HC from ORV to atmosphere at ground level	Leakage of HC from ORV to atmosphere at ground level	HTF Leakage	HTF Leakage	HTF Leakage	Leakage of HC from heat exchanger to atmosphere from Topsides

5.0 ACRONYMS

AAV – Ambient Air Vaporizer

BSCFD – Billion Standard Cubic Feet per Day

CAGR – Compound Annual Growth Rate

CO_x – Carbon oxides

EPA – Environmental Protection Agency

HTF – Heat Transfer Fluid

IFV – Intermediate Fluid Vaporizers

LNG – Liquefied Natural Gas

MMSCFD – Million Standard Cubic Feet per Day

NO_x – Nitrogen oxides

PCHE – Printed Circuit Heat Exchanger

ORV – Open Rack Vaporization

SCV – Submerged Combustion Vaporizer

US – United States

WB – World Bank

6.0 Reference Documents

1. International Gas Union World LNG Report 2015 Edition.
2. BP Statistical Review of World Energy June 2015.
3. Steve Robertson et al., "Global Offshore Prospects Presentation" Douglas-Westwood 2013.
4. Tatiana Morosuk et al., "LNG – Based Cogeneration Systems: Evaluation Using Exergy-Based Analyses" InTechOpen (2012).
5. Dhirav Patel et al., "LNG Vaporizer Selection Based on Site Ambient Conditions." *LNG 17 Conference* 1 (2013): page 1-15.
6. Shinji Egashira, "LNG Vaporizer For LNG Regasification Terminal" Kobelco Technology Review (Dec 2013): page 64-69.
7. Naoya Morimoto et al., "Development And Practical Application of A High Performance Open-Rack LNG Vaporizer (SUPERORV™)" World Gas Conference (2003).

8. H Kelle et al., "Floating Regasification Terminals – Selection & Marinisation of Regasification Equipment for Offshore Use", Offshore Technology Conference (2011): page 1-13.

9. CH-IV International, "OREGON LNG IMPORT TERMINAL - Vaporizer Alternatives Study" Document: 07902-TS-000-107, Dec 2005.

10. Susan Walther, "LNG Smart Air Vaporization Technology Applied to Floating Regasification" Society of Petroleum Engineers India Mumbai Conference (2010).

11. Dynalene HC Technical Data Sheet.

12. Himmelberger et al.,"Optimize Energy Consumption for LNG Vaporization", Hydrocarbon Processing , Jan 2006.

Review of Incumbent and Emerging LNG Regasification Technologies
Part 2 of 2: Offshore Terminals

Chevron, ETC, FE, LNG, MCP
Noma Ogbeifun, Process Engineer
J.C. Kuo, Senior Advisor, Ph.D, P.E.

Prepared for Presentation at the AIChE 2016 Spring Meeting
LNG Section
April 10-14, 2016
Houston, Texas

Abstract:

Recent trends in LNG value chain technology development have pointed away from conventional paths. The emergence of offshore LNG concepts for regasification terminals have been driven by permitting challenges and NIMBY issues experienced with traditional onshore LNG regasification terminals. The LNG regasification technology is designed with consideration for marine motions and limited footprint. Consequently, most traditional onshore regasification technologies are not feasible for offshore applications. This paper provides an overview of the different types of LNG regasification technologies that are currently available and new regasification designs proposed for future terminals. The paper is divided into two parts. Part 1 covers the LNG regasification technologies used in conventional onshore receiving terminal while Part 2 covers the LNG regasification technologies used in offshore receiving terminals.

1.0 Introduction

In 2014, 241.1 million tonnes of LNG was traded globally, a 1.8% increase from 2013[1]. The International Energy Agency in its November 2014 report estimates a sustained annual growth rate of 1.6% in natural gas consumption over several years, positioning natural gas as the second largest energy source behind oil by 2040[2]. With the rise in natural gas consumption, many LNG importing countries are constructing terminals to receive and regasify LNG for local consumption. Since 2000, global LNG receiving capacity increased by 260% to 724 MMTPA in 2014[1]. This increase in global LNG receiving capacity has been driven by the emergence of floating regasification technologies, allowing new countries to become LNG importers by utilizing flexible shipping strategies to access a wide range of LNG supply options.

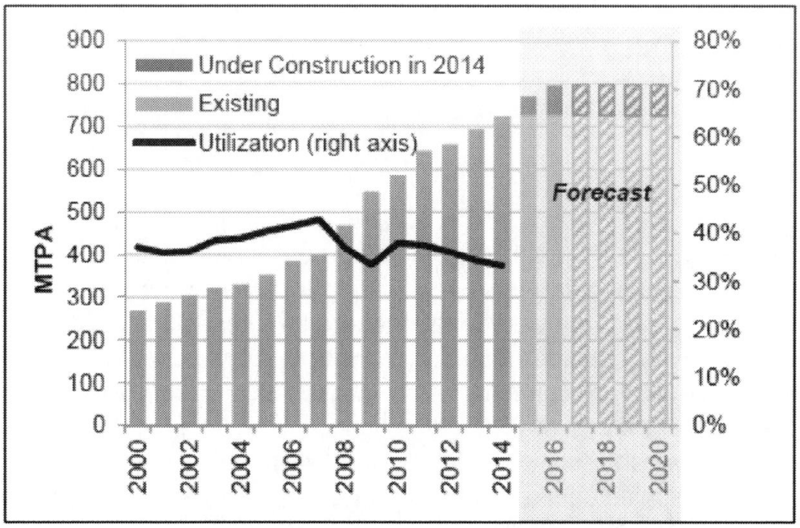

Figure 1: Global LNG Terminal Receiving Capacity[1]

The concept of offshore LNG storage and regasification has become reality given the arduous permitting process, high capital costs, and lengthy construction schedule associated with traditional onshore regasification terminals. With the exception of Adriatic LNG (Gravity Based Structure) and Melaka LNG (Sea Island Jetty), which are fixed offshore LNG terminals, the majority of the new offshore LNG terminals are floating terminals popularly referred to as floating storage and regasification units (FSRUs). According to the International Gas Union, 11 of 30 import markets had floating storage and regasification capacity totalling 54 MMTPA (See Figure 2). The push to pursue floating storage and regasification is mostly driven by the fact that floating terminals, like FSRUs, can be brought online at a lower captial cost and shorter schedule relative to onshore regasification terminals. Floating terminals offer greater flexibility when there is limited access to land for port or onshore regasification terminal siting. The floating terminal can be moored to a offshore buoy (like in the case of the FSRU Toscana). Natural gas is sent onshore via a subsea pipeline. Additional advantages include a possibly easier and shorter permitting process.

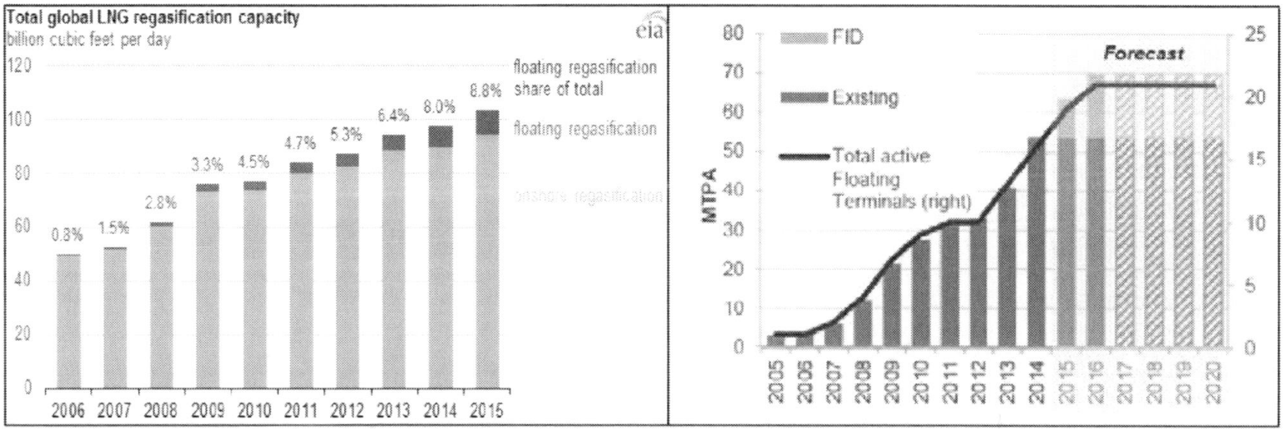

Figure 2: Global LNG Floating Storage and Regasification Capacity[1][3]

The offshore regasification design is integrated into the offshore structure (fixed or floating). The LNG regasification technology is marinised for elements such as marine motions, space constraints, and maintainability. Consequently, traditional onshore regasification designs like submerged combustion vaporizers may not feasible due to ship motion and large plot area. Cost comparisons are subjective and are not part of the scope of this paper. This paper examines the applicability and suitability of different types of LNG regasification technologies that are currently available and new regasification designs proposed for future terminals. The paper is divided into two parts. Part 1 covers the LNG regasification technologies used in conventional onshore receiving terminal while Part 2 covers the LNG regasification technologies used in offshore receiving terminal.

2.0 Closed vs. Open Loop Configurations

LNG regasification, in an offshore terminal, can be achieved through two major methods, open, and closed loop configurations, depending on the severity of environmental regulations. These two main configurations can be applied independently or in combination. Open loop in comparison with closed loop systems significantly reduces the fuel requirement and hence air emissions are reduced as the regasification process uses the natural heat within the seawater rather than heat derived from the combustion of gas. In general, approximately a third of the fuel is required for open loop systems when compared to the closed loop system.

2.1 Open Loop

In locations where seawater temperature is greater than 42°F, warm seawater may be used to supply the LNG regasification duty. Sewater is pumped into the LNG vaporizer and used to vaporize LNG and returned back to the sea in an open loop as shown in Figure 3 below. Typically, the effluent seawater temperature is lowered by approximately 13°F (7°C) from the inlet seawater temperature. In order to satisfy World Bank regulations, the temperature decrease of the ambient seawater must be no more than 37°F per 330 ft from the discharge point. To be effective, the intake and discharge points must be located so that the intake temperature is not affected by the discharge.

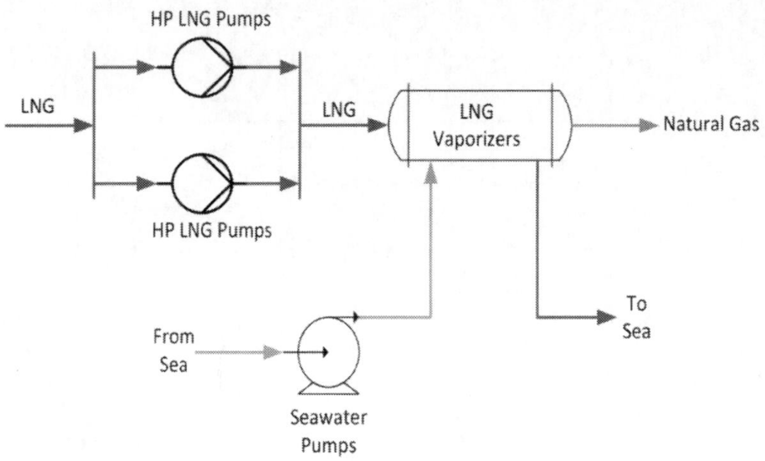

Figure 3: Open Loop Configuration

2.2 Closed Loop

In closed-loop mode, a heating medium, such as steam/hot water from boilers, is used to heat an intermediate heat transfer fluid such as propane or glycol-water and used to vaporize LNG within the LNG vaporizers. The steam condensate/cold water produced is returned to the boilers in a closed loop as shown in Figure 4 below. Steam can be supplied from FSRU boilers and waste heat recovery steam generators in the case of a GBS. This results in minimal use of seawater for LNG regasification.

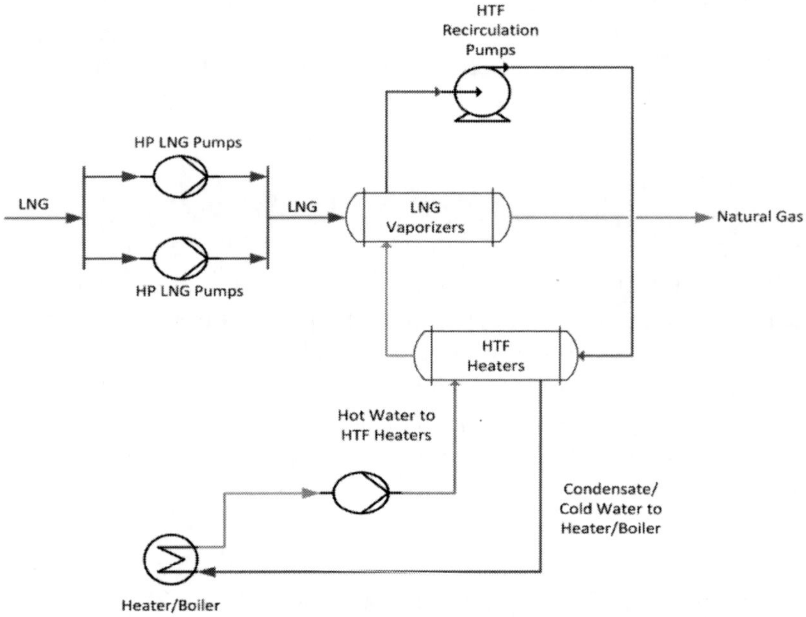

Figure 4: Closed Loop Regasification Configuration

2.3 Combination Mode

In combined mode, LNG regasification duty is supplied using both seawater (in an open loop configuration) and boilers (in a closed loop configuration) as shown in Figure 5 below. Typically, this operating mode usually occurs during peak gas send out.

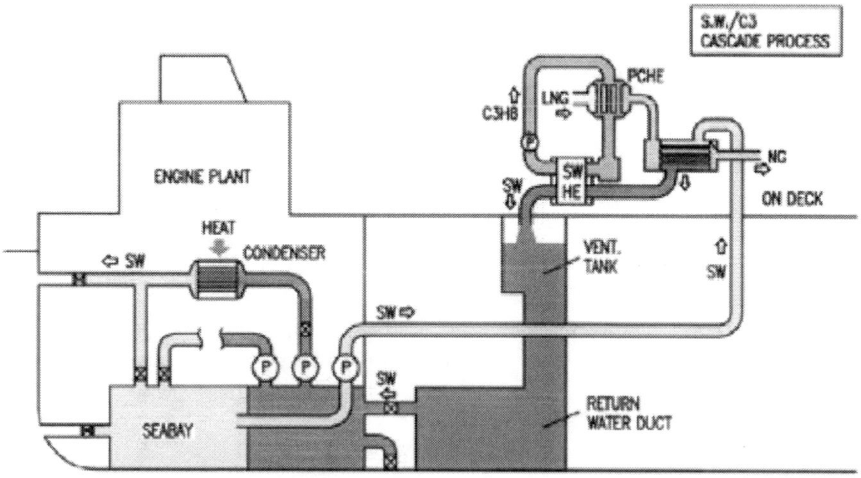

Figure 5: Combined Mode Configuration

3.0 Overview of Offshore LNG Regasification Technologies

The success of offshore LNG regasification terminals is largely dependent on selecting a process design to meet the specified regasification rates and production efficiency. The production efficiency is largely driven by the natural gas end use. In the case of power generation, the desired production efficiency is typically greater or equal to 98% (availability of 358 days per year) with the desired regasification rate upwards of 750 MMSCFD and process turndown around 10%. Operability, cost, and facility type are also important considerations in the selection of a regasification process. Floating terminals are space constrained; requiring compact technologies like open/closed loop seawater regasification using shell and tube LNG vaporizers. Conversely, LNG terminals built on fixed offshore structures (Jetty or GBS) may consider traditional land based regasification technologies like Open Rack Vaporizers (ORV). The Jetty Regasification Unit (JRU) in Melaka, Malaysia and Adriatic LNG GBS are prime example with both terminals using ORVs for LNG regasification. The major offshore LNG regasification technologies are presented in Figure 6. Combustion based technologies are shown in the pink branch of the tree diagram and ambient sourced heat technologies are depicted in the green branch. Specific technologies highlighted in white will be covered in this paper.

Figure 6: Overview of Offshore Regasification Technologies

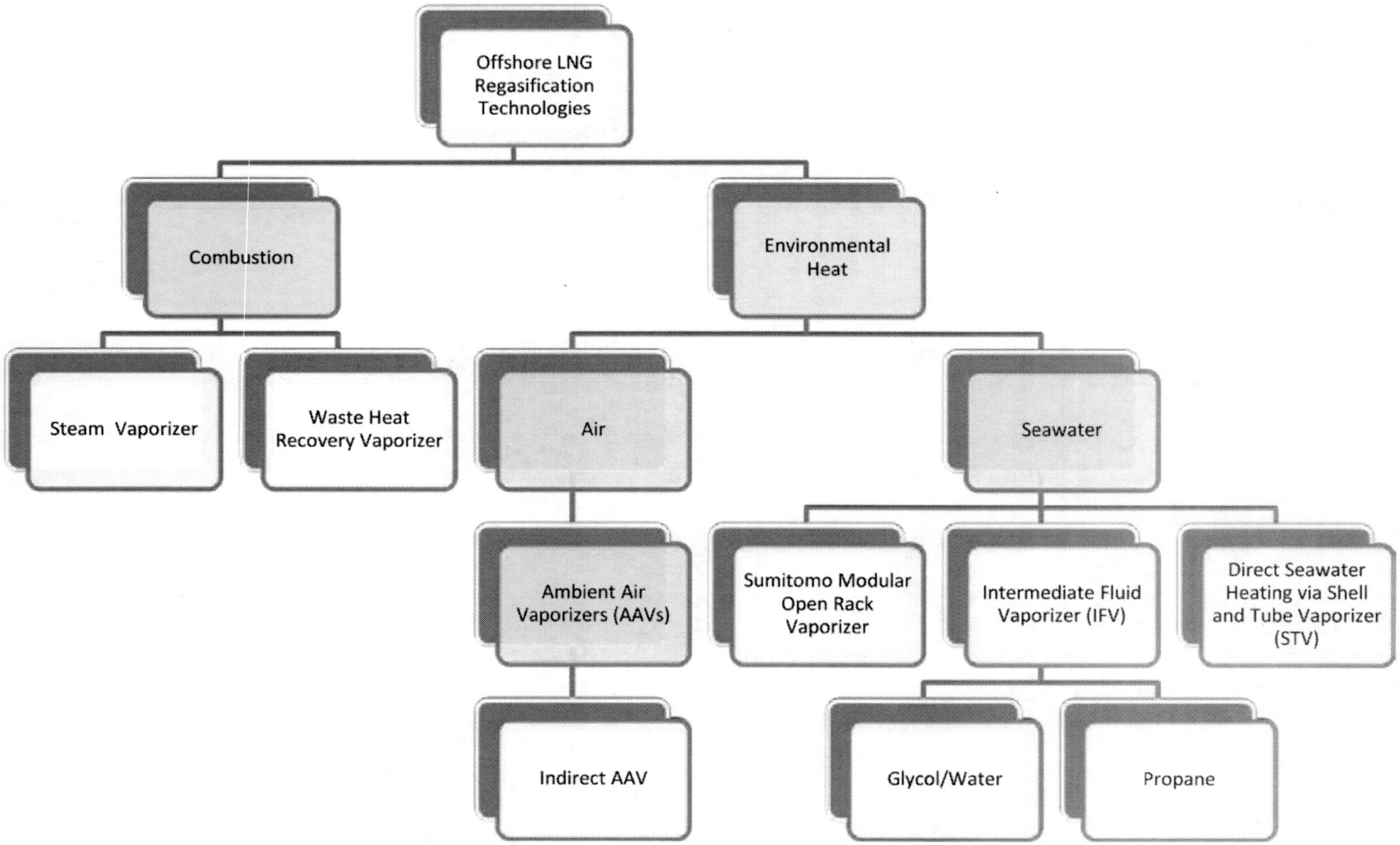

4.0 Environmental Heat

4.1 Ambient Seawater

4.1.1 Sumitomo Modular Open Rack Vaporizer

Modular Open Rack Vaporizers are heat exchangers that use seawater as the primary heat source. It is the same technology as the conventional onshore ORV marinised for offshore applications by Sumitomo Precision Products (see Figure 7). The modular approach minimizes site fabrication and increases constructability, resulting in a reduction in the total cost. Modular ORVs is a proven technology and is currently used by Adriatic LNG.

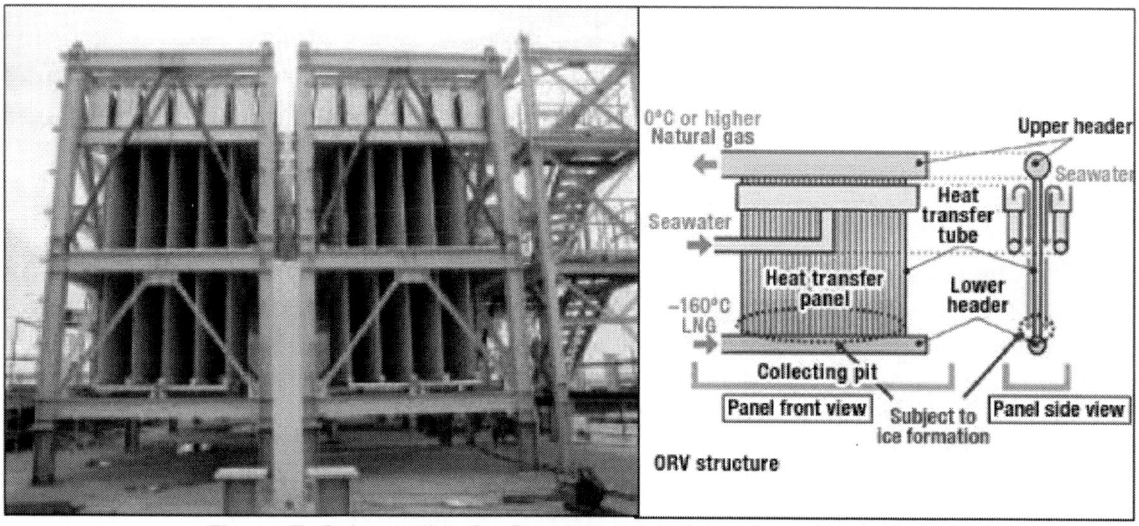

Figure 7: Schematic of a Sumitomo Modular Open Rack Vaporizer (Courtesy of Sumitomo and OSAKA Gas)

Modular ORV technology uses naturally occurring heat in the form of warm seawater flowing over a series of panel coils to vaporize the LNG flowing countercurrent within the panels. Seawater lift pumps bring treated seawater to the distribution troughs at the top of the vaporization panels. Warm seawater flows down the outside of the panels in a falling film. The seawater is cooled by exchanging heat with the LNG. The panels consist of multiple LNG tubes with integral fins. LNG flowing upward through the tubes is vaporized and warmed to approximately 40°F. A 4°F-6°F LNG approach temperature is possible depending on prevalent conditions i.e. summer or winter conditions. The preferred seawater temperature for Modular ORV operation is above 42°F. The cooled seawater from the ORVs is collected in a basin beneath the ORV and routed to the seawater outfall system.

Seawater is pumped into the ORV through intake screens to eliminate debris and marine life. The ORV sea water pumps consume approximately 12.38 BTU/lb_m(LNG)[4] for lifting and driving seawater. Sodium/Calcium hypochlorite is injected into each seawater pump suction continuously to control marine growth in the system. Typically, a water concentration of 0.2 ppm of Sodium/Calcium hypochlorite is desired. In order to shock the system, elevated concentrations of 2.0 ppm would be injected for 20 minutes every 8 hours, during ORV operation. It is expected that dechlorination will naturally occur as the treated water flows over the panels, through the trench, and to the outflow basin, such that no special treatment will be required to meet permitted discharge limits for residual chlorine.

Modular ORV units are generally constructed of aluminum alloy for mechanical strength suitable to operate at the cryogenic temperature. The material has high thermal conductivity which is effective for heat transfer equipment. The tubes are arranged in panels, connected through the LNG inlet and the regasified product outlet piping manifolds are hung from a rack. The panels are coated externally with zinc alloy, providing corrosion protection against seawater. Regular maintenance is required to keep the finned tube surface clean.

The panel arrangement feature provides ease of access for maintenance. Depending on the design of the units, it is also possible to isolate sections of the panels and vary the load on the units. The unit can be turned down to accommodate fluctuations in gas demand, gas outlet temperature and seawater temperature.

4.1.1.1 Environmental Impact

Environmental issues associated with Modular ORVs involve seawater intake, seawater outfall and air emissions. A primary environmental concern with seawater intake for Modular ORVs is marine life in the water column. Screen depth, intake velocity and backwashing are key considerations in the design of the seawater intake screen system. The maximum through slot velocity of the intake screens should be around 0.5 ft/s to minimize impingement and entrainment of marine organisms. An automated air backwash system should be installed to remove impinged organisms and debris on the screen surface.

Seawater from the Modular ORVs is routed to the seawater outfall and returned to the ocean. In order to satisfy World Bank regulations, the temperature decrease of the ambient seawater must be no more than 37°F per 330 ft from the discharge point. Also the residual chlorine levels need to meet the regulation required (0.2mg/l at point of discharge per US EPA standard).

Sodium Hypochlorite is highly corrosive and relatively unstable with the potential to gasifiy, crystallize and foul water system components. Chlorination piping should be designed to maintain Sodium Hypochlorite flow velocity of 0.7 -2 ft/s to avoid gasification and crystallization within flowlines.

4.1.1.2 Safety

ORVs are considered extremely safe. No moving parts are in contact with flammable fluids. ORVs do not require combustion and pose no ignition source.

4.1.2 Intermediate Fluid Vaporizers (IFV)

The concept of LNG regasification using intermediate fluid heat transfer was originally developed by OSAKA GAS in the 1970s in conjunction with the TRI-EX™ heat exchangers [6]. The TRI-EX™ heat exchanger is a type of shell and tube vaporizer where the LNG vaporizer (E1) and trim heater (E3) are conjoined together in a single piece of equipment as shown in Figure 8. LNG is vaporized using a heat transfer fluid (HTF) in a closed intermediate loop.

Typically, the regasification process takes place in a two stage shell and tube configuration. In the first stage (E1), the HTF is heated and vaporized using seawater. The warm HTF vapor exchanges heat with the LNG flowing through the tubes resulting in vaporization. The cold HTF vapor condenses on the tubes and fall back into the shell where it is re-vaporized. The vaporization and condensation of the HTF occurs at the same pressure. The marinised E1 shell and tube exchanger contain liquid level baffles installed perpendicular to the tubes and spaced equidistant from each other on the shell side. This is to maintain HTF level over the tube and prevent the tube exposure during ship motion. In the second stage (E3), the vaporized natural gas is sent to a second shell and tube heat exchanger where seawater is used to bring the natural gas to the specified send out temperature. Periodic cleaning of the seawater tube internals is required to remove marine growth and tube scale buildup.

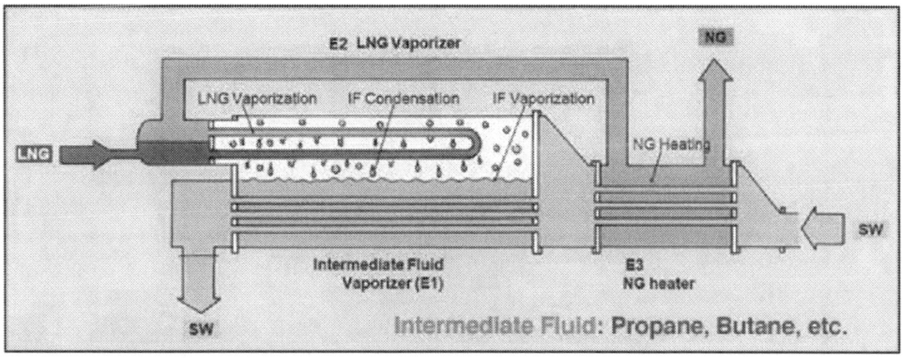

Figure 8: Schematic for IFV Regasification System with Seawater (Courtesy of Kobelco)

Both Propane and Glycol-Water have been used effectively as heat transfer fluids in offshore regasification applications. Propane generally provides better heat transfer properties as it can be easily vaporized using seawater and condensed by LNG thereby requiring a smaller surface area and equipment size. Operating issues related to IFV freezing are eliminated since the thermodynamic freezing point of propane is lower than -162°C. Comparatively, glycol-water mixture requires a larger heat transfer area and higher pumping duty due to its viscosity. Also, poor glycol-water fluid distribution could lead to slosh formation within the vaporizer. The FSRU Toscana (see Figure 9) uses propane as the intermediate fluid and TRI-EX™ heat exchangers in an open loop configuration to vaporize LNG.

Figure 9: IFV Regasification System with Seawater on FSRU Toscana (Courtesy of Kobelco)

Wartsila-Hamworthy IFV technology (see Figure 10) uses cascade heating to supply LNG regasification duty. In the first stage, the HTF (usually propane/glycol-water) is heated using seawater in printed circuit heat exchanger (PCHE). The warm HTF is pumped into the shell side of the shell and tube LNG vaporizer (STV) and used to vaporize LNG. The cold HTF is circulated back to the seawater HTF exchanger for reheating.

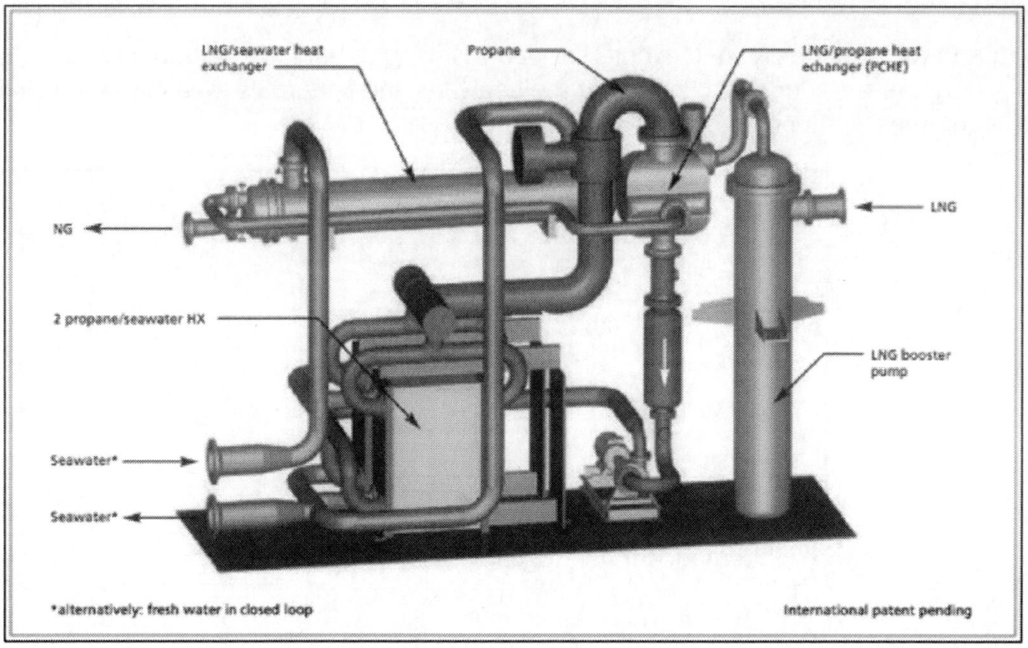

Figure 10: LNG Regasification Using Wartsila Hamworthy IFV Technology (courtesy of Wartsila)

4.1.2.1 Environmental Impact

Environmental issues associated with IFVs involve seawater intake, seawater outfall, air emissions and handling of glycol and propane. The same environmental concerns around seawater handling are equally applicable with the IFV vaporizer option as described in the Modular ORV option. IFVs do not directly produce air emissions. They do indirectly generate air emissions due to the power generation requirement for seawater lift pumps.

Secondary containment will be required around all equipment and vessels. Leaking and spilled liquid will need to be collected in sealable containers, or vacuum (pump) into a storage vessel/container for disposal.

Ethylene glycol is classified as a combustible and will burn producing toxic decomposition products. Use of ethylene glycol will add another toxic, combustible and environmentally regulated chemical to the inventory on the facility. However, this is a controllable and acceptable risk if handled properly. Its use will require reporting to regulating agencies. Ethylene glycol is currently used offshore in closed systems, and therefore is not expected to be denied use by regulators, especially in a closed loop system.

4.1.2.2 Safety

The freezing of glycol/water solution on the shell side of the exchanger in the event of tube rupture is a concern as LNG will enter the shell side of the exchanger impeding the path of the expanding fluid to the relief valve inlet of the shell side of the exchanger.

4.1.3 Direct Heating Shell and Tube Vaporizers

Direct heating Shell and Tube Vaporizers (STV) use seawater as the source of thermal energy. Seawater is pumped into the shell side and used to vaporize LNG in an open loop configuration as shown in Figure 11. Typically, the effluent seawater temperature is lowered by approximately 13°F from the inlet seawater temperature. In order to satisfy World Bank regulations, the temperature decrease of the ambient seawater must be no more than 37°F per 330 ft from the discharge point. The preferred seawater temperature is above 42°F. The potential freezing of seawater on the shell side will lead to reduced heat transfer performance.

Direct Heating STVs are usually of TEMA type NEN or NJN design and have been in service since 1974. TEMA type NEN or NJN STVs are often used in high pressure services and services where minimal joint leakage is required. The NEN designation refers to a single shell pass heat exchanger where seawater is introduced into the heat exchanger through a single shell noozle and cold seawater effluent flows out via single shell noozle. The NJN designation refers to a shell and tube heat exchanger where seawater is introduced into the heat exchanger shell through a single noozles on the shell side and cold seawater effluent flows out via two shell noozles. In both designs, the tubesheets are welded to the shell and front and rear channels minimizing tubesheet thickness . Mechanical cleaning of the tubes is possible through bolted covers on the channels making.

These types of vaporizers can be stacked requiring relatively small spatial footprint when compared to other available vaporization technologies. Another advantage is that they are generally more cost effective than removable bundle STV designs. Therefore, NEN or NJN STVs are frequently used in offshore applications. Typically, a single direct heated STV can deliver up to 56.6 MMBTUs/hr of heating duty at a nominal and maximal LNG capacity of 100 and 115 MSCFD respectively.

Figure 11: Offshore LNG STV Application (Courtesy of Chicago Power & Process, Inc.)

4.1.3.1 Environmental Impact

Environmental issues associated with STVs involve seawater intake, seawater outfall and air emissions. The open-loop technology reduces air emissions, since there is no combustion. A primary environmental concern with seawater intake for STVs is marine life in the water column. Screen depth, intake velocity and backwashing are key considerations in the design of the seawater intake screen system. The maximum through slot velocity of the intake screens should be around 0.5 ft/s to minimize impingement and entrainment of marine organisms. An automated air backwash system should be installed to remove impinged organisms and debris on the screen surface.

4.1.3.2 Safety

Slosh formation and freezing of seawater on the shell side of the exchanger in the event of tube rupture is a concern as LNG will enter the shell side of the exchanger impeding the path of the expanding fluid to the relief valve inlet of the shell side of the exchanger.

4.2 Air

4.2.1 Indirect Ambient Air Vaporizer

Indirect ambient air vaporizers operate by transferring heat from ambient air to a heat transfer fluid (HTF) which in turn exchanges heat with LNG for regasification. The system consists of shell and tube LNG vaporizers, air-fin heaters and a closed HTF loop. Air draft fans drive air flow across the tube bundles maximizing heat transfer to the HTF and minimizing required heat transfer area. The hot HTF is sent to the shell and tube LNG vaporizers for LNG regasification. The cold HTF flows back to the accumulator and is then pumped back to the air fin heaters for heating.

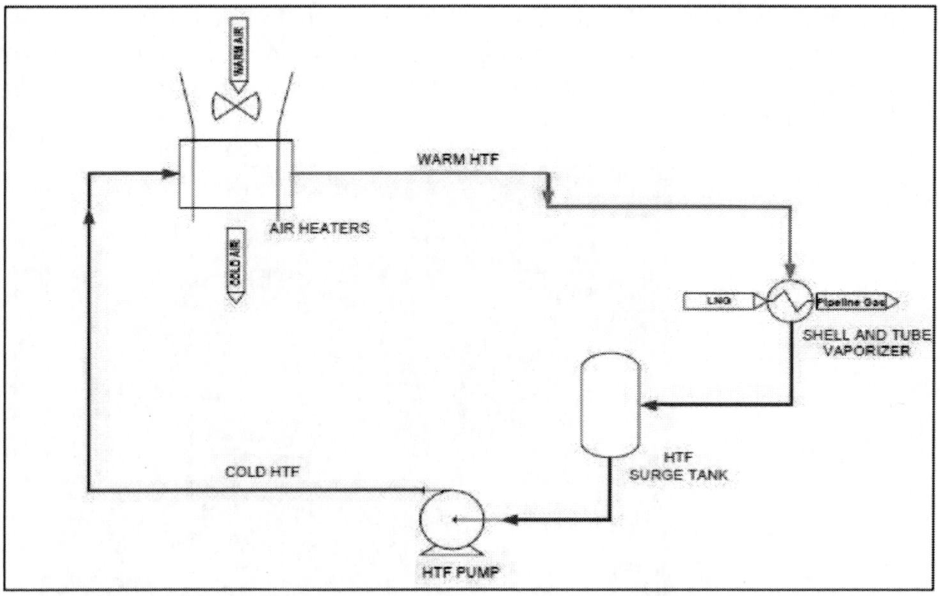

Figure 12: Schematic of Indirect Ambient Air LNG Regasification [9]

The Smart Air® vaporization system (see Figure 13) by Wood Group Mustang has a similar setup to the indirect ambient air regasification system except that the Smart Air® system uses a alkali metal formate/acetate (like Potassium Formate) and water mixture as the heat transfer fluid [10] . The system consists of Air-fin Heaters, accumulator, HTF Pumps, and Shell and Tube LNG vaporizers. The Smart Air® vaporization system Air-fin heaters are proprietary to Wood Group Mustang. Each Air-fin heater contains layers of finned tubes with fans forcing ambient air from the surroundings to flow through the top of the unit downwards, over the tube bundles. The Air-fin heaters are elevated above grade to prevent recirculation of cold dense air from the bottom of the unit to the fan inlet. As with other ambient air vaporization systems, supplementary heat may be required when heating load cannot be met with design ambient conditions.

The alkali metal formate/acetate and water mixture, henceforth referred to as K-HTF, is a water based low temperature heat transfer fluid engineered to deliver excellent heat transfer performance within the temperature ranges of -58°F to 424°F[11]. K-HTF has been gaining popularity due to its good thermophysical properties, low toxicity, non-flammability, and lower pumpimg duty requirements when compared to other water based chemistries such as glycol /water mixtures, alkali chloride/water as well as hydrocarbon and silicone based fluids. However, K-HTFs are corrosive especially towards zinc, aluminium, and galvanized steel.

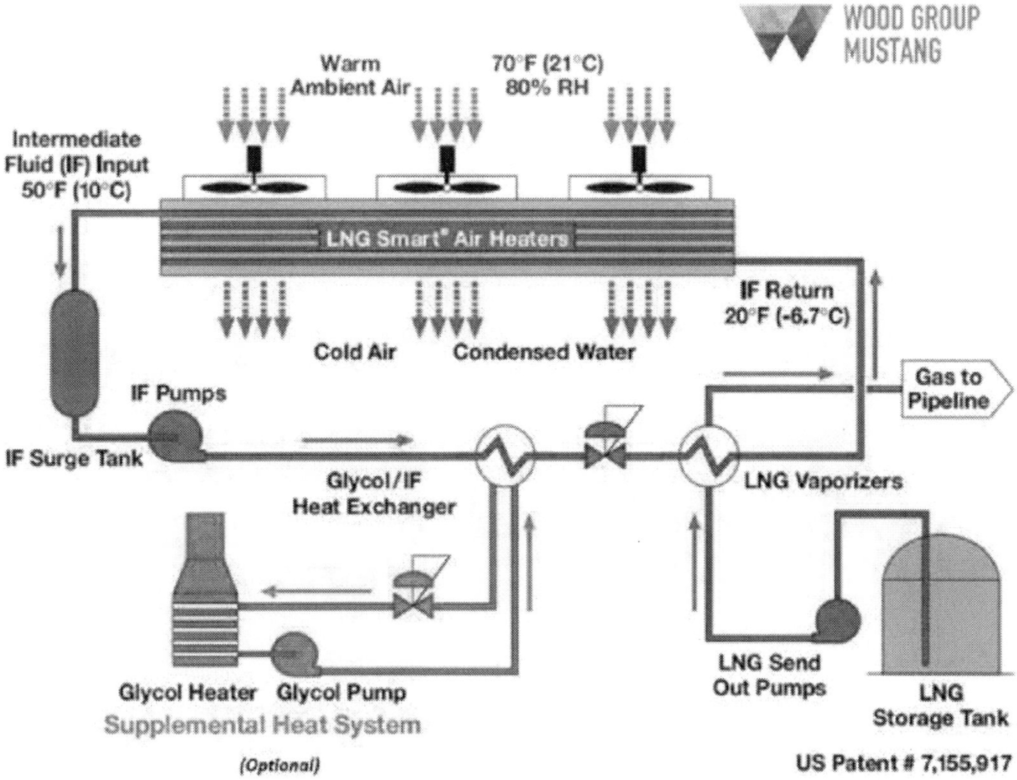

Figure 13: LNG Smart Air® Vaporization Technology (Courtesy of Wood Group Mustang)

4.2.1.1 Environmental Impact

Environmental issues associated with LNG Smart® Air Vaporization System include air emissions from the heat transfer fluid heater (HTF) and handling of the HTF (glycol water mixture or K-HTF). The glycol water closed loop system need to meet the environmental/regulatory implication as listed in the IFV option. Water produced from condensation of water vapor from the surrounding air must be continuously discharged and safely discarded.

4.2.1.2 Safety

The heat transfer fluid heater is source of ignition and the HTF loop has same concern as IFV. K-HTFs are corrosive especially towards zinc, aluminum, and galvanized steel so material selection is key consideration to minimize loss of containment events.

5.0 Combustion

5.1 Steam Vaporizer (SV)

Steam Vaporizers (SV) are typically used when the water temperature is too low for direct heating regasification or environmental regulations prevent seawater use. Steam Vaporizers use condensing steam as a heat source on the shell side of a shell and U-tube heat exchanger (see Figure 14). LNG is vaporized and superheated through the stainless steel tube bundle that is fitted with vortex generators. The steam condensate is recirculated to the boilers for vaporization in a closed loop configuration. Steam flowrate to the SV shell is controlled via temperature control thereby maintaining the send out temperature. The combination of high heat content and large temperature differences lends to a small heat transfer area. Typically, a 125,000 to 130,000 m^3 capacity FSRU has a steam plant which can support a vaporization capacity up to 500 m^3/h of LNG in addition to its own electrical power demand[6]. A design for 260 m^3/h (LNG) would be approximately 6.5 m long x 0.8 m diameter with a dry weight of about 30 tonnes[6].

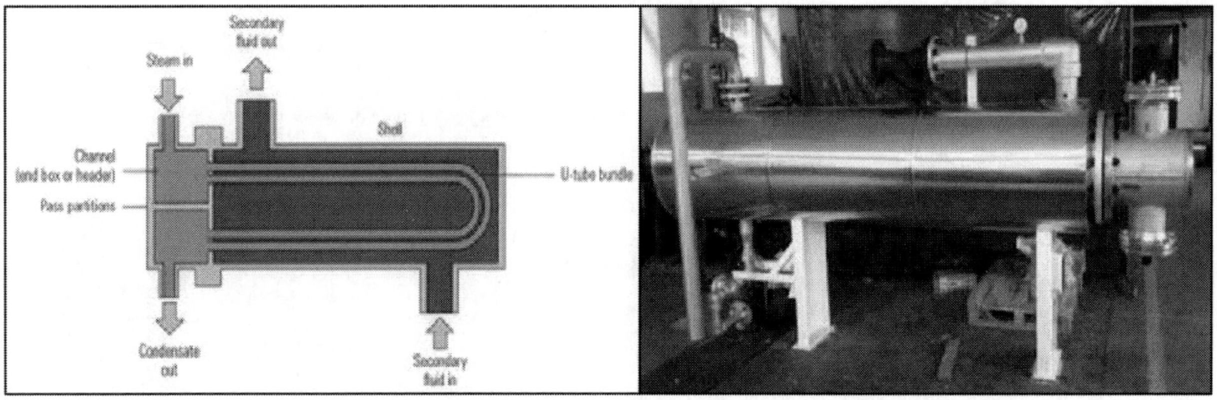

Figure 14: Direct Steam Vaporizer (SV) (Courtesy of Chart Industries)

5.1.1.1 Environmental Impact

Environmental issues associated with SVs involve air emissions from power and steam generation.

5.1.1.2 Safety

Safety measures include combustible gas detection to indicate gas emissions and initiate mitigation measures to prevent ignition of the vapors.

5.2 Waste Heat Recovery LNG Vaporizer (WHRV)

A Waste Heat Recovery LNG Vaporizer (WHRV) is a shell and tube heat exchanger which uses a HTF such as propylene glycol and water mixture circulating in a closed loop to regasify LNG (see Figure 15).

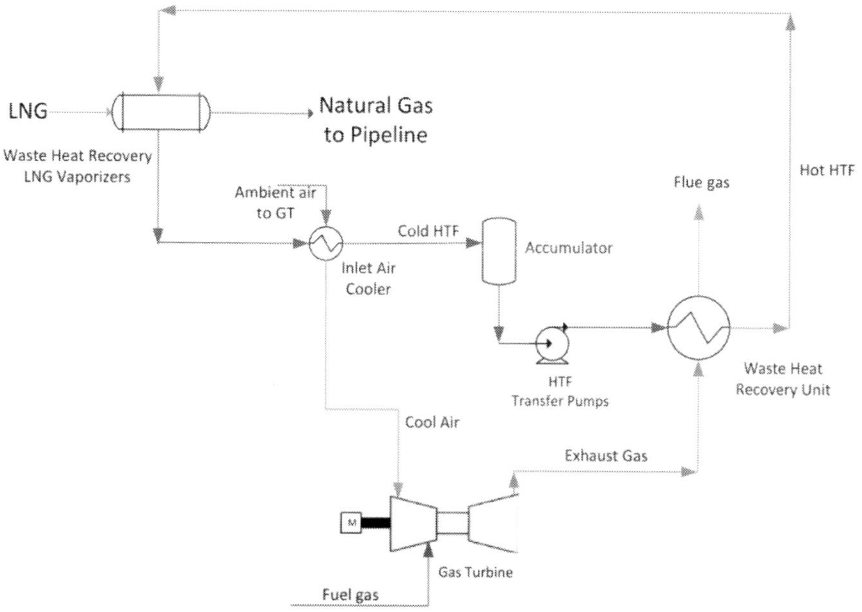

Figure 15: Waste Heat Recovery LNG vaporizer using Thermal Energy from aTurbine Exhaust

Hot turbine exhaust is piped to a waste heat recovery unit and used to heat the HTF. The warm HTF is pumped to the top of the WHRV. The WHRV is installed in the vertical orientation similar to the Modular ORV panels. LNG is pumped from the bottom of the WHRV where it vaporizes as it exchanges heat with the HTF flowing top to bottom. The cold HTF is used to provide inlet air cooling for the gas turbine en route to the waste heat recovery unit for heating. Typically, for each degree Celsius drop in air temperature, power output can be increased by about 1%. However, aero-derivative machines tend to be sensitive to air temperature changes. Currently, WHRVs provide approximately 15% of the regasification duty for the Adriatic LNG terminal.

5.2.1.1 Environmental Impact

Environmental issues associated with waste heat recovery regasification system involve indirect air emissions associated with power generation for driving process equipment like seawater lift pumps and HTF transfer pumps.

5.2.1.2 Safety

In a loss of containment event, gas turbines are a potential source of ignition. Minimizing ignition sources is preferred in any LNG import facility. Safety concerns associated with a technology involving combustion offshore, where paths of egress are limited in the event of an emergency.

Additional vapor control would be required surrounding the waste heat recovery regasification systems. This may mean additional combustible gas detection to indicate gas emissions and initiate mitigation measures to prevent vapor ignition.

6.0 COMPARISON OF ONSHORE LNG REGASIFICATION TECHNOLOGIES

The optimum choice of an LNG vaporization system is determined by the terminal's site selection, the environmental conditions, regulatory limitations and operability considerations. Table 1 compares eight LNG regasification options in term of their applications, operation and maintenance, utility and chemical requirements, and environmental impacts for an offshore LNG regasification terminal.

Table 1: LNG Regasification Type Qualitative Comparison

OFFSHORE								
					IFV			
LNG Regasification Technology	Steam Vaporizer	Waste Heat Recovery Vaporizer	Indirect AAV	Modular ORV (mORV)	Propane	Glycol-Water	Propane	Direct Seawater Heating Shell and Tube Vaporizer
Feature	Indirect LNG vaporization by cooling steam produced from boilers	Indirect LNG vaporization by cooling HTF which is heated by hot exhaust gases	Direct LNG vaporization using air via HTF	Direct LNG vaporization using sea water	Indirect LNG vaporization by condensing propane which is heated by seawater in TRI-EX type Heat Exchanger	Indirect LNG vaporization by cooling hot glycol which is heated by any available heat medium	Indirect LNG vaporization by condensing propane which is heated by seawater	Direct LNG vaporization using sea water
Unit Max Train Throughput, Tons/hr	Design to desired throughput	Design to desired throughput	Design to desired throughput	Design to desired throughput	Design to desired throughput	Design to desired throughput	Design to desired throughput	100 tons per hour per exchanger
Heating Medium	Steam	Hot Exhaust Gas	Air	Seawater	Seawater	Any available heat medium	Seawater	Seawater
Operations and Maintenace	Similar to ORV	Seawater pumps and filtration system Maintenance of vaporizers and cleaning of exchangers	Max heating Duty is based on ambient conditions. Watch out for HTF Corrosion	Seawater pumps and filtration system Maintenance of vaporizers and cleaning of exchangers	Similar to ORV with addition of a propane system	Similar to ORV with addition of a glycol loop system More Complex, requiring coordination with thermal plant.	Similar to ORV with addition of a propane system	Seawater pumps and filtration system Maintenance of vaporizers and cleaning of exchangers
Utilities Required	Seawater, fuel, and electrical power	Fuel gas and electrical power	Electrical power only	Seawater and electrical power	Seawater and electrical power	Seawater and electrical power	Seawater and electrical power	Seawater and electrical power
Chemicals	Sodium/Calcium Hypochlorite	Neutralization required for pH control and NOx reduction by SCR	None	Chlorination for seawater treatment.	Sodium/Calcium Hypochlorite	Sodium/Calcium Hypochlorite	Sodium/Calcium Hypochlorite	Chlorination for seawater treatment.
Environmental Impacts	Flue gas (NOx, CO2, CO) emissions from boiler fuel consumption.	Flue gas (NOx, CO2, CO) emissions and 140 gpm acid water condensate discharge	No significant impact on environment except dense fog Water Condensate	Impacts on marine life from cold seawater and residual chloride content	Impacts on marine life from cold seawater and residual chloride content	Impacts on marine life from cold seawater and residual chloride content	Impacts on marine life from cold seawater and residual chloride content	Impacts on marine life from cold seawater and residual chloride content
Safety	Hot surfaces	HTF Leakage	Leakage of HC from AAV to atmosphere at ground level	Leakage of HC from ORV to atmosphere at ground level	HTF Leakage	HTF Leakage	HTF Leakage	Leakage of HC from heat exchanger to atmosphere from Topsides

7.0 ACRONYMS

AAV – Ambient Air Vaporizer

BSCFD – Billion Standard Cubic Feet per Day

EPA – Environmental Protection Agency

GBS – Gravity Based Structure

HTF – Heat Transfer Fluid

IFV – Intermediate Fluid Vaporizers

JRU - Jetty Regasification Unit

LNG – Liquefied Natural Gas

MMSCFD – Million Standard Cubic Feet per Day

PCHE – Printed Circuit Heat Exchanger

ORV – Open Rack Vaporization

SW – Seawater

US – United States

WB – World Bank

8.0 Reference Documents

1. International Gas Union World LNG Report 2015 Edition.
2. BP Statistical Review of World Energy June 2015.
3. Victoria Zaretskaya et al., "Floating LNG Regasification is used to meet rising natural gas demand in smaller markets" US Energy Information Administration (April 2015).
4. Susan Walther, "LNG Smart Air Vaporization Technology Applied to Floating Regasification" Society of Petroleum Engineers India Mumbai Conference (2010).
5. Shinji Egashira, "LNG Vaporizer For LNG Regasification Terminal" Kobelco Technology Review (Dec 2013): page 64-69.
6. W.S.Wayne et al.," Modification of LNG Carriers for Regasification" Offshore Technology Conference (2003).
7. H Kelle et al., "Floating Regasification Terminals – Selection & Marinisation of Regasification Equipment for Offshore Use", Offshore Technology Conference (2011): page 1-13.
8. CH-IV International, "OREGON LNG IMPORT TERMINAL - Vaporizer Alternatives Study" Document: 07902-TS-000-107, Dec 2005.

9. Dhirav Patel et al., "LNG Vaporizer Selection Based on Site Ambient Conditions." *LNG 17 Conference* 1 (2013): page 1-15.

10. Roger D. Leick, "Adriatic LNG Terminal" ExxonMobil Development Company (March 2005).

11. Dynalene HC Technical Data Sheet.

12. Sumitomo Precision Products Annual Report 2007.

13. Nobuya Himoto et al., "Maximizing Value-added LNG Regasification: Modular ORV for offshore LNG Terminals" AIChE Spring National Meeting, April 2004.

AUTHOR INDEX

AUTHOR INDEX

AIChE
120 Wall Street, FL 23
New York, NY 10005-4020

ISBN 978-1-5108-2485-0

9 781510 824850

CURRAN ASSOCIATES INC.
proceedings
.com